Engineering Damage Mechanics

MOUNT LAUREL BRANCH LIBRARY
BURLINGTON COUNTY COLLEGE
MOUNT LAUREL, NJ

Jean Lemaitre   Rodrigue Desmorat

Engineering Damage Mechanics

Ductile, Creep, Fatigue and Brittle Failures

With 135 Figures

Springer

Professor Jean Lemaitre
Professor Rodrigue Desmorat
Université Paris 6
Ecole Normale Supérieure de Cachan
L.M.T. Cachan
61, Av. du Président Wilson
94230 Cachan, France

ISBN 3-540-21503-4 Springer Berlin Heidelberg New York

Library of Congress Control Number: 2004111141

This work is subject to copyright. All rights are reserved, whether the whole or part of the material is concerned, specifically the rights of translation, reprinting, reuse of illustrations, recitation, broadcasting, reproduction on microfilm or in any other way, and storage in data banks. Duplication of this publication or parts thereof is permitted only under the provisions of the German Copyright Law of September 9, 1965, in its current version, and permission for use must always be obtained from Springer. Violations are liable to prosecution under the German Copyright Law.

Springer is a part of Springer Science+Business Media

springeronline.com

© Springer-Verlag Berlin Heidelberg 2005
Printed in The Netherlands

The use of general descriptive names, registered names, trademarks, etc. in this publication does not imply, even in the absence of a specific statement, that such names are exempt from the relevant protective laws and regulations and therefore free for general use.

Typesetting and Production: LE-TeX Jelonek, Schmidt und Vöckler GbR, Leipzig
Cover design: *design & production* GmbH, Heidelberg

Printed on acid-free paper   55/3141/YL   5 4 3 2 1 0

Foreword

The analysis of stress and deformation of structures and components is not an end in itself. The aim is to predict the serviceability, reliability, and manufacturability of both existing structures and components and of proposed designs. In order to make such assessments, the mode (or modes) of failure need to be known and accounted for. The damage can take many forms, e.g., cracks, voids, chemical attack. In any case, the result is deterioration of the structure or component. Predicting the implications of that deterioration for mechanical integrity is the goal of damage mechanics.

The concept of damage and the realization of its importance for engineering are not new. What is relatively new (within the last 35 years or so) is the development of the framework of Continuum Damage Mechanics. There have been many contributors to this development, but the contributions of the French school and particularly those of Jean Lemaitre stand out. The field of damage mechanics has advanced to the point where it is an engineering tool, with wide applications in industry. In practice, needs arise at several levels. In a preliminary design stage, the need is often for rapid methods of analysis that exhibit trends. In a final design stage or in carrying out a safety assessment the need can be for accurate quantitative predictions. This book provides formulations that span such a range for a variety of technologically important modes of failure, providing a perspective on the advantages and disadvantages of various approaches – from uncoupled, post-processing analyses to fully coupled damage analyses.

After introductory chapters on Continuum Damage Mechanics and numerical analysis of damage, the remaining chapters focus on a mode of damage – ductile failures; low-cycle fatigue; creep, creep-fatigue and dynamic failures; high-cycle fatigue; and failure of brittle and quasi-brittle materials. Each of these chapters appropriately begins with a section on engineering considerations to set the stage and provides a guide to analysis methods and tools. It is quite remarkable that such a wide range of behaviors are incorporated within a unified presentation. The damage mechanics "apple" has blossomed into a tree with many branches!

Since failure is a complex nonlinear process, the predicted behavior can be sensitive to parameter values. Their appropriate identification is key for reliable engineering predictions, as is understanding the sensitivity of predictions to the particular choice of parameter values. The presentation here pays attention to parameter sensitivity as well as to parameter identification.

This book provides a comprehensive guide to Engineering Damage Mechanics. It should appeal to all engineers and students of engineering concerned with lifetime prediction and with the failure resistant design of structures, components, and processes.

Brown University, USA *Alan Needleman*

Introduction

The single apple has become a tree, an apple tree painted by Annie Lemaitre from which two apples fell on the cover page! A decade after "A Course on Damage Mechanics" the topic has grown up to reach the field of applications. Aircraft engines and, more generally, aeronautics, nuclear power plants, metal forming, civil engineering, and the automotive industry have already developed and benifited from damage-based methods to increase performance and security. The time has come to propose simplified or more advanced methods, structured in a unified framework to designers of any mechanical components such as early design with fast calculation of structural failures by closed-form solutions and final validation of solutions by numerical failures analysis. This was the ambition for this book!

This is the reason for having many basic examples and insisting on practical methods such as the difficult problem of the material parameters identification for which a systematic sensitivity analysis is performed for each type of application. Very accurate calculations are too often made with a very poor accuracy of the material parameters! To help, probability concepts are introduced either for random loadings or scatter due to microdefects in the materials. This is done mainly for fatigue failure phenomena and brittle materials but may apply to other cases.

Damage mechanics applies to all materials, including metals and alloys, polymers, elastomers, composites, and concrete, because even if the mechanisms are different on a microscale, they have more or less the same qualitative behaviors on meso- or macroscales. Nevertheless, due to data availability most quantitative examples are related to metals.

- The first chapter reassembles the main concepts of Continuum Damage Mechanics, that is the theoretical tools to apply to specific cases: damage variable, isotropic and anisotropic description, thermodynamics which yields methods of damage measurements, damage laws, coupling with strain behavior, localization, and mesocrack initiation.
- The second chapter is a set of numerical tools for solving the nonlinear problems related to damage evaluation in structures. Post-processing clas-

sical structure analysis, either by the time integration of a damage law, solving a micromechanical two-scale damage model, or when damage is not localized, by solving fully coupled strain-damage structural problems.
- The five following chapters are organized in the same way: four sections from the simplest methods with closed-form solutions to more advanced numerical analyses. The first sections "Engineering Considerations" give the domain of application of the chapter. The second sections "Fast Calculations of Structural Failures" describe some simplified methods to be used in early design. They are applied in the third sections "Basic Engineering Examples" to damage failures of members having stress concentration zones, pressurized cylinders and beams in bending. The fourth sections "Numerical Failure Analysis" describe, using examples, more accurate methods for numerical calculations with computers.
  - The third chapter is devoted to ductile failures involving large deformations for applications in metal forming processes or effects of large overloadings on structures in service.
  - The fourth chapter deals with low-cycle fatigue involving important coupling between damage and plasticity for applications on structures heavily burdened by cyclic loadings.
  - The fifth chapter introduces the effects of temperature-inducing creep and its nonlinear interaction with damage for applications on structures loaded statically or cyclically at elevated temperature, or dynamically.
  - The sixth chapter concerns high-cycle fatigue which uses a two-scale damage model of an elasto-plastic damaged inclusion in an elastic matrix with "elastic fatigue" applications from complex histories of loading to three-dimensional and random loadings.
  - The seventh chapter is devoted to brittle and quasi-brittle materials: quasi-brittle when an irreversible process induces damage, brittle when the fracture occurs without any measurable precursor. Statistical and probabilistic methods are used to represent the large scatter generally observed in the failure of these materials. Their applications concern structures made of concrete, ceramics or composite materials.

How should you use the book? As you like it of course but be aware that each chapter is more or less self-contained, with many referrals to the two first chapters of basic concepts of damage mechanics and its numerical processing. Furthermore, at the end of each chapter on applications, the section "Hierachic Approach" is more or less a summary of the chapter with indications on the domain of validity of each model or method. To help engineers, researchers, students, beginners or not, each section is categorized by the number of apples:

    🍎   means easy to read, easy to apply.
    🍎🍎   means a read with attention and an application with care.
    🍎🍎🍎   means a more advanced theory needing a numerical analysis.

Of course this classification is subjective but it has been checked by some friends working mostly in industry: J. Besson from ENSMP Centre des matériaux (Chap. 1), E. Lorentz from Electricité de France (Chap. 2), F. Moussy from Renault (Chap. 3), J.P. Sermage from E.D.F. SEPTEN (Chap. 4), B. Dambrine from SNECMA (Chap. 5), A. Galtier from ARCELOR for (Chap. 6), B. Bary from C.E.A. (Chap. 7), A. Benallal from C.N.R.S.-LMT Cachan and M. Elgueta from Chili University (overall book), and A. Needleman from Brown University who wrote the foreword. Our thanks to all of them for their expertise and advice. "Merci" also to our friends from "Laboratoire de Mécanique et Technologie" at Cachan who participated in the birth of many parts of this book and particularly to Catherine Génin.

Bon courage pour une lecture fructueuse

LMT Cachan, France
Jean Lemaitre
Rodrigue Desmorat

Contents

1 **Background on Continuum Damage Mechanics** 1
  1.1 Physics and Damage Variables 1
      1.1.1 Definition of a Scalar Damage Variable 3
      1.1.2 Definition of Several Scalar Damage Variables 3
      1.1.3 Definition of a Tensorial Damage Variable 4
      1.1.4 Effective Stress Concept 5
      1.1.5 Effects of Damage 7
  1.2 Thermodynamics of Damage 7
      1.2.1 General Framework 7
      1.2.2 State Potential for Isotropic Damage 10
      1.2.3 State Potential for Anisotropic Damage 11
      1.2.4 Quasi-Unilateral Conditions of Microdefects Closure .. 12
  1.3 Measurement of Damage 16
      1.3.1 Isotropic Elasticity Change 17
      1.3.2 Isotropic Elasticity Change by Ultrasonic Waves 17
      1.3.3 Anisotropic Elasticity Change 18
      1.3.4 Hardness Change 22
      1.3.5 Elasticity Field Change 23
  1.4 Kinetic Laws of Damage Evolution 26
      1.4.1 Damage Threshold and Mesocrack Initiation 27
      1.4.2 Formulation of the Isotropic Unified Damage Law 32
      1.4.3 Formulation of the Anisotropic Damage Law 34
      1.4.4 Fast Identification of Damage Material Parameters .. 35
      1.4.5 Generalization of the Unified Damage Law 41
  1.5 Elasto-(Visco-)Plasticity Coupled with Damage 45
      1.5.1 Basic Equations without Damage Coupling 45
      1.5.2 Coupling with Isotropic Damage 52
      1.5.3 Coupling with Anisotropic Damage 56
      1.5.4 Non-Isothermal Behavior 60
      1.5.5 Two-Scale Model for Damage at Microscale 60

	1.6	Localization and Mesocrack Initiation	65
		1.6.1 Critical Damage Criterion	65
		1.6.2 Strain Damage Localization Criterion	65
		1.6.3 Size and Orientation of the Crack Initiated	73
2	**Numerical Analysis of Damage**		77
	2.1	Uncoupled Analysis	78
		2.1.1 Uniaxial Loading	79
		2.1.2 Proportional Loading	82
		2.1.3 Post-processing a (Visco-)Plastic Computation	85
		2.1.4 Post-processing an Elastic Computation	86
		2.1.5 Jump-in-Cycles Procedure in Fatigue	88
	2.2	Fully-Coupled Analysis	90
		2.2.1 Nonlinear Material Behavior FEA	91
		2.2.2 FE Resolution of the Global Equilibrium	94
		2.2.3 Local Integration Subroutines	96
		2.2.4 Single Implicit Algorithm for Damage Models	97
		2.2.5 Damage Models with Microdefects Closure Effect	105
		2.2.6 Performing FE Damage Computations	111
		2.2.7 Localization Limiters	112
	2.3	Locally-Coupled Analysis	114
		2.3.1 Post-Processing a Reference Structure Calculation	115
		2.3.2 Implicit Scheme for the Two-Scale Model	117
		2.3.3 DAMAGE 2000 Post-Processor	119
	2.4	Precise Identification of Material Parameters	120
		2.4.1 Formulation of an Identification Problem	121
		2.4.2 Minimization Algorithm for Least Squares Problems	123
		2.4.3 Procedure for Numerical Identification	127
		2.4.4 Cross Identification of Damage Evolution Laws	132
		2.4.5 Validation Procedure	133
		2.4.6 Sensitivity Analysis	135
	2.5	Hierarchic Approach and Model Updating	137
	2.6	Table of Material Damage Parameters	138
3	**Ductile Failures**		141
	3.1	Engineering Considerations	142
	3.2	Fast Calculation of Structural Failures	142
		3.2.1 Uniaxial Behavior and Validation of the Damage Law	143
		3.2.2 Case of Proportional Loading	144
		3.2.3 Sensitivity Analysis	146
		3.2.4 Stress Concentration and the Neuber Method	147
		3.2.5 Safety Margin and Crack Arrest	152

	3.3	Basic Engineering Examples		154
		3.3.1	Plates or Members with Holes or Notches	154
		3.3.2	Pressurized Shallow Cylinders	156
		3.3.3	Post-Buckling in Bending	158
		3.3.4	Damage Criteria in Proportional Loading	160
	3.4	Numerical Failure Analysis		166
		3.4.1	Finite Strains	167
		3.4.2	Deep Drawing Limits	170
		3.4.3	Damage in Cold Extrusion Process	172
		3.4.4	Crack Initiation Direction	174
		3.4.5	Porous Materials – the Gurson Model	176
		3.4.6	Frames Analysis by Lumped Damage Mechanics	181
		3.4.7	Predeformed and Predamaged Initial Conditions	184
		3.4.8	Hierarchic Approach up to Full Anisotropy	188
4	**Low Cycle Fatigue**			191
	4.1	Engineering Considerations		192
	4.2	Fast Calculation of Structural Failures		192
		4.2.1	Uniaxial Behavior and Validation of the Damage Law	192
		4.2.2	Case of Proportional Loading	198
		4.2.3	Sensitivity Analysis	199
		4.2.4	Cyclic Elasto-Plastic Stress Concentration	201
		4.2.5	Safety Margin and Crack Growth	208
	4.3	Basic Engineering Examples		208
		4.3.1	Plate or Members with Holes or Notches	208
		4.3.2	Pressurized Shallow Cylinders	210
		4.3.3	Cyclic Bending of Beams	212
	4.4	Numerical Failure Analysis		213
		4.4.1	Effects of Loading History	214
		4.4.2	Multiaxial and Multilevel Fatigue Loadings	216
		4.4.3	Damage and Fatigue of Elastomers	221
		4.4.4	Predeformed and Predamaged Initial Conditions	227
		4.4.5	Hierarchic Approach up to Non-Proportional Effects	228
5	**Creep, Creep-Fatigue, and Dynamic Failures**			233
	5.1	Engineering Considerations		234
	5.2	Fast Calculation of Structural Failures		234
		5.2.1	Uniaxial Behavior and Validation of the Damage Law	235
		5.2.2	Case of Proportional Loading	237
		5.2.3	Sensitivity Analysis	241
		5.2.4	Elasto-Visco-Plastic Stress Concentration	244
		5.2.5	Safety Margin and Crack Growth	247

XIV Contents

    5.3 Basic Engineering Examples 248
        5.3.1 Strain Rate
               and Temperature-Dependent Yield Stress 248
        5.3.2 Plates or Members with Holes or Notches 250
        5.3.3 Pressurized Shallow Cylinder 253
        5.3.4 Adiabatic Dynamics Post-Buckling in Bending 255
    5.4 Numerical Failure Analysis 257
        5.4.1 Hollow Sphere under External Pressure 258
        5.4.2 Effect of Loading History: Creep-Fatigue 262
        5.4.3 Creep-Fatigue
               and Thermomechanical Loadings 263
        5.4.4 Dynamic Analysis of Crash Problems 267
        5.4.5 Ballistic Impact and Penetration of Projectiles 269
        5.4.6 Predeformed and Predamaged Initial Conditions 272
        5.4.7 Hierarchic Approach
               up to Viscous Elastomers 274

6 High Cycle Fatigue 277
    6.1 Engineering Considerations 278
    6.2 Fast Calculation of Structural Failures 279
        6.2.1 Characteristic Effects in High-Cycle Fatigue 279
        6.2.2 Fatigue Limit Criteria 281
        6.2.3 Two-Scale Damage Model
               in Proportional Loading 283
        6.2.4 Sensitivity Analysis 289
        6.2.5 Safety Margin and Crack Growth 292
    6.3 Basic Engineering Examples 294
        6.3.1 Plates or Members with Holes or Notches 294
        6.3.2 Pressurized Shallow Cylinders 295
        6.3.3 Bending of Beams 297
        6.3.4 Random Loadings 297
    6.4 Numerical Failure Analysis 300
        6.4.1 Effects of Loading History 300
        6.4.2 Non-Proportional Loading of a Thinned Shell 303
        6.4.3 Random Distribution of Initial Defects 307
        6.4.4 Stochastic Resolution by Monte Carlo Method 309
        6.4.5 Predeformed and Predamaged Initial Conditions 312
        6.4.6 Hierarchic Approaches
               up to Surface and Gradient Effects 314

7 Failure of Brittle and Quasi-Brittle Materials 321
    7.1 Engineering Considerations 321
    7.2 Fast Calculations of Structural Failures 324
        7.2.1 Damage Equivalent Stress Criterion 324
        7.2.2 Interface Debonding Criterion 327

		7.2.3	The Weibull Model 328

- 7.2.3 The Weibull Model 328
- 7.2.4 Two-Scale Damage Model for Quasi-Brittle Failures 332
- 7.2.5 Sensitivity Analysis 333
- 7.2.6 Safety Margin and Crack Propagation 334

7.3 Basic Engineering Examples 336
- 7.3.1 Plates or Members with Holes and Notches 336
- 7.3.2 Pressurized Shallow Cylinders 337
- 7.3.3 Fracture of Beams in Bending 338

7.4 Numerical Failure Analysis 339
- 7.4.1 Quasi-Brittle Damage Models 339
- 7.4.2 Failure of Pre-stressed Concrete 3D Structures 346
- 7.4.3 Seismic Response of Reinforced Concrete Structures 350
- 7.4.4 Damage and Delamination in Laminate Structures 356
- 7.4.5 Failure of CMC Structures 361
- 7.4.6 Single and Multifragmentation of Brittle Materials 364
- 7.4.7 Hierarchic Approach up to Homogenized Behavior 368

Bibliography 373

Index 375

Notations

It was impossible to avoid using the same letter for different meanings!

Operators

x	scalar
x_i	component of a vector $\vec{x}$
x_{ij}	component of a second order tensor $\boldsymbol{x}$
x_{ijkl}	component of a fourth order tensor $\underline{\boldsymbol{x}}$
$[M]$	matrix
$\dot{x}$	time derivative of x ($\dot{x} = \mathrm{d}x/\mathrm{d}t$)
$x_{i,j}$	gradient of $\vec{x}$ also ∇x
$x_{ij,j}$	divergence of $\boldsymbol{x}$
x_{kk}	trace of $\boldsymbol{x}$
x_H	$\frac{1}{3}$ trace of $\boldsymbol{x}$
x_{ij}^D	component of the deviatoric tensor $\boldsymbol{x}^\mathrm{D}$, $x_{ij}^\mathrm{D} = x_{ij} - x_\mathrm{H}\delta_{ij}$
δ_{ij}	Kronecker delta, $\delta_{ij} = 1$ if $i = j$ and $\delta_{ij} = 0$ if $i \neq j$
$|x|$	absolute value of the scalar x
$|\boldsymbol{x}|_{ij}$	absolute value in terms of principal components of the tensor $\boldsymbol{x}$
$[\![x]\!]$	discontinuity of x
$\langle x \rangle$	Macauley bracket, $\langle x \rangle = x$ if $x \geq 0$ and $\langle x \rangle = 0$ if $x < 0$
$\langle \boldsymbol{x} \rangle^+$ or $\langle \boldsymbol{x} \rangle^+_{ij}$	positive part in terms of principal components of tensor $\boldsymbol{x}$
$\langle \boldsymbol{x} \rangle^-$ or $\langle \boldsymbol{x} \rangle^-_{ij}$	negative part in terms of principal components of tensor $\boldsymbol{x}$
$\bar{x}$	mean value of x
$\bar{\bar{x}}$	standard deviation of the random variable x
Δx	range of x (peak to peak amplitude) or time increment $x_{n+1} - x_n$
$\mathrm{d}, \partial, \delta$	differential operators

$\mathcal{H}$	Heaviside function
$\ln$	neperian logarithm
$\sigma_{ij,j}$	derivative $\frac{\partial}{\partial x_j}\sigma_{ij}$
$\det$	determinant operator
∇	gradient operator
∇^2	laplacian operator

Symbols

$\underline{\boldsymbol{a}}$	rate elasticity tensor
A	damage threshold parameter
A	crack area
$\mathcal{A}_k$	material parameter
A_D	Kachanov law damage parameter
$\boldsymbol{A}$	Almansi strain tensor
b, b_y	isotropic hardening exponents
b_S	Sines criterion parameter
C, C_y	kinematic hardening parameters
C_ϵ	heat capacity
C_h	specific heat
C_MC	Manson–Coffin law parameter
C_P	Paris law parameter
$\underline{\boldsymbol{C}}$	dilatation tensor
D	scalar damage variable
D_{ij}	component of the second order damage tensor $\boldsymbol{D}$
$D_\mathrm{T}, D_\mathrm{S}$	transverse and shear damage variables
D_{ijkl}	component of the fourth order damage tensor
D_c	critical damage parameter
e^p_{ij}	component of the effective plastic strain tensor $\boldsymbol{e}^\mathrm{p}$
E	Young modulus of elasticity
E_{ijkl}	component of the elasticity tensor $\underline{\boldsymbol{E}}$
$\boldsymbol{E}$	Green–Lagrange strain tensor
f	yield function of plastic criterion
f_v	porosity
F	force
F	dissipative potential function
F_X	plastic potential of dissipation
F_D	damage potential of dissipation
$\boldsymbol{F}$	gradient of deformation transformation

g	plastic strain-stress function
G	shear elasticity modulus
G	strain energy release rate
G_c	material toughness parameter
h, h_a	microdefect closure parameters
H	hardness material parameter
H	activation energy parameter
H_N	creep temperature exponent
$H_{ij} = (\mathbf{1} - \mathbf{D})_{ij}^{-1/2}$	component of the damage effective operator $\mathbf{H} = (\mathbf{1} - \mathbf{D})^{-1/2}$
I_{ijkl}	component of the unit fourth order tensor $\mathbf{I}$
[Jac]	Jacobian matrix
k	heat conductivity parameter
K	elastic compressibility modulus
K	stress intensity factor
K_c	cyclic hardening law coefficient
K_T	elastic stress concentration coefficient
k_{Neuber}	Neuber stress concentration correction
K_p, K_p^0, K_p^y, K_p^f	hardening material parameters
K_N, K_N^0	Norton law parameters
K_∞	viscous material parameter
$\mathbf{L}$	strain rate tensor
$\underline{\mathbf{L}}$	fourth order tangent tensorial operator
m	damage threshold exponent
$moon_i$	Mooney parameters
M, M_0, M_y, M_f	isotropic hardening exponents
M_c	cyclic hardening law exponent
M_{ijkl}	component of the effective operator $\mathbf{M}$
n	viscous material parameter exponent
$\vec{n}$	unit normal vector
N, n	number of cycles
N_R	number of cycles to rupture
N, N_0	Norton law viscous exponents
N_{ij}	shape functions

XX   Notations

p	accumulated plastic strain
p_D	damage threshold accumulated plastic strain
p_R	rupture accumulated plastic strain
P	pressure
$\boldsymbol{P}$	ponderation matrix
q_1, q_2	Gurson law parameters
q	state variable of the reversible domain
Q	stress increase of the reversibility domain
$\vec{q}$	thermal flux vector
r	isotropic hardening state variable
r_h	heat source
r_D	Kachanov law damage exponent
R	isotropic hardening stress variable
R_∞, R_∞^y	saturated isotropic hardening parameters
$R_\nu, R_{\nu h}$	triaxiality function
s	specific entropy
s	unified damage law exponent
S	energetic damage law parameter
S	surface
S_D	damage surface
Saf	safety factor
$S_{\mathcal{A}_k}$	sensitivity coefficient
$\boldsymbol{S}$	second Piola–Kirchhoff stress tensor
t	time
t_R	rupture time
T	temperature
T_X	stress triaxiality, $T_X = \sigma_H/\sigma_{eq}$
u	displacement
$\vec{U}$	displacement vector
$\{U\}$	nodal displacement vector
$\{U^e\}$	elementary nodal displacement vector
v	wave speed
V	volume
V_0	reference volume of Weibull law
V_{eff}	effective volume of Weibull law

w	energy density
w_D	damage threshold stored energy density
w_e	elastic strain energy density
w_s	stored energy density
W_e	elastic strain energy
W_1, W_2	hyperelatic energy densities
X	uniaxial kinematic back stress
X_{ij}	component of the back stress deviatoric tensor $\boldsymbol{X}$
X_∞, X_∞^y	saturated kinematic hardening material parameter
Y	energy density release rate
Y_{ij}	component of the energy density release rate tensor $\boldsymbol{Y}$
$\overline{Y}$	effective elastic energy density
Z	necking parameter in pure tension
α	dilatation parameter coefficient
$\boldsymbol{\alpha}$	kinematic hardening state variable
β	Eshelby coefficient
γ, γ_y	kinematic hardening material parameter
γ_{MC}	Manson–Coffin law exponent
$\Gamma(a)$	gamma function: $\Gamma(a) = \int_0^\infty t^{a-1} \exp(-t) dt$
$\gamma(a, x)$	incomplete gamma function: $\gamma(a, x) = \int_0^x t^{a-1} \exp(-t) dt$
δ_0	size of the mesocrack initiated
δ_{GTN}	Gurson law coalescence parameter
$\boldsymbol{\Delta}$	hypoelastic strain rate tensor
$\epsilon, \epsilon_{ij}, \boldsymbol{\epsilon}$	uniaxial and tensorial total strains
$\epsilon_e, \epsilon_{ij}^e, \boldsymbol{\epsilon}^e$	uniaxial and tensorial elastic strains
$\epsilon_p, \epsilon_{ij}^p, \boldsymbol{\epsilon}^p$	uniaxial and tensorial plastic strains
ϵ_{pD}	damage threshold plastic strain in pure tension
$\epsilon_{pR}, \epsilon_{pR}^\star$	rupture plastic strains in pure tension
ϵ_{pu}	plastic strain for ultimate stress in pure tension
ϵ_R	rupture strain in pure tension
$\epsilon_{p\Sigma}$	signed equivalent plastic strain
$\boldsymbol{\epsilon}^\pi$	irreversible strain
η	hydrostatic sensitivity damage parameter
η_P	Paris law exponent

ϕ_D	damage dissipated energy density
ϕ_p	plastic dissipated energy density
$\phi_\text{Dp}, \phi_\text{F}$	fracture dissipated energy density
λ	Lamé elastic parameter
λ	elongation
$\dot{\lambda}$	plastic multiplier
μ	Lamé elastic parameter in shear
$\dot{\mu}$	internal sliding multiplier
ν	Poisson ratio of elastic contraction
$\vec{\nu}$	unit reference vector
ν_{ij}	anisotropic contraction ratio
π	cumulative internal sliding
π_D	damage threshold
θ	numerical parameter of the θ-method
ρ	mass density
ρ	radius of curvature
$\sigma, \sigma_{ij}, \boldsymbol{\sigma}$	uniaxial and tensorial Cauchy stresses
$\tilde{\sigma}, \tilde{\sigma}_{ij}, \tilde{\boldsymbol{\sigma}}$	uniaxial and tensorial effective stresses
$\sigma^\mu, \sigma_{ij}^\mu, \boldsymbol{\sigma}^\mu$	stresses at microscale
σ_H	hydrostatic stress, $\sigma_\text{H} = \frac{1}{3}\sigma_\text{kk}$
σ_eq	von Mises equivalent stress
σ_Σ	signed von Mises stress
$\sigma^\star$	damage equivalent stress, $\sigma^\star = \sigma_\text{eq} R_\nu^{1/2}$
σ_n	nominal stress
σ_v	viscous stress
σ_R	rupture stress
σ_u	ultimate stress
σ_y	yield stress
σ_y02	engineering yield stress for $\epsilon_\text{p} = 0.2 \cdot 10^{-2}$
σ_f	engineering fatigue limit at 10^6 or 10^7 cycles
σ_f^∞	asymptotic fatigue limit
σ_s	reversibility threshold
Σ_{ij}	component of normalized stress tensor $\boldsymbol{\Sigma}$
ψ	Helmholtz specific free energy

ψ_e	elastic state potential
ψ_p	plastic state potential
ψ_T	thermal state potential
$\psi^\star$	Gibbs specific free enthalpy
$\psi_\mathrm{e}^\star$	elastic specific free enthalpy

1
Background on Continuum Damage Mechanics

This chapter gives the reader the necessary and sufficient information to deal with practical applications of Continuum Damage Mechanics (the term was first introduced by J. Hult in 1972) to predict the crack initiation in structures subjected to heavy loadings. This ranges from the early definition of a **scalar** damage variable by L.M. Kachanov in 1958 to the anisotropic **tensorial** damage variable, from classical thermodynamics aspects to the effective stress that take into account the **closure of microdefects** in compression-like loadings, from simple measurement of local damage to fields of damage, from a simple damage law to induced **anisotropy** and a generalization to quasi-brittle materials, from elasto-(visco-)plasticity coupled with damage to the phenomenon of **localization**, the ultimate stage before crack initiation.

The basic concepts are written in a concise and sufficiently general way in order to be applied to any material: metals and alloys, polymers, elastomers, composites, concretes, ceramics... For more basic details please refer to the book *A Course on Damage Mechanics* written by Jean Lemaitre and published by Springer-Verlag in 1992 and 1996. Other (good!) references are given at the end of the book.

1.1 Physics and Damage Variables

Damage, in its mechanical sense in solid materials is the creation and growth of microvoids or microcracks which are discontinuities in a medium considered as continuous at a larger scale. In engineering, the mechanics of continuous media introduces a Representative Volume Element (RVE) on which all properties are represented by homogenized variables. To give an order of magnitude, its size can vary from about $(0.1 \text{ mm})^3$ for metals and ceramics to about $(100 \text{ mm})^3$ for concrete. The damage discontinuities are "small" with respect to the size of the RVE but of course large compared to the atomic spacing (Fig. 1.1).

1 Background on Continuum Damage Mechanics

From a physical point of view, damage is always related to plastic or irreversible strains and more generally to a strain dissipation either on the mesoscale, the scale of the RVE, or on the microscale, the scale of the discontinuities:

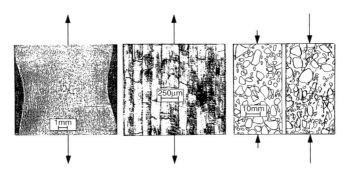

Fig. 1.1. Examples of damage in a metal (*left*, micro-cavities in copper from L. Engel and H. Klingele), in a composite (*middle*, microcracks in carbon-fiber/epoxy resin laminate from O. Allix), and in a concrete (*right*, crack pattern in a ASTM I cement, 4.8 and 9.5 mm aggregates, petrography from S.P. Shah)

- In the first case (mesolevel), the damage is called ductile damage if it is nucleation and growth of cavities in a mesofield of plastic strains under static loadings; it is called creep damage when it occurs at elevated temperature and is represented by intergranular decohesions in metals; it is called low cycle fatigue damage when it occurs under repeated high level loadings, inducing mesoplasticity.
- In the second case (microlevel), it is called brittle failure, or quasi-brittle damage, when the loading is monotonic; it is called high cycle fatigue damage when the loading is a large number of repeated cycles. Ceramics, concrete, and metals under repeated loads at low level below the yield stress are subjected to quasi-brittle damage.

In all cases, these are volume defects such as microcavities, or surface defects such as microcracks. This is the reason to have several definitions of a damage variable:

- If only ductile damage is considered, it may be defined as the volume density of microvoids,

$$D_{\rm v} = \frac{\delta V_{\rm voids}}{\delta V_{\rm RVE}} = f_{\rm v}. \tag{1.1}$$

This is the starting point of the Gurson model described in Sect. 3.4.5.

- For a larger generality where microcavities and microcracks may exist, the damage variable is physically defined by the surface density of microcracks and intersections of microvoids lying on a plane cutting the RVE of cross section δS (Fig. 1.2). For the plane with normal $\vec{n}$ where this density is maximum, we have

$$D_{(\vec{n})} = \frac{\delta S_D}{\delta S} \,. \qquad (1.2)$$

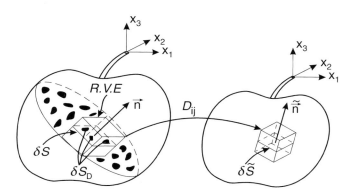

Fig. 1.2. Physical damage and mathematical continuous damage

1.1.1 Definition of a Scalar Damage Variable⁰

If the damage is **isotropic**, the scalar variable $D_{(\vec{n})}$ does not depend on the normal. The intrinsic variable is a scalar (L.M. Kachanov 1958):

$$\boxed{D = \frac{\delta S_D}{\delta S}} \qquad (1.3)$$

It can be used as such for one-dimensional problems. It can also be used as an easy evaluation of the approximate damage in three-dimensional problems (particularly in proportional loading).

1.1.2 Definition of Several Scalar Damage Variables⁰⁰

If several mechanisms of damage occur, simultaneously or not, each of them may be represented by a scalar variable (D_k) with the same physical meaning as above. This is the case in classical composites where delamination of fibers and matrix cracking may occur. Two or three independent variables are used (P. Ladevèze 1983), for instance:

- D_F for quasi-brittle fibers breakage,
- D_T for transverse matrix cracking, and
- D_S acting on shear for splitting.

1.1.3 Definition of a Tensorial Damage Variable♂♂♂

Damage is often **non-isotropic** due to microcracking more or less perpendicular to the largest positive principal stress. Then, the surface density of microdefects in a plane with normal $\vec{n}$ acts through an operator which transforms the surface δS and $\vec{n}$ of Fig. 1.2 into a smaller but continuous area, $\delta \tilde{S} = \delta S - \delta S_\mathrm{D}$, and into another normal $\tilde{\vec{n}}$ (S. Murakami 1981).

To keep the same physical meaning as above the damage acts through the operator $(\mathbf{1} - \boldsymbol{D})$ and

$$(\delta_{ij} - D_{ij})\,n_j\delta S = \tilde{n}_i \delta \tilde{S}, \tag{1.4}$$

where δ_{ij} is the Kronecker delta and $\boldsymbol{D}$ is a second order tensor. As there is no distorsion of the surface δS, it induces the property of orthotropy consistent with the fact that the damage is governed by the plastic strain represented by the second order tensor $\boldsymbol{\epsilon}^\mathrm{p}$.

Remark – *In fact, the largest generality for a damage variable is a representation by a fourth order tensor, as it can be shown in several ways (J.L. Chaboche 1978, D. Krajcinovic 1981, F.A. Leckie and E.T. Onat 1981, C.L. Chow 1987). Such a tensor is difficult to use and is not necessary for damage-induced by meso- or microplasticity.*

- *As in the previous section, consider a damage plane area, δS, with normal $\vec{n}$ and a reference vector $\vec{\nu}$ such that the tensor $\nu_i n_j \delta S$ defines the geometrical reference configuration. Continuum Damage Mechanics defines the effective continuous configuration by a modified area $\delta \tilde{S}$ and a modified normal $\tilde{\vec{n}}$, as shown in Fig. 1.3.*
  The damage $\boldsymbol{D}$ is the operator which transforms the second order tensor $\nu_i n_j \delta S$ of the reference configuration into the tensor of the effective configuration $\nu_i \tilde{n}_j \delta \tilde{S}$. This is a fourth order tensor, where

$$(I_{ijkl} - D_{ijkl})\nu_k n_l \delta S = \nu_i \tilde{n}_j \delta \tilde{S}, \tag{1.5}$$

with the following symmetries: $D_{ijkl} = D_{ijlk} = D_{jikl} = D_{klij}$.

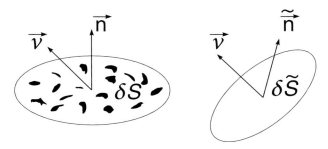

Fig. 1.3. Reference and effective configurations

- Another way to define a fourth order tensor $\underline{\boldsymbol{D}}$ is to consider the operator which changes the initial elasticity tensor E_{ijkl} into the actual elasticity tensor $\tilde{E}_{ijkl}$ softened by damage:

$$(I_{ijrs} - D_{ijrs})E_{rskl} = \tilde{E}_{ijkl}. \qquad (1.6)$$

From a purely theoretical point of view, this definition does not yield a real state variable because it requires the knowledge of a particular behavior (elasticity here).

1.1.4 Effective Stress Concept^{♂♂}

The constitutive equations for strain and damage characterize the plain material itself without the volume or the surface discontinuities. An interesting concept to introduce next is the stress acting on the resisting area, $\delta \tilde{S} = \delta S - \delta S_D$, as shown in Fig. 1.2 – the effective stress.

- In the uniaxial case of **isotropic damage** without the closure effect of microcracks in compression, this mean value of the microstresses is simply given by the force equilibrium (Y.N. Rabotnov 1968):

$$\tilde{\sigma}\delta\tilde{S} = \sigma\delta S \quad \text{with} \quad D = \frac{\delta S_D}{\delta S} = \frac{\delta S - \delta \tilde{S}}{\delta S} \qquad (1.7)$$

or

$$\boxed{\tilde{\sigma} = \frac{\sigma}{1-D}.} \qquad (1.8)$$

- In the multiaxial case of isotropic damage, all the stress components act on the same effective area. The effective stress tensor is simply

$$\boxed{\tilde{\sigma}_{ij} = \frac{\sigma_{ij}}{1-D}.} \qquad (1.9)$$

- The case of **anisotropic damage** is much more complicated^{♂♂♂} to ensure a good representation of the physics as well as compatibility with thermodynamics. In fact, the effective stress with a second order damage tensor representation is an approximation of the exact effective stress deduced from the general representation of the damage by the fourth order tensor $\underline{\boldsymbol{D}}$, as defined in the remark in the previous section. This effective stress is defined as previously, but, here, by the projection of the stress vector on the reference vector $\vec{\nu}$ in Fig. 1.3:

$$\nu_i \tilde{\sigma}_{ij} \tilde{n}_j \delta\tilde{S} = \nu_i \sigma_{ij} n_j \delta S \qquad (1.10)$$

or

$$\tilde{\sigma}_{ij}(I_{ijkl} - D_{ijkl})\nu_k n_l \delta S = \sigma_{kl}\nu_k n_l \delta S \qquad (1.11)$$

using the preceeding equation (1.5).

With the symmetry properties of the damage tensor D_{ijkl} given in Sect. 1.1.3, the effective stress is the symmetric tensor,

$$\tilde{\sigma}_{ij} = \sigma_{kl}\left(\boldsymbol{I} - \boldsymbol{D}\right)^{-1}_{klij}, \qquad (1.12)$$

which may also be written as

$$\tilde{\sigma}_{ij} = M_{ijkl}\sigma_{kl}, \quad \text{with} \quad M_{ijkl} = \left(\boldsymbol{I} - \boldsymbol{D}\right)^{-1}_{klij}. \qquad (1.13)$$

Back to a second order damage tensor, there are several possibilities to write a restriction of the fourth order formulation but only one fulfills the following conditions:

- Symmetry of the effective stress (note that $\tilde{\sigma}_{ij} = \sigma_{ik}(\boldsymbol{1} - \boldsymbol{D})^{-1}_{kj}$ is not symmetric!).
- An effective stress independent of the strain behavior and, in particular, independent from the Poisson ratio ($\tilde{\sigma}_{ij} = E_{ijkl}\tilde{E}^{-1}_{klrs}\sigma_{rs}$ may depend on the Poisson ratio).
- Compatibility with the thermodynamics framework: existence of strain potentials and principle of strain equivalence. The symmetrization $\tilde{\sigma}_{ij} = \frac{1}{2}\left[\sigma_{ik}(\boldsymbol{1}-\boldsymbol{D})^{-1}_{kj} + (\boldsymbol{1}-\boldsymbol{D})^{-1}_{ik}\sigma_{kj}\right]$ is not derived from a potential (J.P. Cordebois and F. Sidoroff 1979).
- Different effect of the damage on the hydrostatic behavior represented by the stress (σ_H) and on the deviatoric part (σ^D_{ij}) by means of a hydrostatic sensitivity parameter (η) that is easy to identify.

This restriction is represented by

$$\boxed{\tilde{\sigma}_{ij} = (H_{ik}\sigma^D_{kl}H_{lj})^D + \frac{\sigma_H}{1 - \eta D_H}\delta_{ij}, \quad \text{with } H_{ij} = (\boldsymbol{1}-\boldsymbol{D})^{-1/2}_{ij},} \qquad (1.14)$$

where $D_H = \frac{1}{3}D_{kk}$ is the hydrostatic damage and $\boldsymbol{H}$ is the effective damage tensor. The corresponding fourth order tensor M_{ijkl} is

$$M_{ijkl} = H_{ik}H_{lj} - \frac{1}{3}\left[H^2_{kl}\delta_{ij} + H^2_{ij}\delta_{kl}\right] + \frac{1}{9}H^2_{pp}\delta_{ij}\delta_{kl} + \frac{1}{3(1-\eta D_H)}\delta_{ij}\delta_{kl}. \qquad (1.15)$$

For the uniaxial case, the effective von Mises stress ($\tilde{\sigma}_{eq}$) differs from the effective tensile stress ($\tilde{\sigma}_1$) (assumed here in direction 1) as follows:

$$\begin{aligned}\tilde{\sigma}_{eq} &= \frac{2}{3}\frac{\sigma}{1-D_1} + \frac{1}{3}\frac{\sigma}{1-D_2} \\ \tilde{\sigma}_1 &= \frac{4}{9}\frac{\sigma}{1-D_1} + \frac{2}{9}\frac{\sigma}{1-D_2} + \frac{1}{3}\frac{\sigma}{1-\eta D_H},\end{aligned} \qquad (1.16)$$

where $D_3 = D_2$. The coefficient η does not depend very much on the materials as most often $\eta \approx 3$. The isotropic damage case corresponds to $D_{ij} = D\delta_{ij}$ and $\eta = 1$.

The effective stress taking into account the quasi-unilateral conditions of microdefects closure is formulated in Sect. 1.2.4.

1.1.5 Effects of Damage♂

Damage by the creation of free surfaces of discontinuities reduces the value of many properties:

- It decreases the elasticity modulus
- It decreases the yield stress before or after hardening
- It decreases the hardness
- It *increases* the creep strain rate
- It decreases the ultrasonic waves velocity
- It decreases the density
- It *increases* the electrical resistance

Some of these effects are used to evaluate the damage by inverse methods (see Sect. 1.3). Furthermore, the effects on mechanical strength and stiffness are different in tension and in compression due to microcracks opening under tension and their closure under compression.

1.2 Thermodynamics of Damage

The thermodynamics of irreversible processes allows for the modelling of different materials' behavior in three steps:

1. Definition of state variables, the actual value of each defining the present state of the corresponding mechanism involved
2. Definition of a state potential from which derive the state laws such as thermo-elasticity and the definition of the variables associated with the internal state variables
3. Definition of a dissipation potential from which derive the laws of evolution of the state variables associated with the dissipative mechanisms

These three steps offer several choices for the definitions, each chosen in accordance with experimental results and purpose of use. Then, the second principle of the thermodynamics must be checked for any evolution.

The two potential functions introduce parameters which depend on the material and the temperature. They must be identified from experiments in each case.

1.2.1 General Framework♂♂

The **state variables**, observable and internal, are chosen in accordance with the physical mechanisms of deformation and degradation of the material (Table 1.1).

Taking small deformations into consideration, the total strain is split into a thermo-elastic part ϵ^e and a plastic part ϵ^p:

$$\epsilon_{ij} = \epsilon_{ij}^e + \epsilon_{ij}^p . \tag{1.17}$$

1 Background on Continuum Damage Mechanics

Table 1.1. State and associated variables (J. Lemaitre 1978)

Mechanisms	Type	State variables		Associated variables
		Observable	Internal	
Thermoelasticity	Tensor	ϵ_{ij}		σ_{ij}
Entropy	Scalar	T		s
Plasticity	Tensor		$\epsilon^{\mathrm{p}}_{ij}$	$-\sigma_{ij}$
Isotropic hardening	Scalar		r	R
Kinematic hardening	Tensor		α_{ij}	X_{ij}
Damage	Scalar (isotropic)		D	$-Y$
	Tensor (anisotropic)		D_{ij}	$-Y_{ij}$

The Helmholtz specific free energy taken as the **state potential** of the material is a function of all the state variables. Written as $\psi(\epsilon^{\mathrm{e}}_{ij}, D \text{ or } D_{ij}, r, \alpha_{ij}, T)$, some qualitative experimental results on the possibility of couplings show that the state potential is the sum of thermo-elastic (ψ_{e}), plastic (ψ_{p}), and purely thermal (ψ_{T}) contributions. Here, it is more convenient to consider the potential as the Gibbs specific free enthalpy ($\psi^{\star}$) deduced from the Helmholtz free energy by a partial Legendre transform on the strain:

$$\psi^{\star} = \sup_{\epsilon}\left[\frac{1}{\rho}\sigma_{ij}\epsilon_{ij} - \psi\right] \tag{1.18}$$

or

$$\psi^{\star} = \sup_{\epsilon^{\mathrm{e}}}\left[\frac{1}{\rho}\sigma_{ij}\epsilon^{\mathrm{e}}_{ij} - \psi_{\mathrm{e}}\right] + \frac{1}{\rho}\sigma_{ij}\epsilon^{\mathrm{p}}_{ij} - \psi_{\mathrm{p}} - \psi_{\mathrm{T}}, \tag{1.19}$$

where ρ is the density and where ψ_{p} and ψ_{T} do not depend on the total strain.

It is finally expressed as

$$\psi^{\star} = \psi^{\star}_{\mathrm{e}} + \frac{1}{\rho}\sigma_{ij}\epsilon^{\mathrm{p}}_{ij} - \psi_{\mathrm{p}} - \psi_{\mathrm{T}}, \tag{1.20}$$

where

- The elastic contribution $\psi^{\star}_{\mathrm{e}}$ is affected by damage to model the experimentally-observed coupling between elasticity and damage through the effective stress concept $\tilde{\sigma}$ associated with the **principle of strain equivalence** (J. Lemaitre 1971). It states that the **strain** constitutive equations of a damaged material is derived from the same formalism as for a non-damaged material except that the stress is replaced by the effective stress.
- The part $\psi_{\mathrm{p}} = \frac{1}{\rho}\left(\int_0^r R dr + \frac{1}{3}C\alpha_{ij}\alpha_{ij}\right)$ is the contribution due to plastic hardening. When multiplied by ρ, it is the **energy stored** (w_{s}) in the RVE. The material parameter C accounts for the linear part of the kinematic hardening.
- The extra contribution ψ_{T} is a function of the temperature only. It partially defines the heat capacity of the material.

1.2 Thermodynamics of Damage

First, the state laws are derived from the state potential. They are the laws of thermoelasticity,

$$\epsilon_{ij} = \rho \frac{\partial \psi^\star}{\partial \sigma_{ij}} = \rho \frac{\partial \psi_e^\star}{\partial \sigma_{ij}} + \epsilon_{ij}^{\mathrm{p}} \longrightarrow \epsilon_{ij}^{\mathrm{e}} = \rho \frac{\partial \psi^\star}{\partial \sigma_{ij}}$$

$$s = \frac{\partial \psi^\star}{\partial T}.$$

(1.21)

The other derivatives define the associated variables as follows:

$$R = -\rho \frac{\partial \psi^\star}{\partial r},$$

$$X_{ij} = -\rho \frac{\partial \psi^\star}{\partial \alpha_{ij}},$$

$$-Y = -\rho \frac{\partial \psi^\star}{\partial D} \quad \text{or} \quad -Y_{ij} = -\rho \frac{\partial \psi^\star}{\partial D_{ij}}.$$

(1.22)

The **second principle** of thermodynamics, written as the Clausius–Duhem inequality is satisfied when the damage rate is positive:

$$\sigma_{ij} \dot{\epsilon}_{ij}^{\mathrm{p}} - \dot{w}_{\mathrm{s}} + Y_{ij} \dot{D}_{ij} - \frac{q_i T_{,i}}{T} \geq 0$$

(1.23)

This means the dissipation sum of the dissipation due to plastic power $(\sigma_{ij}\dot{\epsilon}_{ij}^{\mathrm{p}})$, minus the **stored energy density rate** $(\dot{w}_{\mathrm{s}} = R\dot{r} + X_{ij}\dot{\alpha}_{ij})$, plus the dissipation due to damage $(Y_{ij}\dot{D}_{ij})$, and plus the thermal energy ($\vec{q}$ is the thermal flux), is transformed into heat.

Finally, the kinetic laws governing the evolution of the internal variables are derived from a **dissipation potential** (F) which is a convex function of the associated variables to ensure fulfillment of the second principle:

$$F = F(\sigma, R, X_{ij}, Y \text{ or } Y_{ij}; D \text{ or } D_{ij}, T).$$

(1.24)

The state variables representing the temperature and the damage may act, but only as parameters. Introducing the plastic criterion function (f), the nonlinear kinematic hardening term (F_X) and the damage potential (F_D), according to qualitative experiments on the possibilities of coupling, F is the sum $F = f + F_\mathrm{X} + F_\mathrm{D}$. The evolution laws are formally written as

$$\dot{\epsilon}_{ij}^{\mathrm{p}} = -\dot{\lambda} \frac{\partial F}{\partial (-\sigma_{ij})} = \dot{\lambda} \frac{\partial F}{\partial \sigma_{ij}},$$

$$\dot{r} = -\dot{\lambda} \frac{\partial F}{\partial R},$$

$$\dot{\alpha}_{ij} = -\dot{\lambda} \frac{\partial F}{\partial X_{ij}},$$

$$\dot{D} = -\dot{\lambda} \frac{\partial F}{\partial (-Y)} = \dot{\lambda} \frac{\partial F}{\partial Y} \quad \text{or} \quad \dot{D}_{ij} = -\dot{\lambda} \frac{\partial F}{\partial (-Y_{ij})} = \dot{\lambda} \frac{\partial F}{\partial Y_{ij}}.$$

(1.25)

This is the normality rule of generalized standard materials.

For phenomena which do not depend explicitly on time, such as plasticity, the potential F is not differentiable and $\dot{\lambda}$ is the plastic multiplier, calculated from the consistency condition, $f = 0$, $\dot{f} = 0$. The first condition, $f = 0$, means that the state of stress is on the actual yield condition; the second, $\dot{f} = 0$, means that an increase of the state of stress induces an increase of the yield stress. Elastic unloading occurs when $f < 0$ or $\dot{f} < 0$, the internal variables then keeping a constant value ($\dot{\epsilon}^{\rm p}_{ij} = \dot{r} = \dot{\alpha}_{ij} = \dot{D}_{ij} = 0$). The complete loading or unloading conditions are called the Kuhn–Tucker conditions: $\dot{\lambda} \geq 0$, $f \leq 0$, $\dot{\lambda} f = 0$.

For phenomena like visco-plasticity which depends explicitly upon time, $\dot{\lambda}$ is the viscosity function $\dot{\lambda}(f)$.

To complete the formal description of elasto-(visco-)plasticity coupled with damage, the **accumulated plastic strain rate** ($\dot{p}$) is defined in accordance with the yield criterion considered. For the von Mises criterion it is

$$\dot{p} = \sqrt{\frac{2}{3}\dot{\epsilon}^{\rm p}_{ij}\dot{\epsilon}^{\rm p}_{ij}}\,. \tag{1.26}$$

In 1D (uniaxial tension-compression case), it simply means $\dot{p} = |\dot{\epsilon}_{\rm p}|$.

1.2.2 State Potential for Isotropic Damage

According to the principle of strain equivalence, the strain potential for linear isotropic thermo-elasticity and isotropic damage (D) is

$$\rho \psi_{\rm e}^{\star} = \frac{1+\nu}{2E}\frac{\sigma_{ij}\sigma_{ij}}{1-D} - \frac{\nu}{2E}\frac{\sigma_{kk}^2}{1-D} + \alpha(T - T_{\rm ref})\sigma_{kk}\,, \tag{1.27}$$

where E is the Young's modulus, ν the Poisson ratio, α the thermal expansion coefficient, and $T_{\rm ref}$ a reference temperature. The thermo-elasticity law is derived from this potential as

$$\epsilon^{\rm e}_{ij} = \rho\frac{\partial \psi_{\rm e}^{\star}}{\partial \sigma_{ij}} = \frac{1+\nu}{E}\tilde{\sigma}_{ij} - \frac{\nu}{E}\tilde{\sigma}_{kk}\delta_{ij} + \alpha(T - T_{\rm ref})\delta_{ij}\,, \tag{1.28}$$

where the effective stress is $\tilde{\sigma}_{ij} = \dfrac{\sigma_{ij}}{1-D}$.

The **energy density release rate** (Y), the associated variable with the damage variable, is also derived from the state potential and may be written as (J.L. Chaboche 1976)

$$Y = \rho\frac{\partial \psi^{\star}}{\partial D} = \frac{\tilde{\sigma}_{\rm eq}^2 R_\nu}{2E}\,, \tag{1.29}$$

introducing the triaxiality function

$$R_\nu = \frac{2}{3}(1+\nu) + 3(1-2\nu)\left(\frac{\sigma_\mathrm{H}}{\sigma_\mathrm{eq}}\right)^2 \quad (1.30)$$

where $\sigma_\mathrm{H} = \sigma_{kk}/3$ is the **hydrostatic stress**, $\sigma_\mathrm{eq} = \sqrt{\frac{3}{2}\sigma^D_{ij}\sigma^D_{ij}}$ the **von Mises equivalent stress**, and $\sigma^D_{ij} = \sigma_{ij} - \sigma_\mathrm{H}\delta_{ij}$ the stress deviator. The ratio $\sigma_\mathrm{H}/\sigma_\mathrm{eq}$ is the **stress triaxiality** denoted by T_X.

If w_e is the elastic strain energy density defined by $\mathrm{d}w_\mathrm{e} = \sigma_{ij}\mathrm{d}\epsilon^\mathrm{e}_{ij}$, then

$$Y = \frac{w_\mathrm{e}}{1-D} \quad \text{and} \quad Y = \frac{1}{2}\frac{\mathrm{d}w_\mathrm{e}}{\mathrm{d}D}\bigg|_{\substack{\sigma=const \\ T=const}} . \quad (1.31)$$

At this level, an interesting concept is the **damage equivalent stress** ($\sigma^\star$) (J. Lemaitre 1981) defined as the uniaxial stress which gives the same amount of elastic strain energy (w_e) as a multiaxial state of stress. For linear isothermal isotropic elasticity, $w_\mathrm{e} = \rho\psi^\star_\mathrm{e}$ or in terms of the stress deviator ($\boldsymbol{\sigma}^D$) for shear energy and the hydrostatic stress (σ_H) for the hydrostatic energy,

$$\rho\psi^\star_\mathrm{e} = \frac{1+\nu}{2E}\frac{\sigma^D_{ij}\sigma^D_{ij}}{1-D} + \frac{3(1-2\nu)}{2E}\frac{\sigma^2_\mathrm{H}}{1-D} = \frac{\sigma^2_\mathrm{eq}R_\nu}{2E(1-D)} . \quad (1.32)$$

Writing the equality between the uniaxial case, for which $R_\nu = 1$, and the multiaxial case,

$$\frac{\sigma^{\star 2}}{2E(1-D)} = \frac{\sigma^2_\mathrm{eq}R_\nu}{2E(1-D)} , \quad (1.33)$$

yields

$$\boxed{\sigma^\star = \sigma_\mathrm{eq} R_\nu^{1/2} .} \quad (1.34)$$

1.2.3 State Potential for Anisotropic Damage♂♂

In accordance with the principle of strain equivalence, with the definition of the effective stress (1.14) of Sect. 1.1.4 and with P. Ladevèze's (1983) framework for the description of an anisotropic state of damage, the state potential $\rho\psi^\star_\mathrm{e}$ represented by the tensor $\boldsymbol{D}$ is

$$\rho\psi^\star_\mathrm{e} = \frac{1+\nu}{2E}H_{ij}\sigma^D_{jk}H_{kl}\sigma^D_{li} + \frac{3(1-2\nu)}{2E}\frac{\sigma^2_\mathrm{H}}{1-\eta D_\mathrm{H}} + \alpha(T-T_\mathrm{ref})\sigma_{kk} , \quad (1.35)$$

where

$$H_{ij} = (\mathbf{1}-\boldsymbol{D})^{-1/2}_{ij} \quad (1.36)$$

The law of elasticity is derived from the Gibbs potential and as expressed in (1.28) still introduces the symmetric effective stress tensor defined in Sect. 1.1.4:

$$\tilde{\sigma}_{ij} = (H_{ik}\sigma^{D}_{kl}H_{lj})^{D} + \frac{\sigma_{H}}{1-\eta D_{H}}\delta_{ij}. \quad (1.37)$$

The associated variable with $\boldsymbol{D}$ is the energy density release rate tensor, $\boldsymbol{Y}$, which is also derived from the Gibbs energy, $Y_{ij} = \rho\partial\psi^{\star}/\partial D_{ij}$. This derivative needs some care due to possible variation of the damage principal directions. Using $H_{ik}\dot{H}_{kj} + \dot{H}_{ik}H_{kj} = H^{2}_{ik}\dot{D}_{kl}H^{2}_{lj}$ and

$$A_{ijkl}\dot{H}_{kl} = H^{2}_{ip}\dot{D}_{pq}H^{2}_{qj}, \quad \text{with} \quad A_{ijkl} = \frac{1}{2}(H_{ik}\delta_{jl}+H_{jl}\delta_{ik}+H_{il}\delta_{jk}+H_{jk}\delta_{il}), \quad (1.38)$$

where the symmetrization is due to $H_{kl} = H_{lk}$, one gets:

$$Y_{ij} = \frac{1+\nu}{E}\sigma^{D}_{kp}H_{pq}\sigma^{D}_{ql}A^{-1}_{klmn}H^{2}_{mi}H^{2}_{jn} + \frac{\eta(1-2\nu)}{2E}\frac{\sigma^{2}_{H}}{(1-\eta D_{H})^{2}}\delta_{ij}. \quad (1.39)$$

It is possible to verify that the dissipation $Y_{ij}\dot{D}_{ij}$ is positive or zero at least for practical cases.

By chance, the law of damage evolution in Sect. 1.4 will not be a function of $\boldsymbol{Y}$ but it will be a function of the **effective elastic energy density**, the scalar $\overline{Y} = \int \tilde{\sigma}_{ij}d\epsilon^{e}_{ij}$, as in the isotropic case. $\overline{Y}$ can be written as a function of the effective stress,

$$\boxed{\overline{Y} = \frac{1}{2}E_{ijkl}\epsilon^{e}_{kl}\epsilon^{e}_{ij} = \frac{1}{2}\tilde{\sigma}_{ij}\epsilon^{e}_{ij} = \frac{\tilde{\sigma}^{2}_{eq}\tilde{R}_{\nu}}{2E},} \quad (1.40)$$

with the effective triaxiality function

$$\boxed{\tilde{R}_{\nu} = \frac{2}{3}(1+\nu) + 3(1-2\nu)\left(\frac{\tilde{\sigma}_{H}}{\tilde{\sigma}_{eq}}\right)^{2},} \quad (1.41)$$

where

$$\boxed{\tilde{\sigma}_{eq} = (\boldsymbol{H}\,\boldsymbol{\sigma}^{D}\boldsymbol{H})_{eq} = \left[\frac{3}{2}(\boldsymbol{H}\,\boldsymbol{\sigma}^{D}\boldsymbol{H})^{D}_{ij}(\boldsymbol{H}\,\boldsymbol{\sigma}^{D}\boldsymbol{H})^{D}_{ij}\right]^{1/2}} \quad (1.42)$$

$$\boxed{\tilde{\sigma}_{H} = \frac{\sigma_{H}}{1-\eta D_{H}}.}$$

1.2.4 Quasi-Unilateral Conditions of Microdefects Closure◊◊◊

For most materials under certain conditions of loading, the microdefects may close during compression. This is more often the case for very brittle materials. The phenomenon of partial closure of microcracks increases the area which effectively carries the load in compression and the stiffness may then be partially or fully recovered. For most materials, considering $\tilde{E}^+$ and $\tilde{E}^-$ as the effective elasticity modulus in tension and in compression, the ratio $(E - \tilde{E}^-)/(E - \tilde{E}^+)$ has a constant value close to 0.2 when the damage models of Sects. 1.2.2 and 1.2.3 give a value equal to 1. This phenomenon is referred to as quasi-unilateral conditions, the term "quasi" standing for "initial elasticity partially recovered in compression."

From a theoretical point of view, the damage state due to the presence of microdefects is represented by the internal variable D or $\boldsymbol{D}$. Being a thermodynamical state, it is independent of both the intensity and the sign of the loading at constant internal variables (metals, concretes, composites, polymers ... do not physically recover from their wounds!). This means that

- **No extra damage variable has to be introduced** to model the microdefects closure effect
- D or $\boldsymbol{D}$ act differently in tension and in compression

For unidimensional states of stress, a solution is to define an effective stress such as

- $\dfrac{\sigma}{1-D}$ in tension with D the relative reduction of the resisting area in tension identified by $D = 1 - \tilde{E}^+/E$ and
- $\dfrac{\sigma}{1-hD}$ in compression with hD, the relative reduction of the resisting area in compression identified by $hD = 1 - \tilde{E}^-/E$ or $h = (E - \tilde{E}^-)/(E - \tilde{E}^+)$,

where h is a **microdefects closure parameter** that is material-dependent (but most often $h \approx 0.2$).

For 3D states of stress, the difficulty is to recognize what is compression and what is tension! The theoretical background needed to handle such a problem mostly concerns the definition and use of positive and negative parts of tensors built with their three principal values.

The positive part $\langle \boldsymbol{s} \rangle^+$, also written as $\langle \boldsymbol{s} \rangle^+_{ij}$ (with negative part $\langle \boldsymbol{s} \rangle^-$ or $\langle \boldsymbol{s} \rangle^-_{ij}$ respectively), of a symmetric tensor $\boldsymbol{s}$ is built from its positive (negative) eigenvalues s_K and the corresponding normalized eigenvectors $\vec{q}^K$, as in

$$\langle \boldsymbol{s} \rangle^+_{ij} = \sum_{K=1}^{3} \langle s_K \rangle q_i^K q_j^K \quad \text{and} \quad \langle \boldsymbol{s} \rangle^-_{ij} = s_{ij} - \langle \boldsymbol{s} \rangle^+_{ij}, \tag{1.43}$$

where $\langle \cdot \rangle$ stands for the positive part of a scalar: $\langle x \rangle = \begin{cases} x & \text{if } x \geq 0, \\ 0 & \text{if } x < 0. \end{cases}$

Using the property $\langle \boldsymbol{s} \rangle^+_{ij} \langle \boldsymbol{s} \rangle^-_{ij} = 0$, it is easy to demonstrate that

$$\begin{aligned} \langle s_{kk} \rangle^2 &= \langle s_{kk} \rangle^2 + \langle -s_{kk} \rangle^2 \\ s_{ij} s_{ij} &= \langle \boldsymbol{s} \rangle^+_{ij} \langle \boldsymbol{s} \rangle^+_{ij} + \langle \boldsymbol{s} \rangle^-_{ij} \langle \boldsymbol{s} \rangle^-_{ij}. \end{aligned} \tag{1.44}$$

The products $\langle s_{kk}\rangle^2$, $\langle \boldsymbol{s}\rangle^+_{ij}\langle \boldsymbol{s}\rangle^+_{ij}$, and $\langle \boldsymbol{s}\rangle^-_{ij}\langle \boldsymbol{s}\rangle^-_{ij}$ may be continuously differentiated as follows:

$$\frac{\partial}{\partial s_{ij}}\left(\frac{1}{2}\langle s_{kk}\rangle^2\right) = \langle s_{kk}\rangle\delta_{ij},$$

$$\frac{\partial}{\partial s_{ij}}\left(\frac{1}{2}\langle \boldsymbol{s}\rangle^+_{rs}\langle \boldsymbol{s}\rangle^+_{rs}\right) = \langle \boldsymbol{s}\rangle^+_{ij}, \quad (1.45)$$

$$\frac{\partial}{\partial s_{ij}}\left(\frac{1}{2}\langle \boldsymbol{s}\rangle^-_{rs}\langle \boldsymbol{s}\rangle^-_{rs}\right) = \langle \boldsymbol{s}\rangle^-_{ij}.$$

1.2.4.1 Quasi-Unilateral Conditions for Isotropic Damage
(P. Ladevèze and J. Lemaitre 1984)

The microdefects closure parameter h operates by $(1-hD)^{-1}$ on the negative part of the principal stresses and by $(1-D)^{-1}$ on the positive principal stresses. For isotropic damage, the isothermal state potential $\rho\psi_e^\star$ is

$$\rho\psi_e^\star = \frac{1+\nu}{2E}\left[\frac{\langle\boldsymbol{\sigma}\rangle^+_{ij}\langle\boldsymbol{\sigma}\rangle^+_{ij}}{1-D} + \frac{\langle\boldsymbol{\sigma}\rangle^-_{ij}\langle\boldsymbol{\sigma}\rangle^-_{ij}}{1-hD}\right] - \frac{\nu}{2E}\left[\frac{\langle\sigma_{kk}\rangle^2}{1-D} + \frac{\langle-\sigma_{kk}\rangle^2}{1-hD}\right]. \quad (1.46)$$

The elasticity law is still derived from $\epsilon^e_{ij} = \rho\,\partial\psi_e^\star/\partial\sigma_{ij}$,

$$\boxed{\epsilon^e_{ij} = \frac{1+\nu}{E}\left[\frac{\langle\boldsymbol{\sigma}\rangle^+_{ij}}{1-D} + \frac{\langle\boldsymbol{\sigma}\rangle^-_{ij}}{1-hD}\right] - \frac{\nu}{E}\left[\frac{\langle\sigma_{kk}\rangle}{1-D} - \frac{\langle-\sigma_{kk}\rangle}{1-hD}\right]\delta_{ij}.} \quad (1.47)$$

It defines an effective stress as

$$\boxed{\begin{aligned}\tilde{\sigma}_{ij} &= \frac{\langle\boldsymbol{\sigma}\rangle^+_{ij}}{1-D} + \frac{\langle\boldsymbol{\sigma}\rangle^-_{ij}}{1-hD} \\ &\quad + \frac{\nu}{1-2\nu}\left(\frac{\delta_{kl}\langle\boldsymbol{\sigma}\rangle^+_{kl} - \langle\sigma_{kk}\rangle}{1-D} + \frac{\delta_{kl}\langle\boldsymbol{\sigma}\rangle^-_{kl} + \langle-\sigma_{kk}\rangle}{1-hD}\right)\delta_{ij}.\end{aligned}} \quad (1.48)$$

The energy density release rate $Y = \rho\partial\psi^\star/\partial D$ now strongly depends on h as

$$\boxed{Y = \frac{1+\nu}{2E}\left[\frac{\langle\boldsymbol{\sigma}\rangle^+_{ij}\langle\boldsymbol{\sigma}\rangle^+_{ij}}{(1-D)^2} + h\frac{\langle\boldsymbol{\sigma}\rangle^-_{ij}\langle\boldsymbol{\sigma}\rangle^-_{ij}}{(1-hD)^2}\right] - \frac{\nu}{2E}\left[\frac{\langle\sigma_{kk}\rangle^2}{(1-D)^2} + h\frac{\langle-\sigma_{kk}\rangle^2}{(1-hD)^2}\right].}$$
$$(1.49)$$

1.2.4.2 Quasi-Unilateral Conditions for Anisotropic Damage
(P. Ladevèze 1983, R. Desmorat 1999)

As for the isotropic case, the scalar parameter h_a, different from h in general, is introduced in order to model the different effects of $\boldsymbol{D}$ on tension and compression. We define

$$H^{\text{p}}_{ij} = (\mathbf{1} - \boldsymbol{D})^{-1/2}_{ij}, \qquad H^{\text{n}}_{ij} = (\mathbf{1} - h_{\text{a}}\boldsymbol{D})^{-1/2}_{ij}, \qquad (1.50)$$

where p stands for *positive* and n for *negative*, as it will be exhibited by their use.

The isothermal state potential that can model quasi-unilateral anisotropic damage is

$$\rho\psi^{\star}_{\text{e}} = \frac{1+\nu}{2E}\left(H^{\text{p}}_{ij}\sigma^{\text{D}}_{+jk}H^{\text{p}}_{kl}\sigma^{\text{D}}_{+li} + H^{\text{n}}_{ij}\sigma^{\text{D}}_{+jk}H^{\text{n}}_{kl}\sigma^{\text{D}}_{+li}\right)$$
$$+ \frac{3(1-2\nu)}{2E}\left(\frac{\langle\sigma_{\text{H}}\rangle^2}{1-\eta D_{\text{H}}} + \frac{\langle-\sigma_{\text{H}}\rangle^2}{1-h_{\text{a}}\eta D_{\text{H}}}\right), \qquad (1.51)$$

where $\boldsymbol{\sigma}^{\text{D}}_{+}$ is built with the eigenvalues λ_K and the corresponding eigenvectors $(\vec{T}^K)$ of the non-symmetric matrix $(\boldsymbol{H}^{\text{p}}\boldsymbol{\sigma}^{\text{D}})$ in such a way that $(\boldsymbol{H}^{\text{p}}\boldsymbol{\sigma}^{\text{D}}_{+}\boldsymbol{H}^{\text{p}}\boldsymbol{\sigma}^{\text{D}}_{+})$ may be continuously differentiated. This means

$$d\left(\frac{1}{2}H^{\text{p}}_{ij}\sigma^{\text{D}}_{+jk}H^{\text{p}}_{kl}\sigma^{\text{D}}_{+li}\right) = H^{\text{p}}_{ij}\sigma^{\text{D}}_{+jk}H^{\text{p}}_{kl}d\sigma^{\text{D}}_{li} + \sigma^{\text{D}}_{+li}H^{\text{p}}_{ij}\sigma^{\text{D}}_{+jk}dH^{\text{p}}_{kl}, \qquad (1.52)$$

where $\vec{T}^K$ and λ_K are given by the eigenvalue problem

$$\sigma^{\text{D}}_{ij}T_j = \lambda\left(\boldsymbol{H}^{\text{p}}\right)^{-1}_{ij}T_j \quad \text{(without summation)} \qquad (1.53)$$

and the normalization $T^I_i(\boldsymbol{H}^{\text{p}})^{-1}_{ij}T^J_j = \delta^{IJ}$ is made. Be careful: the scalar product $\vec{T}^I \cdot \vec{T}^J \neq \delta_{IJ}$ as $(\boldsymbol{H}^{\text{p}}\boldsymbol{\sigma}^{\text{D}})$ is non-symmetric. The eigenvalues λ_K are real because $\boldsymbol{\sigma}^{\text{D}}$ is symmetric and $(\boldsymbol{H}^{\text{p}})^{-1}$ is positive-defined and symmetric. The deviatoric stress tensor may then be rewritten as

$$\sigma^{\text{D}}_{ij} = \sum_{K=1}^{3}\lambda_K\left[(\boldsymbol{H}^{\text{p}})^{-1}_{ik}T^K_k\right]\left[(\boldsymbol{H}^{\text{p}})^{-1}_{jl}T^K_l\right] \qquad (1.54)$$

and the special positive part $\boldsymbol{\sigma}^{\text{D}}_{+}$ is therefore defined as

$$\sigma^{\text{D}}_{+ij} = \sum_{K=1}^{3}\langle\lambda_K\rangle\left[(\boldsymbol{H}^{\text{p}})^{-1}_{ik}T^K_k\right]\left[(\boldsymbol{H}^{\text{p}})^{-1}_{jl}T^K_l\right]. \qquad (1.55)$$

The same procedure is followed for the negative part $\boldsymbol{\sigma}^{\text{D}}_{-}$ with $\boldsymbol{H}^{\text{n}}$ and with the eigenvectors and the negative eigenvalues of $(\boldsymbol{H}^{\text{n}}\boldsymbol{\sigma}^{\text{D}})$.

The elasticity law is derived from the state potential,

$$\epsilon^{\text{e}}_{ij} = \rho\frac{\partial\psi^{\star}_{\text{e}}}{\partial\sigma_{ij}}$$
$$= \frac{1+\nu}{E}\left[(H^{\text{p}}_{ik}\sigma^{\text{D}}_{+kl}H^{\text{p}}_{lj})^{\text{D}} + (H^{\text{n}}_{ik}\sigma^{\text{D}}_{-kl}H^{\text{n}}_{lj})^{\text{D}}\right]$$
$$+ \frac{1-2\nu}{E}\left[\frac{\langle\sigma_{\text{H}}\rangle}{1-\eta D_{\text{H}}} - \frac{\langle-\sigma_{\text{H}}\rangle}{1-\eta h_{\text{a}} D_{\text{H}}}\right]\delta_{ij} \qquad (1.56)$$
$$\epsilon^{\text{e}}_{ij} = \frac{1+\nu}{E}\tilde{\sigma}_{ij} - \frac{\nu}{E}\tilde{\sigma}_{kk}\delta_{ij},$$

and defines the effective stress,

$$\tilde{\sigma}_{ij} = (H^{\mathrm{p}}_{ik}\sigma^{\mathrm{D}}_{+kl}H^{\mathrm{p}}_{lj})^{\mathrm{D}} + (H^{\mathrm{n}}_{ik}\sigma^{\mathrm{D}}_{-kl}H^{\mathrm{n}}_{lj})^{\mathrm{D}} + \left[\frac{\langle\sigma_{\mathrm{H}}\rangle}{1-\eta D_{\mathrm{H}}} - \frac{\langle-\sigma_{\mathrm{H}}\rangle}{1-\eta h_{\mathrm{a}} D_{\mathrm{H}}}\right]\delta_{ij}, \tag{1.57}$$

which takes into account the quasi-unilateral effect. The effective energy density is also a function of the microdefects closure parameter. It is expressed as

$$\overline{Y} = \frac{1+\nu}{2E}\mathrm{tr}\left[\left(\boldsymbol{H}^{\mathrm{p}}\boldsymbol{\sigma}^{\mathrm{D}}_{+}\boldsymbol{H}^{\mathrm{p}}\right)^2 + h_{\mathrm{a}}\left(\boldsymbol{H}^{\mathrm{n}}\boldsymbol{\sigma}^{\mathrm{D}}_{-}\boldsymbol{H}^{\mathrm{n}}\right)^2\right] \\ + \frac{3(1-2\nu)}{2E}\left[\frac{\langle\sigma_{\mathrm{H}}\rangle^2}{(1-\eta D_{\mathrm{H}})^2} + h_{\mathrm{a}}\frac{\langle-\sigma_{\mathrm{H}}\rangle^2}{(1-\eta h_{\mathrm{a}} D_{\mathrm{H}})^2}\right]. \tag{1.58}$$

In order to keep the possibility to differentiate the state potential, the positive and negative parts are taken with respect to the deviatoric stress tensor. This means that the parameter h_{a} affects both tension and compression ($\boldsymbol{\sigma}^{\mathrm{D}}$ has positive and negative components in any uniaxial case), therefore the model is quite complex even for the tensile test. Furthermore, the value of the microdefects closure parameter depends on the model itself: for a given material, h for the isotropic damage model is different from h_{a} identified for the anisotropic damage model. The good thing is that the common value for many materials, $h \approx 0.2$, for the isotropic damage model corresponds to $h_{\mathrm{a}} \approx 0$ for the anisotropic damage model.

Finally, the law of elasticity may be written as ($h_{\mathrm{a}} = 0$)

$$\boxed{\begin{aligned}\epsilon^{\mathrm{e}}_{ij} &= \frac{1+\nu}{E}\left[(H^{\mathrm{p}}_{ik}\sigma^{\mathrm{D}}_{+kl}H^{\mathrm{p}}_{lj})^{\mathrm{D}} + \left(\langle\boldsymbol{\sigma}^{\mathrm{D}}\rangle_{-}\right)^{\mathrm{D}}_{ij}\right] \\ &+ \frac{1-2\nu}{E}\left[\frac{\langle\sigma_{\mathrm{H}}\rangle}{1-\eta D_{\mathrm{H}}} - \langle-\sigma_{\mathrm{H}}\rangle\right]\delta_{ij}\end{aligned}} \tag{1.59}$$

and the effective energy density is

$$\boxed{\overline{Y} = \frac{1+\nu}{2E}\mathrm{tr}\left(\boldsymbol{H}^{\mathrm{p}}\boldsymbol{\sigma}^{\mathrm{D}}_{+}\boldsymbol{H}^{\mathrm{p}}\right)^2 + \frac{3(1-2\nu)}{2E}\frac{\langle\sigma_{\mathrm{H}}\rangle^2}{(1-\eta D_{\mathrm{H}})^2}.} \tag{1.60}$$

1.3 Measurement of Damage

The direct measurement of damage as the surface density of microdefects is difficult to perform and is used only in laboratories well equipped for micrography from both the human and microscopy points of view. It is easier to take advantage of the coupling between damage and elasticity (or plasticity) to evaluate the damage by inverse methods.

1.3.1 Isotropic Elasticity Change$^{\male}$

In the case of isotropic damage, the uniaxial law of elasticity in tension coupled with damage reduces to

$$\epsilon_e = \frac{\sigma}{E(1-D)} = \frac{\sigma}{\tilde{E}}, \qquad (1.61)$$

where E is the Young's modulus of undamaged elasticity and $\tilde{E}$ is the actual modulus of damaged elasticity. The damage is then expressed as the loss of stiffness,

$$\boxed{D = 1 - \frac{\tilde{E}}{E}.} \qquad (1.62)$$

To evaluate the damage as a function of the accumulated plastic strain, $p = \int_0^t |\dot{\epsilon}_p| dt$, it is advised to perform a very low cycle test at a constant strain range on a tension-compression specimen (rupture for 10 to 100 cycles) and to measure the elastic strain by means of small-strain gauges during unloadings (Fig. 1.4, J. Dufailly 1976). If $\Delta\epsilon_p$ is the plastic strain range over a cycle, the accumulated plastic strain for N cycles is simply $p \approx 2\Delta\epsilon_p N$.

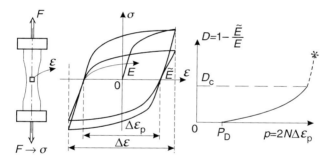

Fig. 1.4. Measurement of damage by means of elasticity change

The evaluation of the elasticity modulus needs much care because an absolute accuracy of about 10^{-6} is needed on the measurement of the strain. Furthermore the damage is almost always very localized on a small volume and it looses its continuous signification as soon as a mesocrack appears. In tension, this is before the whole fracture, when the elasticity modulus decreases rapidly, an instant which corresponds to a quick change of the curvature in the graph of $D = 1 - \tilde{E}/E$ as a function of the accumulated plastic strain p (Fig. 1.4).

1.3.2 Isotropic Elasticity Change by Ultrasonic Waves$^{\male\male}$

A wave's speed is related to the density ρ and to the elastic properties of the considered medium. Considering the longitudinal waves in a linear, isotropic, elastic, long cylindrical specimen with Young's modulus E and a Poisson ratio ν, the wave speed is determined by

$$v = \sqrt{\frac{E}{\rho} \frac{1-\nu}{(1+\nu)(1-2\nu)}}. \tag{1.63}$$

If the material is damaged mainly by microcracks, the density remains almost unchanged and if they are randomly oriented to make the damage isotropic, the Poisson ratio also remains constant. The wave speed becomes

$$\tilde{v} = \sqrt{\frac{\tilde{E}}{\rho} \frac{1-\nu}{(1+\nu)(1-2\nu)}}, \tag{1.64}$$

from which the damage may be determined as

$$D = 1 - \frac{\tilde{E}}{E} \quad \longrightarrow \quad \boxed{D = 1 - \frac{\tilde{v}^2}{v^2}}. \tag{1.65}$$

This method requires the measurement of the propagation time of the waves. The accuracy is higher when the distance covered by the waves is large. Unfortunately, due to the localization of the damage, the distance available for an uniform field of damage is always small, often too small for metals, but the method works nicely for concrete with a frequency of 0.1 to 1 MHz (Y. Berthaud 1988).

1.3.3 Anisotropic Elasticity Change♂♂♂

When anisotropic damage is considered, the damage tensor has six components or three if the principal orthotropic frame is known, through, for example, an independent measurement of the plastic strains. Considering the case

$$\boldsymbol{D} = \begin{bmatrix} D_1 & 0 & 0 \\ 0 & D_2 & 0 \\ 0 & 0 & D_3 \end{bmatrix}, \quad \boldsymbol{H} = \begin{bmatrix} H_1 = \frac{1}{\sqrt{1-D_1}} & 0 & 0 \\ 0 & H_2 = \frac{1}{\sqrt{1-D_2}} & 0 \\ 0 & 0 & H_3 = \frac{1}{\sqrt{1-D_3}} \end{bmatrix}, \tag{1.66}$$

with elastic strains corresponding to an uniaxial tension σ_1 in the direction 1 of a previously damaged material, the elasticity law (1.28) with the effective stress (1.37) of Sect. 1.2.3 can be expanded as

$$\begin{bmatrix} \epsilon_1^e & 0 & 0 \\ 0 & \epsilon_2^e & 0 \\ 0 & 0 & \epsilon_3^e \end{bmatrix} = \frac{1+\nu}{E} \left(\begin{bmatrix} H_1 & 0 & 0 \\ 0 & H_2 & 0 \\ 0 & 0 & H_3 \end{bmatrix} \begin{bmatrix} \frac{2\sigma_1}{3} & 0 & 0 \\ 0 & \frac{-\sigma_1}{3} & 0 \\ 0 & 0 & \frac{-\sigma_1}{3} \end{bmatrix} \begin{bmatrix} H_1 & 0 & 0 \\ 0 & H_2 & 0 \\ 0 & 0 & H_3 \end{bmatrix} \right)^D$$
$$+ \frac{1-2\nu}{3E} \frac{\sigma_1}{1-\eta D_\mathrm{H}} \begin{bmatrix} 1 & 0 & 0 \\ 0 & 1 & 0 \\ 0 & 0 & 1 \end{bmatrix}. \tag{1.67}$$

The damaged elasticity modulus in the direction 1 and the associated contraction ratios also measured by strain gauges are defined by

$$\tilde{E}_1 = \frac{\sigma_1}{\epsilon_1^e}, \qquad \tilde{\nu}_{12} = -\frac{\epsilon_2^e}{\epsilon_1^e}, \quad \text{and} \quad \tilde{\nu}_{13} = -\frac{\epsilon_3^e}{\epsilon_1^e}. \qquad (1.68)$$

This leads to three expressions for the elastic properties as functions of the damage:

$$\frac{E}{\tilde{E}_1} = \frac{1+\nu}{9}\left(\frac{4}{1-D_1} + \frac{1}{1-D_2} + \frac{1}{1-D_3}\right) + \frac{1-2\nu}{3(1-\eta D_\mathrm{H})}, \qquad (1.69)$$

$$\tilde{\nu}_{12}\frac{E}{\tilde{E}_1} = \frac{1+\nu}{9}\left(\frac{2}{1-D_1} + \frac{2}{1-D_2} - \frac{1}{1-D_3}\right) - \frac{1-2\nu}{3(1-\eta D_\mathrm{H})}, \qquad (1.70)$$

$$\tilde{\nu}_{13}\frac{E}{\tilde{E}_1} = \frac{1+\nu}{9}\left(\frac{2}{1-D_1} - \frac{1}{1-D_2} + \frac{2}{1-D_3}\right) - \frac{1-2\nu}{3(1-\eta D_\mathrm{H})}. \qquad (1.71)$$

For a tensile loading in direction 2 (or 3), the equations remain the same with subscripts 1 and 2 (or 1 and 3) inverted. Then uniaxial tensions applied in the directions 2 and 3 give 6 additional equations to determine the 3 components of the damage (D_1, D_2, D_3), plus the coefficient η if it is unknown, with a verification of the symmetries:

$$\frac{\tilde{\nu}_{ij}}{\tilde{E}_i} = \frac{\tilde{\nu}_{ji}}{\tilde{E}_j} \quad \text{(no summation)}. \qquad (1.72)$$

If a uniformly damaged cube is available to machine 3 tensile specimens in the 3 principal directions, 3 elasticity tests (with measurement of the elasticity modulus and contraction ratios) allow for the calculation of the damage components,

$$\boxed{\begin{aligned}
D_1 &= 1 - \frac{\tilde{E}_1}{E}(1+\nu)\left[2 + \tilde{\nu}_{12} - \frac{\tilde{E}_1}{\tilde{E}_2}\right]^{-1}, \\
D_2 &= 1 - \frac{\tilde{E}_2}{E}(1+\nu)\left[2 - (1-\tilde{\nu}_{12})\frac{\tilde{E}_2}{\tilde{E}_1}\right]^{-1}, \\
D_3 &= 1 - \frac{\tilde{E}_3}{E}(1+\nu)\left[2 + \tilde{\nu}_{32} - \frac{\tilde{E}_3}{\tilde{E}_2}\right]^{-1}, \\
\text{and} \quad \eta D_\mathrm{H} &= 1 - \frac{\tilde{E}_1}{E}\frac{1-2\nu}{1-2\tilde{\nu}_{12}},
\end{aligned}} \qquad (1.73)$$

with $D_\mathrm{H} = \frac{1}{3}(D_1 + D_2 + D_3)$.

Due to the required size of the specimens, this procedure is not possible most of the time. Nevertheless, a cube smaller than the classical tension specimens may be analyzed by ultrasonic waves in the three principal directions. In this case, the transversal wave speed must be also measured.

1.3.3.1 Damage Induced by Uniaxial Tension

From the previous general expressions, approximate formulae may be derived for damage induced by uniaxial tension (in direction 1). Using experiments of ductile damage with repeated unloadings for which only the elasticity modulus ($\tilde{E}_1$) and the contraction coefficient ($\tilde{\nu}_{12}$) are measured,

$$D_1 \approx 1 - \frac{\tilde{E}_1}{E} \tag{1.74}$$

$$D_2 = D_3 \approx 1 - \frac{\tilde{E}_1}{E} \frac{1+\nu}{1+3\tilde{\nu}_{12}-2\nu}. \tag{1.75}$$

For the extremal values $\frac{1}{2} \leq \frac{\tilde{\nu}}{\nu} \leq 1$ and $\frac{3}{4} \leq \frac{\tilde{E}}{E} \leq 1$, this corresponds to a maximal relative error of 15% when compared with the exact formulae. The last expression of (1.73) for ηD_H remains valid as only $\tilde{E}_1$ and $\tilde{\nu}_{12}$ are needed.

1.3.3.2 Damage in Thin Sheets

Practical applications often concern thin sheets in which it is possible to machine small tension specimens in two perpendicular directions and in the direction at 45 degrees, as shown in Fig. 1.5.

Careful measurements made on the small specimens give $\tilde{E}_1$, $\tilde{E}_2$, $\tilde{E}_{\frac{\pi}{4}}$ and $\tilde{\nu}_{12}$, $\tilde{\nu}_{21}$, $\tilde{\nu}_{\frac{\pi}{4}}$, from which D_1 and D_2 are determined. The parameter η is obtained as the slope of $\left(1 - \frac{\tilde{E}_1}{E}\frac{1-2\nu}{1-2\tilde{\nu}_{12}}\right)$ as a function of the hydrostatic

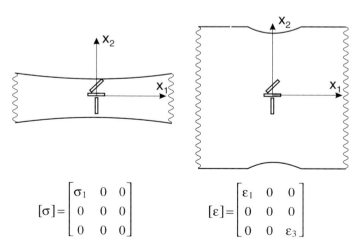

Fig. 1.5. Uniaxial tension and plane tension in three specimens machined in anisotropic damaged sheets

damage $D_H = \frac{1}{3}(D_1+D_2+D_3)$. A hypothesis concerning the damage created in direction 3 has to be made for the calculation of D_H. It is $D_2 = D_3$ for damage induced by uniaxial tension loading; $D_2 = 0$ and $D_3 = D_1$ for damage induced by plane tension loading (large plates). Such choices have an influence on the measurement of η only. They are consistent with the consideration of the damage evolution law of Sect. 1.4.

The data at 45 degrees may be used as complementary data for the identification of the damage or to check the quality of the measurements,

$$\tilde{E}_{\frac{\pi}{4}} = 4\left[\frac{1}{\tilde{E}_1} + \frac{1}{\tilde{E}_2} + \frac{1}{\tilde{G}_{12}} - 2\frac{\tilde{\nu}_{12}}{\tilde{E}_1}\right]^{-1} \qquad (1.76)$$

$$\tilde{\nu}_{\frac{\pi}{4}} = \frac{\tilde{E}_{\frac{\pi}{4}}}{4}\left[2\frac{\tilde{\nu}_{12}}{\tilde{E}_1} + \frac{1}{\tilde{G}_{12}} - \frac{1}{\tilde{E}_1} - \frac{1}{\tilde{E}_2}\right], \qquad (1.77)$$

with the damaged shear modulus being

$$\tilde{G}_{12} = G\sqrt{(1-D_1)(1-D_2)}, \quad \text{and} \quad G = \frac{E}{2(1+\nu)}. \qquad (1.78)$$

Two examples are given for anisotropic ductile damage created in uniaxial tension and plane tension in the direction 1 of thin sheets of ARCELOR steel SOLDUR 355. The results are given in Fig. 1.6 for two values of the plastic strain applied on the big samples in order to produce the damage. In both cases, the ratio D_2/D_1 is close to the ratio $|\epsilon_2^P|/|\epsilon_1^P|$ as $D_2/D_1 \approx 1/2$ in the case of uniaxial tension and $D_2/D_1 \approx 0$ in the case of plane tension. The value of η is in the range of many other experiments for which it has been found: $2.1 \leq \eta \leq 3.5$.

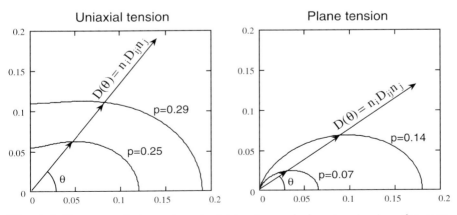

Fig. 1.6. Damage anisotropy in uniaxial tension and plane tension in polar coordinates (M. Sauzay 2000)

1.3.4 Hardness Change

Hardness or, better for damage evaluation, microhardness is influenced by the softening effect of damage. The process of inserting a diamond indenter in the material where the load (F) and the projected indented area (S) are measured defines the hardness as

$$H = \frac{F}{S}. \tag{1.79}$$

The material around the mark is loaded in 3D, proportional plasticity which avoids the consideration of a kinematic hardening (see Sect. 2.1.2). In the case of isotropic damage with the same effect on tension and compression, the plastic potential is written with the effective von Mises equivalent stress, the isotropic hardening stress R, and the yield stress σ_y,

$$f = \frac{\sigma_{eq}}{1-D} - R - \sigma_y = 0, \tag{1.80}$$

which means that the actual plastic yield stress corresponding to a plastic strain in the material is

$$\sigma_s = (R + \sigma_y)(1-D). \tag{1.81}$$

Theoretical analyses and many experiments have proved that this actual yield stress is related to the hardness H linearly,

$$H = k\sigma_s. \tag{1.82}$$

- If a microhardness test is performed on a virgin part of a piece of material where there is no hardening (or only the small hardening due to the micro-indentation),

$$H = k\sigma_y. \tag{1.83}$$

- If a microhardness test is performed on a damaged part of the same piece of material, as the damage occurs for quasi-saturated hardening $R + \sigma_y \approx \sigma_u$, one has

$$\tilde{H} = k\sigma_u(1-D). \tag{1.84}$$

with σ_u being the ultimate stress.

From the two last equations the damage is evaluated as

$$\boxed{D = 1 - \frac{\tilde{H}}{H}\frac{\sigma_y}{\sigma_u}.} \tag{1.85}$$

This method gives about the same results as the elasticity changes but, with some care, it allows for the measurement of surface damage fields with **in situ measurements** (R. Billardon and J. Dufailly 1987). Nevertheless the method does not take into consideration eventual internal stresses, which sometimes exist in zones subjected to high gradients of plastic strains and which may artificially increase or decrease the hardness.

The damage here is measured in compression by use of the coupling between plasticity and damage. Good comparisons with damage measurements in tension prove that the yield criterion does not depend much on the microdefects closure phenomenon. This non-dependency is due to the mechanism of plasticity itself produced by slips in shear.

1.3.5 Elasticity Field Change♂♂♂
(D. Claise, F. Hild and S. Roux 2002)

The microhardness technique allows for the measurement of damage fields, but in discrete values for each indentation. More powerful is the measurement, by **digital image correlation**, of the displacement field of the damaged structure loaded in its elastic range, associated with a finite element analysis (FEA) in which the loading and the displacement fields are given and the local elastic properties related to the damage field are unknown. This technique is described for plane stress or plane strain problems but its extension to 3D problems is possible.

Digital image correlation consists of correlating a small image of the surface of a deformed body with the image of the same zone in its initial state, by a translation that represents the displacement of its center. It needs:

- Some preparation (or not) of the surface in order to obtain a texture heterogeneity that can be observed by a camera.
- A CDD camera which is the key point for the resolution of the method. A 12-bit camera allows for a displacement resolution of 10^{-2} pixel or 5×10^{-5} as the absolute value of the strain, but a 16-bit camera is likely to allow for the measurement of strains with absolute values as low as 10^{-6}, with a relative accuracy of 5% and a gauge length of 1024 pixels.
- A fast acquisition set up to record the amount of information defining the images.
- A standard Personal Computer.
- Software for the image data processing by multiscale correlation (for example, the software CORRELI developed by F. Hild at LMT-Cachan).
- A special finite element code giving directly the damage field whose principle is explained below. This technique is applied to real damaged structures loaded in such a way that the strain in the most damaged zone remains elastic, with the largest possible value.

1.3.5.1 Damage Field from Displacements Measurements

The equilibrium condition $\sigma_{ij,j} = 0$ for elasticity coupled with damage constitutive equation (λ and μ are Lamé elasticity parameters of the undamaged material),

24   1 Background on Continuum Damage Mechanics

$$\sigma_{ij} = 2\mu(1-D)\epsilon^{\text{e}}_{ij} + \lambda(1-D)\epsilon^{\text{e}}_{kk}\delta_{ij}\,, \tag{1.86}$$

is

$$\left(2\mu\epsilon^{\text{e}}_{ij} + \lambda\epsilon^{\text{e}}_{kk}\delta_{ij}\right)(1-D)_{,j} + \left(2\mu\epsilon^{\text{e}}_{ij} + \lambda\epsilon^{\text{e}}_{kk}\delta_{ij}\right)_{,j}(1-D) = 0 \tag{1.87}$$

or

$$\left(2\mu\epsilon^{\text{e}}_{ij} + \lambda\epsilon^{\text{e}}_{kk}\delta_{ij}\right)\frac{(1-D)_{,j}}{1-D} + \left(2\mu\epsilon^{\text{e}}_{ij} + \lambda\epsilon^{\text{e}}_{kk}\delta_{ij}\right)_{,j} = 0\,, \tag{1.88}$$

which is rewritten with $\dfrac{(1-D)_{,j}}{1-D} = [\ln(1-D)]_{,j}$:

$$\left(2\mu\epsilon^{\text{e}}_{ij} + \lambda\epsilon^{\text{e}}_{kk}\delta_{ij}\right)[\ln(1-D)]_{,j} + \left(2\mu\epsilon^{\text{e}}_{ij} + \lambda\epsilon^{\text{e}}_{kk}\delta_{ij}\right)_{,j} = 0\,. \tag{1.89}$$

The continuity of the stress vector across interfaces or element boundaries for finite element analyses leads to (with here a boundary oriented by a normal $\vec{e}_x$)

$$\begin{aligned}
[\![(1-D)\{\lambda(\epsilon^{\text{e}}_{xx} + \epsilon^{\text{e}}_{yy} + \epsilon^{\text{e}}_{zz}) + 2\mu\epsilon^{\text{e}}_{xx}\}]\!] &= 0\,, \\
[\![(1-D)\mu\epsilon^{\text{e}}_{xy}]\!] &= 0\,, \\
[\![(1-D)\mu\epsilon^{\text{e}}_{xz}]\!] &= 0\,,
\end{aligned} \tag{1.90}$$

where $[\![\star]\!]$ denotes the jump of a quantity $\star$. Similar expressions are obtained for normals $\vec{e}_y$, $\vec{e}_z$, or $\vec{n}$.

In equations (1.89) and (1.90), the displacement field $\vec{u}(\mathbf{x})$ and therefore the elastic strain field $\boldsymbol{\epsilon}^{\text{e}}(\mathbf{x})$ are known (assumed equal to the strain $\boldsymbol{\epsilon}(\mathbf{x})$ determined by image correlation in the elastic range). These equations allow for the determination of the field $(1-D(\mathbf{x}))$ up to a multiplicative factor only since no load measures are considered. If the critical damage D_{c} is known, the observed crack initiation condition corresponds to $(1-D) = (1-D_{\text{c}})$ and gives the multiplicative factor.

Based on the FEA, a weak formulation of the problem may be built, assuming a spatial discretization by 2D quadratic elements and a damage constant per element ($D = D^{(p)}$ for element p). The strain energy $W_{\text{e}}^{(p)}$ of an element p classically depends on the elastic properties and of the nodal displacements $\{U^{\text{e}}\}$ (see Sect. 2.2). It is linear in $(1-D)$. In the absence of applied load on each middle node k between two elements p and q, the derivative of the strain energy $W_{\text{e}}^{(pq)} = W_{\text{e}}^{(p)} + W_{\text{e}}^{(q)}$ of both elements with respect to the displacement $\vec{u}^k$ of node k vanishes, so that

$$\frac{\partial W_{\text{e}}^{(pq)}}{\partial u_i^k} = 0\,. \tag{1.91}$$

Due to the linearity with respect to $(1-D^{(p)})$ in element p and to $(1-D^{(q)})$ in element q, one obtains two equations per middle node. These equations are written in the synthetic form

$$g_{ik}^{(p)}(\{U^{(p)}\})(1 - D^{(p)}) = g_{ik}^{(q)}(\{U^{(q)}\})(1 - D^{(q)}), \quad (1.92)$$

where the functions $g_{ik}^{(p \text{ or } q)}$ are generic and only depend on the nodal displacements $\{U^{(p)}\}$, $\{U^{(q)}\}$ of elements p and q.

Taking the logarithm of (1.92) yields

$$\ln(1 - D^{(p)}) - \ln(1 - D^{(q)}) = \ln|g_{ik}^{(p)}(\{U^{(p)}\})| - \ln|g_{ik}^{(q)}(\{U^{(q)}\})| \quad (1.93)$$

and leads to the overdetermined linear system

$$[M]\{d\} = \{q\}, \quad (1.94)$$

where the vector $\{d\}$ of components the logarithm of $(1-D)$ over each element is the unknown,

$$\{d\} = \begin{bmatrix} \ln(1 - D^{(1)}) \\ \vdots \\ \ln(1 - D^{(p)}) \\ \vdots \\ \ln(1 - D^{(N)}) \end{bmatrix}, \quad (1.95)$$

and where the matrix $[M]$ is filled up with 0, 1 and -1 only. The second member $\{q\}$ is due to the terms $\ln|g_{ik}^{(p)}| - \ln|g_{ik}^{(q)}|$ and is function of the measured nodal displacements. The system is solved by the use of the least square method:

$$([M]\{d\} - \{q\})^{\mathrm{T}}[W_{\mathrm{cor}}]([M]\{d\} - \{q\}) \quad \text{minimum} \quad \text{gives } \{d\}, \quad (1.96)$$

where $[W_{\mathrm{cor}}]$ is a diagonal weight matrix. We specify one component of $\{d\}$ and define $\{\delta\}$ as the new unknown vector

$$\{\delta\} = \{d\} - \{d\}_0 \quad \text{where} \quad \{d\}_0 = \begin{bmatrix} 0 \\ \vdots \\ 0 \\ \ln(1 - D)_{i_0} \\ 0 \\ \vdots \\ 0 \end{bmatrix}. \quad (1.97)$$

The damage field is finally obtained as the solution of

$$\boxed{([M]^{\mathrm{T}}[W_{\mathrm{cor}}][M])\{\delta\} = ([M]^{\mathrm{T}}[W_{\mathrm{cor}}])\{q\} - ([M]^{\mathrm{T}}[W_{\mathrm{cor}}][M])\{d\}_0.} \quad (1.98)$$

1.3.5.2 Example of a Cross-Shaped Specimen

As an example, Fig. 1.7 shows a cross-shaped specimen that is loaded biaxially in the triaxial testing machine ASTREE of LMT Cachan and also the lens of the camera. The material is a vinylester-glass fibers composite considered as isotropic since a random distribution of the fibers is observed (Fig. 1.7). The size of the analyzed square is 36.5 × 36.5 mm and it is 2.5 mm thick. The displacement measured just before failure by a mesocrack is also shown in the figure and the corresponding damage field is given in the third picture. Three highly damaged elements are visible on the top left corner and correspond to the initiation of a mesocrack precursor of the final failure of the specimen.

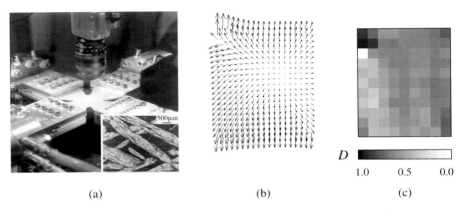

Fig. 1.7. Sample in the ASTREE machine with its microstructure (**a**), displacement field (**b**), damage field of the central zone (**c**) (D. Claise, F. Hild and S. Roux 2002)

1.4 Kinetic Laws of Damage Evolution

According to the thermodynamics framework of Sect. 1.2.1, the evolution law for damage derives from the **potential of dissipation** and particularly from the function F_D:

$$\dot{D} = \dot{\lambda}\frac{\partial F_D}{\partial Y} \quad \text{or} \quad \dot{D}_{ij} = \dot{\lambda}\frac{\partial F_D}{\partial Y_{ij}}. \tag{1.99}$$

There are many possible choices for the analytical form of the function F_D, depending on the knowledge of experimental results, the purpose of use and the ability of the model maker! The best is the simplest with the domain of validity required, where simplest means the smallest possible number of material parameters.

1.4.1 Damage Threshold♂♂ and Mesocrack Initiation♂

All measurements of damage during plastic loadings, creep, or fatigue show that no mechanical damage occurs before a certain irreversible or accumulated plastic strain (p_D) is reached, on a meso- or microscale. This **threshold** p_D depends on the material but also encounters strong variations with the type of loading. This is because the damage initiation is in fact related to the amount of energy needed for the incubation of defects, as the **stored energy threshold** (w_D) of the material.

According to the formalism of elasto-(visco-)plasticity with isotropic and kinematic hardening of Sect. 1.2, the stored energy is

$$w_s = \int_0^t (R\dot{r} + X_{ij}\dot{\alpha}_{ij})\,dt. \qquad (1.100)$$

The w_s vs p curve obtained is drawn in Fig. 1.8. It is also shown that the isotropic hardening saturates at the value $R = R_\infty$ for large p and recall that $p = r$ as long as there is no damage. With the consideration of the state law $X_{ij} = \frac{2}{3}C\alpha_{ij}$, one has then

$$w_s = \rho\psi_p \approx R_\infty p + \frac{3}{4C}X_{ij}X_{ij} \quad \text{as long as } D = 0. \qquad (1.101)$$

For nonlinear kinematic hardening, the contribution $\frac{3}{4C}X_{ij}X_{ij}$ reaches a saturation value for monotonic loadings and becomes a positive periodic function of time for cyclic loadings. This last expression then exhibits a linear dependency in p for large values of the accumulated plastic strain and values for w_s much larger than what can be measured when the observed tendency is to reach a constant value asymptotically.

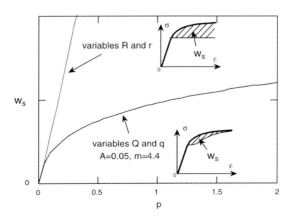

Fig. 1.8. Stored energy and damage threshold in pure tension (exponential isotropic hardening with $b = 80$)

For the above purpose, the classical framework needs to be corrected (A. Chrysochoos 1987). A new set of thermodynamics variables, (Q,q), is considered to represent isotropic hardening which introduces the correction function $z(r) = \frac{A}{m} r^{(1-m)/m}$ (A and m are material parameters), leading to new expressions such as

$$w_\mathrm{s} = \int_0^t (Q\dot{q} + X_{ij}\dot{\alpha}_{ij})\,\mathrm{d}t = \int_0^t (R(r)z(r)\dot{r} + X_{ij}\dot{\alpha}_{ij})\,\mathrm{d}t. \tag{1.102}$$

Within the thermodynamics framework, the change of variables is written as

$$Q(q) = R(r) \quad \text{and} \quad \mathrm{d}q = z(r)\mathrm{d}r \tag{1.103}$$

so that w_s increases much less with p, as shown in Fig. 1.8:

$$w_\mathrm{s} \approx AR_\infty p^{1/m} + \frac{3}{4C} X_{ij} X_{ij} \quad \text{as long as } D = 0. \tag{1.104}$$

Nothing else in the constitutive equations is modified.

One considers then that damage is initiated when the corrected stored energy reaches the threshold value and w_D is considered as a material parameter. This allows us to represent the loading dependency of the damage threshold in terms of accumulated plastic strain p_D (Fig. 1.9).

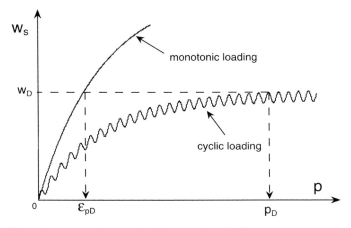

Fig. 1.9. Stored energy defining the damage thresholds in monotonic and in cyclic loadings

For practical applications, approximate expressions for p_D may be derived. In the monotonic case, the kinematic hardening may be considered as an additional isotropic hardening (see Sect. 2.1.2) and with $p = \epsilon_\mathrm{p}$ and $r = p$ as long as there is no damage. The threshold is defined by reference to the true

1.4 Kinetic Laws of Damage Evolution

limit of irreversibility taken as the **asymptotic fatigue limit** σ_f^∞,

$$w_D = \int_0^{\epsilon_{pD}} (\sigma_u - \sigma_f^\infty) \frac{A}{m} \epsilon_p^{\frac{1-m}{m}} d\epsilon_p \quad \text{or} \quad \boxed{w_D = A(\sigma_u - \sigma_f^\infty) \epsilon_{pD}^{1/m}}, \quad (1.105)$$

with σ_u as the ultimate stress and ϵ_{pD} as the **damage threshold in pure tension**.

- **For monotonic loadings**,

$$\boxed{p_D = \epsilon_{pD}}. \quad (1.106)$$

- **For cyclic or fatigue loadings** between $\sigma_{min} \leq 0$ and $\sigma_{max} > 0$, the stored energy due to kinematic hardening is small and may be neglected. An approximation of perfect plasticity at the maximum stress in tension and at the minimum stress in compression allows us to write

$$w_D = \frac{1}{2} \int_0^{p_D} (\sigma_{max} - \sigma_f^\infty) \frac{A}{m} p^{\frac{1-m}{m}} dp + \frac{1}{2} \int_0^{p_D} (|\sigma_{min}| - \sigma_f^\infty) \frac{A}{m} p^{\frac{1-m}{m}} dp$$

$$= A \left(\frac{\sigma_{max} + |\sigma_{min}|}{2} - \sigma_f^\infty \right) p_D^{1/m}, \quad (1.107)$$

with a stress range, $\Delta\sigma = \sigma_{max} - \sigma_{min} = \sigma_{max} + |\sigma_{min}| > 2\sigma_f^\infty$, corresponding to a plastic strain range, $\Delta\epsilon_p$. Equation (1.107) gives the threshold p_D as a loading-dependent function of the threshold in pure tension (ϵ_{pD}):

$$p_D = \epsilon_{pD} \left(\frac{\sigma_u - \sigma_f^\infty}{\frac{\sigma_{max} + |\sigma_{min}|}{2} - \sigma_f^\infty} \right)^m = \epsilon_{pD} \left(\frac{\sigma_u - \sigma_f^\infty}{\frac{\Delta\sigma}{2} - \sigma_f^\infty} \right)^m. \quad (1.108)$$

For 3D cyclic loadings such as $(\Delta\boldsymbol{\sigma})_{eq} = \sigma_{eq\,max} + \sigma_{eq\,min} > 2\sigma_f^\infty$, it can be written as

$$\boxed{p_D = \epsilon_{pD} \left(\frac{\sigma_u - \sigma_f^\infty}{\frac{(\Delta\boldsymbol{\sigma})_{eq}}{2} - \sigma_f^\infty} \right)^m}. \quad (1.109)$$

But the most accurate expression for any loading is of course to calculate the stored energy w_s and to write

$$\boxed{\max w_s = w_D \text{ at damage initiation}}. \quad (1.110)$$

One needs then to identify the correction parameters A, m, and the stored energy at damage initiation (w_D). Two experiments with damage measurements give the threshold of monotonic tension (ϵ_{pD}) and in fatigue

(p_D) for a known plastic strain range ($\Delta\epsilon_\text{p}$). The stored energy in monotonic tension is

$$w_\text{s} = \frac{A}{m}\int_0^{\epsilon_\text{pD}} R(p)p^{(1-m)/m}\mathrm{d}p + \frac{X^2(\epsilon_\text{pD})}{2C} = w_\text{D} \qquad (1.111)$$

and is

$$w_\text{s} = \frac{A}{m}\int_0^{p_\text{D}} R(p)p^{(1-m)/m}\mathrm{d}p + \frac{X_\text{max}^2}{2C} = w_\text{D} \qquad (1.112)$$

in fatigue with (see Sect. 1.5.1):
- for exponential isotropic hardening: $R(p) = R_\infty(1-\exp(-bp))$
- for linear kinematic hardening: $X(\epsilon_\text{pD}) = C\epsilon_\text{pD}$ in monotonic loading, $X_\text{max} = C\epsilon_\text{p max} = C\Delta\epsilon_\text{p}/2$ in alternate fatigue loading $\sigma_\text{min} = -\sigma_\text{max}$
- for nonlinear kinematic hardening: $X(\epsilon_\text{pD}) = X_\infty(1-\exp(-\gamma\epsilon_\text{pD}))$ in monotonic loading, $X_\text{max} = X_\infty \tanh\frac{\gamma\Delta\epsilon_\text{p}}{2}$ in alternate fatigue loading $\sigma_\text{min} = -\sigma_\text{max}$

where R_∞, b, C, γ and $X_\infty = C/\gamma$ are the hardening parameters.
The exponent m is determined by (1.108) such that

$$\boxed{m = \frac{\ln\dfrac{p_\text{D}}{\epsilon_\text{pD}}}{\ln\left(\dfrac{\sigma_\text{u}-\sigma_\text{f}^\infty}{\dfrac{\Delta\sigma}{2}-\sigma_\text{f}^\infty}\right)}} \qquad (1.113)$$

and A is identified by a comparison between (1.111) and (1.112).
For nonlinear kinematic hardening,

$$\boxed{A = \frac{mX_\infty}{2\gamma R_\infty}\frac{[1-\exp(-\gamma\epsilon_\text{pD})]^2 - \tanh^2\frac{\gamma\Delta\epsilon_\text{p}}{2}}{\displaystyle\int_{\epsilon_\text{pD}}^{p_\text{D}}(1-\exp(-bp))\,p^{\frac{1}{m}-1}\mathrm{d}p}} \qquad (1.114)$$

or, with the hypothesis of saturated hardening at damage initiation in monotonic tension,

$$A \approx \frac{X_\infty}{2\gamma R_\infty}\frac{1-\tanh^2\frac{\gamma\Delta\epsilon_\text{p}}{2}}{p_\text{D}^{1/m}-\epsilon_\text{pD}^{1/m}}. \qquad (1.115)$$

If possible, take A as the average value given by different tests where the plastic strain range $\Delta\epsilon_\text{p}$ and the damage threshold p_D are measured.

Finally, still for nonlinear kinematic hardening,

$$w_D = \frac{A R_\infty}{m} \int_0^{\epsilon_{pD}} (1 - \exp(-bp)) \, p^{(1-m)/m} dp + \frac{X_\infty^2 \left[1 - \exp(-\gamma \epsilon_{pD})\right]^2}{2C}.$$

(1.116)

- **For cyclic or fatigue loadings that are periodic by blocks**, each level or block i is a cyclic loading between $\sigma_{\text{eq min}}^{(i)}$ and $\sigma_{\text{eq max}}^{(i)}$. Damage will be initiated during the n_B-th block when the stored energy density reaches its threshold, i.e. as soon as $w_s = w_D$, or in an equivalent manner when the accumulated plastic strain reaches the loading-dependent threshold p_D deduced from:

$$\left\langle \frac{\sigma_{\text{eq min}}^{(n_B)} + \sigma_{\text{eq max}}^{(n_B)}}{2} - \sigma_f^\infty \right\rangle \left[p_D^{1/m} - (p_{n_B-1})^{1/m} \right]$$

$$+ \sum_{i=1}^{n_B-1} \left\langle \frac{\sigma_{\text{eq min}}^{(i)} + \sigma_{\text{eq max}}^{(i)}}{2} - \sigma_f^\infty \right\rangle \left[(p_i)^{1/m} - (p_{i-1})^{1/m} \right]$$

$$= (\sigma_u - \sigma_f^\infty) \epsilon_{pD}^{\frac{1}{m}}$$

(1.117)

where p_i is the value of the accumulated plastic strain reached at the end of the i-th block. This equation generalizes equation (1.109) to multilevel fatigue loadings. An example of application is given in Sect. 4.4.1.

At the other end of the damage evolution, a **mesocrack is initiated** when the density of defects reaches the value for which the process of localization and instability develops (see Sect. 1.6), that is $D = D_c$ in the plane where $D(\vec{n})$ is maximum. The critical damage D_c is a material parameter. A way to evaluate D_c is to apply the concept of effective stress at fracture: in a pure tensile test when damage develops at saturated hardening, the stress decreases from the ultimate stress σ_u to the rupture stress σ_R in such a way that for isotropic damage:

$$\tilde{\sigma}_R = \sigma_u \quad \text{or} \quad \frac{\sigma_R}{1 - D_c} = \sigma_u$$

(1.118)

from which

$$D_c = 1 - \frac{\sigma_R}{\sigma_u}$$

(1.119)

leading to values of the critical damage between 0.2 and 0.5 for many materials.

For anisotropic damage, $\tilde{\sigma}_{eq} = \frac{2}{3}\frac{\sigma_1}{1-D_1} + \frac{1}{3}\frac{\sigma_1}{1-D_2}$ with $D_2 \approx D_1/2$ in tension. Writing again $\tilde{\sigma}_{eqR} = \sigma_u$ and $D_1 = D_c$ gives values of the critical damage of the same order of magnitude.

1.4.2 Formulation of the Isotropic Unified Damage Law

The thermodynamics approach ensures that the main variable governing the damage evolution or the damage rate ($\dot{D}$) is its associate variable (Y), the **energy density release rate**. Then the dissipative damage potential function (F_D) is primarily a function of Y. Many observations and experiments show that the damage is also governed by the **plastic strain** which is introduced through the plastic multiplier ($\dot{\lambda}$) as

$$\dot{D} = \dot{\lambda}\frac{\partial F_\mathrm{D}}{\partial Y} \quad \text{if} \quad p > p_\mathrm{D} \quad \text{or} \quad \max w_\mathrm{s} > w_\mathrm{D}, \tag{1.120}$$

with $\dot{\lambda}$ calculated from the constitutive equations of (visco-)plasticity coupled with the damage deduced from the dissipative potential function (F),

$$F = f + F_\mathrm{X} + F_\mathrm{D}. \tag{1.121}$$

The (visco-)plasticity loading function f is determined by the von Mises criterion (see also Sect. 1.5),

$$f = \left(\frac{\boldsymbol{\sigma}}{1-D} - \boldsymbol{X}\right)_\mathrm{eq} - R - \sigma_\mathrm{y} = \sigma_\mathrm{v}, \tag{1.122}$$

where

- $\left(\dfrac{\boldsymbol{\sigma}}{1-D} - \boldsymbol{X}\right)_\mathrm{eq} = \sqrt{\dfrac{3}{2}\left(\dfrac{\sigma^\mathrm{D}_{ij}}{1-D} - X_{ij}\right)\left(\dfrac{\sigma^\mathrm{D}_{ij}}{1-D} - X_{ij}\right)}$
- σ_v is the viscous stress for viscoplasticity, $\sigma_\mathrm{v} = 0$ for plasticity

The normality rule,

$$\dot{\epsilon}^\mathrm{P}_{ij} = \dot{\lambda}\frac{\partial F}{\partial \sigma_{ij}} = \dot{\lambda}\frac{\partial f}{\partial \sigma_{ij}} = \frac{3}{2}\frac{\dfrac{\sigma^\mathrm{D}_{ij}}{1-D} - X_{ij}}{\left(\dfrac{\boldsymbol{\sigma}}{1-D} - \boldsymbol{X}\right)_\mathrm{eq}}\frac{\dot{\lambda}}{1-D}, \tag{1.123}$$

coupled with the definition of the accumulated plastic strain rate, $\dot{p} = \sqrt{\frac{2}{3}\dot{\epsilon}^\mathrm{P}_{ij}\dot{\epsilon}^\mathrm{P}_{ij}}$, and of the evolution law for the variable r,

$$\dot{r} = -\dot{\lambda}\frac{\partial F}{\partial R} = -\dot{\lambda}\frac{\partial f}{\partial R} = \dot{\lambda}, \tag{1.124}$$

leads to

$$\dot{p} = \frac{\dot{\lambda}}{1-D}. \tag{1.125}$$

1.4 Kinetic Laws of Damage Evolution

Many experimental results show also that F_D must be a nonlinear function of Y. Then a simple (good !) choice is

$$F_D = \frac{S}{(s+1)(1-D)} \left(\frac{Y}{S}\right)^{s+1}, \qquad (1.126)$$

from which

$$\dot{D} = \left(\frac{Y}{S}\right)^s \dot{p}. \qquad (1.127)$$

S and s are two material parameters that are functions of the temperature.

The full damage constitutive equation is (J. Lemaitre 1987)

$$\boxed{\begin{aligned} \dot{D} &= \left(\frac{Y}{S}\right)^s \dot{p} \quad \text{if} \quad \max w_s > w_D \text{ or } p > p_D, \\ \dot{D} &= 0 \quad \text{if not}, \\ D &= D_c \quad \rightarrow \quad \text{mesocrack initiation}, \end{aligned}} \qquad (1.128)$$

where

- $\begin{cases} Y = \dfrac{\tilde{\sigma}_{\text{eq}}^2 R_\nu}{2E} \\ R_\nu = \dfrac{2}{3}(1+\nu) + 3(1-2\nu)\left(\dfrac{\sigma_H}{\sigma_{\text{eq}}}\right)^2 \end{cases}$ in the simplest case

- $Y = \dfrac{1+\nu}{2E}\left[\dfrac{\langle\boldsymbol{\sigma}\rangle_{ij}^+\langle\boldsymbol{\sigma}\rangle_{ij}^+}{(1-D)^2} + h\dfrac{\langle\boldsymbol{\sigma}\rangle_{ij}^-\langle\boldsymbol{\sigma}\rangle_{ij}^-}{(1-hD)^2}\right] - \dfrac{\nu}{2E}\left[\dfrac{\langle\sigma_{kk}\rangle^2}{(1-D)^2} + h\dfrac{\langle-\sigma_{kk}\rangle^2}{(1-hD)^2}\right]$
  if quasi-unilateral conditions are considered

- $w_s = \int_0^t (R(r)z(r)\dot{r} + X_{ij}\dot{\alpha}_{ij})\,dt$ and w_D is a material parameter (related to ϵ_{pD})

- $p_D = \epsilon_{pD}$ for monotonic loading,

  $p_D = \epsilon_{pD}\left(\dfrac{\sigma_u - \sigma_f^\infty}{\dfrac{\sigma_{\text{eq max}} + \sigma_{\text{eq min}}}{2} - \sigma_f^\infty}\right)^m$ for cyclic loading

- D_c is a material parameter

This law **unifies** many particular models:

- Ductile damage if $\dot{p}$ is governed by plasticity (see Chap. 3)
- Creep damage if $\dot{p}$ is given by a viscosity law such as Norton power law (see Chap. 5)
- Fatigue damage if $\dot{p}$ is calculated from cyclic plasticity (see Chaps. 4 and 6)
- Quasi-brittle damage if $\dot{p}$ is at a microscopic level (see Chaps. 6 and 7)

1.4.3 Formulation of the Anisotropic Damage Law
(J. Lemaitre, R. Desmorat and M. Sauzay 2000)

It is a "simple" extension of the isotropic case if the following damage dissipation potential is considered:

$$F_{\rm D} = \left(\frac{\overline{Y}(\epsilon^{\rm e})}{S}\right)^s Y_{ij} \left|\frac{{\rm d}\epsilon^{\rm p}}{{\rm d}r}\right|_{ij}, \tag{1.129}$$

where $|.|$ applied to a tensor means the absolute value in terms of the principal components. The term $\overline{Y}$ is the effective elastic energy density which can be written as a function of the effective stress,

$$\overline{Y} = \frac{1}{2} E_{ijkl} \epsilon^{\rm e}_{kl} \epsilon^{\rm e}_{ij} = \frac{1}{2} \tilde{\sigma}_{ij} \epsilon^{\rm e}_{ij} = \frac{\tilde{\sigma}_{\rm eq}^2 \tilde{R}_\nu}{2E}, \tag{1.130}$$

introducing the effective stress triaxiality function,

$$\tilde{R}_\nu = \frac{2}{3}(1+\nu) + 3(1-2\nu)\left(\frac{\tilde{\sigma}_{\rm H}}{\tilde{\sigma}_{\rm eq}}\right)^2, \tag{1.131}$$

where

$$\tilde{\sigma}_{\rm eq} = (\boldsymbol{H}\,\boldsymbol{\sigma}^{\rm D}\boldsymbol{H})_{\rm eq} \quad \text{and} \quad \tilde{\sigma}_{\rm H} = \frac{\sigma_{\rm H}}{1-\eta D_{\rm H}}. \tag{1.132}$$

Then

$$\dot{D}_{ij} = \dot{\lambda}\frac{\partial F}{\partial Y_{ij}} \quad \text{with} \quad \dot{\lambda} = \dot{r} \tag{1.133}$$

$$\dot{D}_{ij} = \dot{r}\left(\frac{\overline{Y}}{S}\right)^s \left|\frac{{\rm d}\epsilon^{\rm p}}{{\rm d}r}\right|_{ij}. \tag{1.134}$$

Or finally,

$$\boxed{\begin{aligned} \dot{D}_{ij} &= \left(\frac{\overline{Y}}{S}\right)^s |\dot{\epsilon}^{\rm p}|_{ij} \quad &\text{if} \quad \max w_{\rm s} > w_{\rm D} \quad \text{or} \quad p > p_{\rm D} \\ \dot{D}_{ij} &= 0 \quad &\text{if not}, \end{aligned}} \tag{1.135}$$

an equation written in the principal frame of the plastic strain rate tensor $\dot{\epsilon}^{\rm p}$ which shows that the **principal directions of the damage rate tensor coincide with those of the plastic strain rate**. As an example, the pure tensile test reads

$$\boldsymbol{\sigma} = \sigma \begin{bmatrix} 1 & 0 & 0 \\ 0 & 0 & 0 \\ 0 & 0 & 0 \end{bmatrix}, \quad \boldsymbol{\epsilon}^{\rm p} = \epsilon_{\rm p} \begin{bmatrix} 1 & 0 & 0 \\ 0 & -\frac{1}{2} & 0 \\ 0 & 0 & -\frac{1}{2} \end{bmatrix}, \quad \text{and} \quad \boldsymbol{D} = D \begin{bmatrix} 1 & 0 & 0 \\ 0 & \frac{1}{2} & 0 \\ 0 & 0 & \frac{1}{2} \end{bmatrix}. \tag{1.136}$$

And as in the isotropic case,

$$\dot{D}_{ij} = 0 \quad \text{if} \quad w_{\rm s} < w_{\rm D} \quad \text{or} \quad p < p_{\rm D}. \tag{1.137}$$

The condition of a mesocrack initiation is fulfilled when the intensity of the damage in one plane reaches the critical value $D_{\rm c}$. According to the physical definition of the damage, this occurs when the norm of the damage vector $D_{ij}n_j$ or when the largest principal value of the damage D_I reaches D_c:

$$\boxed{\max D_I = D_c} \quad \rightarrow \quad \boxed{\text{mesocrack initiation.}} \qquad (1.138)$$

$\vec{n}_I(\max D_I)$ gives the orientation of the plane on which lies the mesocrack.

1.4.4 Fast Identification of Damage Material Parameters♂

As already stated in Sect. 1.3, the best set of data on damage evolution comes from very low cycle tests, at constant strain range on tension-compression specimens, from which the damage is deduced using the elasticity modulus change. Unfortunately, these tests are not commonly available. We will describe here a method to obtain the values of the material parameters for a specific material from usual properties listed in catalogues or handbooks, or derivations using only a few usual tests as tensile and fatigue data:

- The damage law itself needs E, ν, S, s
- The damage threshold needs ϵ_{pD}, σ_f^∞, σ_u and m
- The condition of mesocrack initiation needs D_c
- The quasi-unilateral conditions need h for isotropic damage or h_a for anisotropic damage but most often $h = 0.2$ or $h_a = 0$
- anisotropic damage needs η but a good candidate is $\eta = 3$

1.4.4.1 Partial Identification from a Simple Tensile Test

A simple uniaxial tension test, as shown in Fig. 1.10, gives:

- The Young modulus E and the Poisson ratio ν
- The yield stress σ_y, the conventional yield stress $\sigma_{y02} = \sigma_{(\epsilon_p = 0.2\, 10^{-2})}$ and the ultimate stress σ_u
- The rupture stress σ_R and the rupture plastic strain ϵ_{pR}
- The necking parameter $Z = \dfrac{S_0 - S_R}{S_0}$ with S_0 the initial section of the specimen and S_R its rupture minimal section

Considering the uniaxial, isotropic, unified damage law in monotonic loading with the hardening saturated at σ_u gives

$$\dot{D} = \left(\frac{\sigma_u^2}{2ES}\right)^s \dot{\epsilon}_p \qquad (1.139)$$

or, by integration with $D = 0$ if $\epsilon_p \leq \epsilon_{pD}$,

$$D = \left(\frac{\sigma_u^2}{2ES}\right)^s (\epsilon_p - \epsilon_{pD}) . \qquad (1.140)$$

36   1 Background on Continuum Damage Mechanics

The damage usually starts to grow when the plastic strain reaches the value corresponding to the ultimate stress σ_u,

$$\boxed{\epsilon_{pD} \approx \epsilon_p(\sigma = \sigma_u).} \tag{1.141}$$

As introduced in Sect. 1.4.1, the critical value of the damage is given by

$$\boxed{D_c = 1 - \frac{\sigma_R}{\sigma_u}} \tag{1.142}$$

and $0.2 \leq D_c \leq 0.5$ in general.

Upon fracture, the local rupture strain in the necking region ($\epsilon_{pR}^\star$) is again estimated from the necking parameter $Z = (S_0 - S_R)/S_0$ considering plastic incompressibility ($\epsilon_{kk}^P = 0$):

$$\epsilon_{pR}^\star = 2\left(1 - \sqrt{1 - Z}\right). \tag{1.143}$$

Then,

$$D_c = \left(\frac{\sigma_u^2}{2ES}\right)^s \left(\epsilon_{pR}^\star - \epsilon_{pD}\right). \tag{1.144}$$

This relation allows us to determine the 3 remaining parameters: S, s, and m. To go further, another type of test is needed.

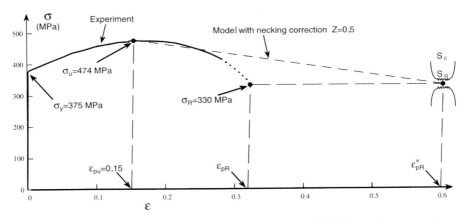

Fig. 1.10. Material parameters from a tension test on ferritic steel at room temperature

1.4.4.2 Final Identification from Fatigue Tests

Usually it is not too difficult to obtain some information about fatigue! That is at least the case for the engineering fatigue limit (σ_f) corresponding to a number of cycles to rupture of 10^6 (or 10^7), or better, the asymptotic fatigue limit (σ_f^∞) and some low fatigue tests results (Fig. 1.11).

The unified damage law can be easily applied to low cycle fatigue if we make the assumption that the material is perfectly plastic for each level of stress, i.e.,

$$\dot{D} = \left(\frac{\sigma_{max}^2}{2ES(1-D)^2}\right)^s \dot{p}. \tag{1.145}$$

As $\dot{p} = |\dot{\epsilon}_p|$, the integration of (1.145) over one cycle N, neglecting the variation of D during the cycle, is simply

$$\frac{\delta D}{\delta N} = \int_{1 \text{ cycle}} \dot{D} dt = \left(\frac{\sigma_{max}^2}{2ES(1-D)^2}\right)^s 2\Delta\epsilon_p, \tag{1.146}$$

where $\Delta\epsilon_p$ is the plastic strain range corresponding to σ_{max} or the stress range $\Delta\sigma = 2\sigma_{max}$ through the cyclic stress-strain behavior.

Using N_D as the number of cycles corresponding to the damage threshold and N_R as the number of cycles corresponding to a mesocrack initiation when $D = D_c$, a second integration of (1.145) over the whole fatigue process gives

$$\int_0^{D_c} (1-D)^{2s} \delta D = \left(\frac{\sigma_{max}^2}{2ES}\right)^s 2\Delta\epsilon_p \int_{N_D}^{N_R} \delta N \tag{1.147}$$

or

$$\frac{1}{2s+1}\left[1 - (1-D_c)^{2s+1}\right] = \left(\frac{\sigma_{max}^2}{2ES}\right)^s 2\Delta\epsilon_p (N_R - N_D), \tag{1.148}$$

with N_D given by

$$p_D = 2\Delta\epsilon_p N_D = \epsilon_{pD}\left(\frac{\sigma_u - \sigma_f^\infty}{\sigma_{max} - \sigma_f^\infty}\right)^m. \tag{1.149}$$

The number of cycles needed to initiate a mesocrack is finally expressed as

$$N_R = \frac{\epsilon_{pD}}{2\Delta\epsilon_p}\left(\frac{\sigma_u - \sigma_f^\infty}{\sigma_{max} - \sigma_f^\infty}\right)^m + \frac{1 - (1-D_c)^{2s+1}}{2(2s+1)\Delta\epsilon_p}\left(\frac{2ES}{\sigma_{max}^2}\right)^s, \tag{1.150}$$

with $(2ES)^s$ taken from (1.144) of the tensile test results

$$(2ES)^s = \sigma_u^{2s}\left(\frac{\epsilon_{pR}^\star - \epsilon_{pD}}{D_c}\right) \tag{1.151}$$

so that

$$\boxed{N_R = \frac{\epsilon_{pD}}{2\Delta\epsilon_p}\left(\frac{\sigma_u - \sigma_f^\infty}{\sigma_{max} - \sigma_f^\infty}\right)^m + \frac{1-(1-D_c)^{2s+1}}{2(2s+1)D_c\Delta\epsilon_p}\left(\frac{\sigma_u}{\sigma_{max}}\right)^{2s}(\epsilon_{pR}^\star - \epsilon_{pD}).} \tag{1.152}$$

This equation written for two "good" results of low cycle fatigue tests, $N_{R1}(\sigma_{\max 1}, \Delta\epsilon_{p1})$ and $N_{R2}(\sigma_{\max 2}, \Delta\epsilon_{p2})$ leads to two equations for the unknowns s and m which can be solved by substitution by any mathematical software, for example.

If the threshold is written in terms of stored energy, we also need the value of A to calculate w_D according to (1.114) or (1.115) of Sect. 1.4.1:

$$A \approx \frac{X_\infty}{2\gamma R_\infty} \frac{1 - \tanh^2 \frac{\gamma \Delta\epsilon_p}{2}}{p_D^{1/m} - \epsilon_{pD}^{1/m}} \quad \text{and}$$

$$w_D = \frac{AR_\infty}{m} \int_0^{\epsilon_{pD}} (1 - \exp(-bp)) \, p^{(1-m)/m} dp + \frac{X_\infty^2 (1 - \exp(-\gamma\epsilon_{pD}))^2}{2C},$$
(1.153)

where R_∞, C, γ, $X_\infty = C/\gamma$ are the hardening parameters (see Sect. 1.5.1).

The word "fast" in the title of this section means a procedure of a few hours once one has the tensile curve and the Wöhler curve, but obtaining these experimental results can take a few days or even months!

For anisotropic damage, the same procedure gives material parameters close to those obtained for the isotropic law when using damage data with $D < 0.2$.

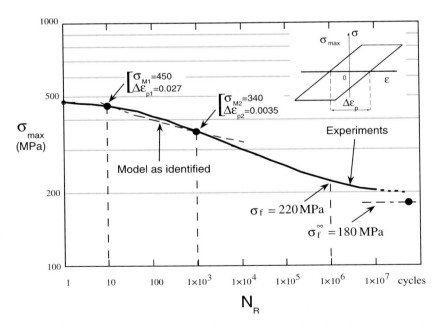

Fig. 1.11. Material parameters from a Wöhler curve of ferritic steel at room temperature

1.4.4.3 Example of Identification for Ferritic Steel

The experimental tensile and Wöhler curves are shown in Figs. 1.10 and 1.11 but the identification is made only with the coordinates of the 6 points marked by "•" on the figures, that is: $\sigma_u(\epsilon_{pu})$, $\sigma_R(\epsilon_{pR})$, $\sigma_R(\epsilon_{pR}^\star)$ and σ_f^∞, $N_{R1}(\sigma_{\max 1}, \Delta\epsilon_{p1})$ and $N_{R2}(\sigma_{\max 2}, \Delta\epsilon_{p2})$.

1. From the tensile curve you can obtain the following:
   - Elasticity modulus $E = 200000$ MPa.
   - Poisson ratio measured $\nu = 0.3$.
   - Yield stress $\sigma_y = 375$ MPa.
   - Conventional yield stress $\sigma_{y02} = 380$ MPa.
   - Ultimate stress: $\sigma_u = 474$ MPa.
   - $\epsilon_p(\sigma_u) = 0.15$, then $\epsilon_{pD} = 0.15$.
   - $\sigma_R = 330$ MPa then $D_c = 0.3$.
   - $Z = 0.5$ then $\epsilon_{pR}^\star = 0.6$.

2. The Wöhler curve yields
   - Asymptotic fatigue limit $\sigma_f^\infty = 180$ MPa. If the asymptotic fatigue limit cannot be estimated take the engineering fatigue limit σ_f instead.
   - Low cycle fatigue tests corresponding to $N_R \approx 10$ and $N_R \approx 1000$ cycles:
     1. $\sigma_{M1} = 450$ MPa, $\Delta\epsilon_{p1} = 0.027$, mean value $N_{R1} = 10$ cycles,
     2. $\sigma_{M2} = 340$ MPa, $\Delta\epsilon_{p2} = 0.0035$, mean value $N_{R2} = 984$ cycles,

   from which $s = 2.4$, $m = 6$. The value of S can then be calculated using (1.144):

$$S = \frac{\sigma_u^2}{2E}\left(\frac{\epsilon_{pR}^\star - \epsilon_{pD}}{D_c}\right)^{1/s} = 0.665 \text{ MPa}. \qquad (1.154)$$

3. Finally, take $h = 0.2$, $h_a = 0$, and $\eta = 3$. If the two scale damage model is applied (see Sect. 1.5.5), $C_y = \dfrac{\sigma_u - \sigma_y}{\epsilon_{pu}} = 660$ MPa and be happy.

4. Last, but most important, is to check how close the above values return the starting curves. The graphs given by the model are shown by the dotted lines in Figs. 1.10 and 1.11:
   - for the tensile curve with $D \approx \left(\dfrac{\sigma_u^2}{2ES}\right)^s \langle\epsilon_p - \epsilon_{pD}\rangle$,
   - for the Wöhler curve in the range of low cycle fatigue.

   Considering the small number of input information, the comparison with the experimental data is "not bad." The range of high cycle fatigue is the domain of application of the two-scale damage model of Sect. 1.5.5.

1.4.4.4 Case of Anisotropic Damage

If now the same work is performed with the anisotropic formulae, the only change is in the damage law where $\tilde{R}_\nu$ is now different from unity, even in the uniaxial case:

$$\boldsymbol{\sigma} = \begin{bmatrix} \sigma & 0 & 0 \\ 0 & 0 & 0 \\ 0 & 0 & 0 \end{bmatrix}, \quad \boldsymbol{D} = D \begin{bmatrix} 1 & 0 & 0 \\ 0 & \frac{1}{2} & 0 \\ 0 & 0 & \frac{1}{2} \end{bmatrix}, \tag{1.155}$$

and

$$\tilde{R}_\nu = \frac{2}{3}(1+\nu) + 3(1-2\nu)\left(\frac{\tilde{\sigma}_H}{\tilde{\sigma}_{eq}}\right)^2, \tag{1.156}$$

with

$$\tilde{\sigma}_H = \frac{\sigma_H}{1 - \eta D_H} = \frac{\sigma}{1 - \frac{2\eta}{3}D}$$

$$\tilde{\sigma}_{eq} = \left[\frac{3}{2}(\boldsymbol{H}\boldsymbol{\sigma}^D\boldsymbol{H})^D : (\boldsymbol{H}\boldsymbol{\sigma}^D\boldsymbol{H})^D\right]^{1/2} = \frac{2}{3}\frac{\sigma}{1-D} + \frac{1}{3}\frac{\sigma}{1 - \frac{D}{2}}. \tag{1.157}$$

Then,

$$\tilde{R}_\nu = \frac{2}{3}(1+\nu) + 3(1-2\nu)\left[\left(1 - \frac{2\eta}{3}D\right)\left(\frac{2}{1-D} + \frac{1}{1 - \frac{D}{2}}\right)\right]^{-2} \tag{1.158}$$

in tension.

With $\nu = 0.3$ and $\eta = 3$, the maximum value of $\tilde{R}_\nu$ for $D \leq 0.2$ is 1.12, which is 12% more than the isotropic damage value $R_\nu = 1$. As the uniaxial law turns into

$$\dot{D} = \left(\frac{\sigma_{max}^2 \tilde{R}_\nu}{2ES}\right)^s \dot{p}, \tag{1.159}$$

this changes the value of S and D_c by an amount which remains in the order of magnitude of the experimental discrepancy or the hypothesis of perfect plasticity. In practice, keep s as identified for the isotropic damage model and update S and D_c on the monotonic tensile curve.

For more accurate predictions, the "fast identification" results may be used as the starting values of an optimization process to fit all available data that can be obtained on a specific material. This numerical optimization process, the least square method, consists of finding the set of parameters to minimize the difference between experiments and predictions. Numerical methods for precise identification are described in Sect. 2.4 with an exampley.

The end of Sect. 1.5.5 offers a more precise identification of the asymptotic fatigue limit.

1.4.5 Generalization of the Unified Damage Law

For materials which do not exhibit meso-plasticity, it is not possible to consider the damage as governed by the accumulated plastic strain rate. The damage law $\dot{D} = (Y/S)^s \dot{p}$ is then useless. A damage law $D = D(Y)$ cannot be considered either as it does not lead to damage growth in fatigue. For such materials the idea is to relate the damage rate to the main dissipative mechanism, often **internal sliding and friction**, and to consider damage as governed by a cumulative measure of the internal sliding. This may apply to metals for which internal slips are mainly due to dislocations creation and evolution, as well as many non-metallic materials such as elastomers (Sect. 4.4.3), concretes, and composites.

We then use a framework that is similar to elasto-(visco-)plasticity coupled with damage and define $\boldsymbol{\epsilon}^\pi$, $\boldsymbol{\alpha}$, q, and D as internal variables associated with $-\boldsymbol{\sigma}^\pi$, $\boldsymbol{X}$, Q, and $-Y$ (see Table 1.2). The physical meaning of the thermodynamic variables depends on the type of material and of the physical dissipative mechanisms. It will be discussed for the case of metals, elastomers, and concrete later (Sect. 1.4.5.3). Nevertheless,

- $\boldsymbol{\epsilon}^\pi$ is an internal inelastic strain which can be recovered
- $\boldsymbol{X}$ is a residual microstress of kinematic nature
- Q defines the size increase of the reversibility domain in the $\boldsymbol{\sigma}^\pi$ stress space
- D is the damage variable ($\boldsymbol{D}$ if induced anisotropy is considered)

Table 1.2. State and associated variables

Mechanisms	Type	State variables		Associated
		Observable	Internal	variables
Elasticity	Tensor	$\boldsymbol{\epsilon}$		$\boldsymbol{\sigma}$
Internal sliding	Tensor		$\boldsymbol{\epsilon}^\pi$	$-\boldsymbol{\sigma}^\pi$
	Tensor		$\boldsymbol{\alpha}$	$\boldsymbol{X}$
	Scalar		q	Q
Damage	Scalar		D	$-Y$

Next we consider the small strain hypothesis with **isotropy** of the dissipative mechanisms, including damage. For the finite strain framework see Sect. 4.4.3.

The general form of the **state potential** can be expressed as

$$\rho\psi = (1-D)\left[w_1(\boldsymbol{\epsilon}) + w_2(\boldsymbol{\epsilon} - \boldsymbol{\epsilon}^\pi)\right] + w_s(q, \boldsymbol{\alpha}), \qquad (1.160)$$

where w_1 and w_2 define the strain energy density and w_s is the stored energy density as a function of the scalar variable q and the tensorial variable $\boldsymbol{\alpha}$. The state laws are

$$\begin{aligned}
\sigma_{ij} &= \rho \frac{\partial \psi}{\partial \epsilon_{ij}} = (1-D)\frac{\partial(w_1+w_2)}{\partial \epsilon_{ij}}, \\
\sigma_{ij}^\pi &= -\rho \frac{\partial \psi}{\partial \epsilon_{ij}^\pi} = (1-D)\frac{\partial w_2}{\partial \epsilon_{ij}}, \\
X_{ij} &= \rho \frac{\partial \psi}{\partial \alpha_{ij}} = \frac{\partial w_s}{\partial \alpha_{ij}}, \\
Q &= \rho \frac{\partial \psi}{\partial q} = \frac{\partial w_s}{\partial q}, \\
Y &= -\rho \frac{\partial \psi}{\partial D} = w_1 + w_2.
\end{aligned} \qquad (1.161)$$

They naturally define the effective stresses ($\tilde{\boldsymbol{\sigma}}$, $\tilde{\boldsymbol{\sigma}}^\pi$) such that the elasticity law written in terms of strains and effective stresses does not depend on D (strain equivalence principle):

$$\begin{aligned}
\tilde{\boldsymbol{\sigma}} &= \frac{\partial(w_1+w_2)}{\partial \epsilon}, \\
\tilde{\boldsymbol{\sigma}}^\pi &= \frac{\partial w_2}{\partial \epsilon}.
\end{aligned} \qquad (1.162)$$

The **dissipation potential** is defined as

$$F = f + F_X + F_D, \qquad (1.163)$$

where

- $f = \left\| \dfrac{\boldsymbol{\sigma}^\pi}{1-D} - \boldsymbol{X} \right\| - Q - \sigma_s < 0$ defines the reversibility domain, $\|.\|$ is a norm in the stresses space (not necessarily von Mises norm), and σ_s is the reversibility limit
- the functions $F_X = \dfrac{\gamma}{2C_X} X_{ij} X_{ij}$ and $Q = Q(q)$ model the internal sliding nonlinearity
- $F_D = \dfrac{S}{(s+1)(1-D)} \left(\dfrac{Y}{S} \right)^{s+1}$ is the damage potential

The evolutions laws can then be derived from the dissipation potential (normality rule)

$$\begin{aligned}
\dot{\epsilon}_{ij}^\pi &= \dot{\mu} \frac{\partial F}{\partial \sigma_{ij}^\pi}, \\
\dot{\alpha}_{ij} &= -\dot{\mu} \frac{\partial F}{\partial X_{ij}}, \\
\dot{q} &= -\dot{\mu} \frac{\partial F}{\partial Q}, \\
\dot{D} &= \dot{\mu} \frac{\partial F}{\partial Y},
\end{aligned} \qquad (1.164)$$

where $\dot{\mu}$ is the internal sliding multiplier equal to a norm of the inelastic strain rate:

$$\frac{\dot{\mu}}{1-D} = \frac{\dot{r}}{1-D} = \|\dot{\boldsymbol{\epsilon}}^{\pi}\|. \qquad (1.165)$$

This defines the **cumulative measure of the internal sliding** as

$$\pi = \int_0^t \|\dot{\boldsymbol{\epsilon}}^{\pi}\| \mathrm{d}t. \qquad (1.166)$$

The multiplier $\dot{\mu}$ is given by the consistency condition, $f = 0$ and $\dot{f} = 0$, for non-viscous materials or by a viscosity law such as the generalized Norton law, $\dot{\pi} = \langle f/K_N \rangle^N = \langle \sigma_v/K_N \rangle^N$, for viscous materials undergoing a viscous stress σ_v.

The generalized damage evolution law is derived from (1.164), leading to

$$\boxed{\begin{aligned}\dot{D} &= \left(\frac{Y}{S}\right)^s \dot{\pi} \quad \text{if} \quad \pi > \pi_\mathrm{D} \\ D &= D_\mathrm{c} \quad \longrightarrow \quad \text{mesocrack initiation,}\end{aligned}} \qquad (1.167)$$

which corresponds to the damage governed by the main dissipative mechanisms through $\dot{\pi}$ and where S and s are the damage material parameters and π_D is the damage threshold.

1.4.5.1 Positivity of the Intrinsic Dissipation

As in the previous thermodynamics framework, the generalized damage model satisfies the positivity of the intrinsic dissipation, $\mathcal{D} = \sigma_{ij}\dot{\epsilon}_{ij} - \rho\dot{\psi}$, as

$$\begin{aligned}\mathcal{D} &= \sigma_{ij}^{\pi}\dot{\epsilon}_{ij}^{\pi} - Q\dot{q} - X_{ij}\dot{\alpha}_{ij} + Y\dot{D} \\ &= \left(\sigma_{ij}^{\pi}\frac{\partial F}{\partial \sigma_{ij}^{\pi}} + Q\frac{\partial F}{\partial Q} + X_{ij}\frac{\partial F}{\partial X_{ij}} + Y\frac{\partial F}{\partial Y}\right)\dot{\lambda} \geq 0\end{aligned} \qquad (1.168)$$

at constant temperature when the dissipation potential $F(\boldsymbol{\sigma}^{\pi}, Q, \boldsymbol{X}, Y; D)$ is a non-negative convex function of its arguments $\boldsymbol{\sigma}^{\pi}$, R, $\boldsymbol{X}$, and Y, where $F(\boldsymbol{0}, 0, \boldsymbol{0}, 0; D) = 0$ and the damage D acts as a parameter. Using the evolution laws, the intrinsic dissipation may also be rewritten as

$$\mathcal{D} = \left(\sigma_\mathrm{s} + \sigma_\mathrm{v} + \frac{\gamma}{C_\mathrm{X}}X_{ij}X_{ij}\right)(1-D)\|\dot{\boldsymbol{\epsilon}}^{\pi}\| + Y\dot{D} \geq 0. \qquad (1.169)$$

This condition requires the damage rate to be positive or zero ($\dot{D} \geq 0$, an eventual recovery of damage needs a new variable) and the stored energy rate $\dot{w}_\mathrm{s} = X_{ij}\dot{\alpha}_{ij} + Q\dot{q}$ to remain smaller or equal to the inelastic strain power ($\sigma_{ij}^{\pi}\dot{\epsilon}_{ij}^{\pi}$).

1.4.5.2 Extension to Induced Anisotropy

Note that the generalization of previous damage evolution extension to induced damage anisotropy is also possible. It is

$$\dot{D}_{ij} = \left(\frac{\overline{Y}}{S}\right)^s |\dot{\boldsymbol{\epsilon}}^\pi|_{ij} , \qquad (1.170)$$

with $|.|$ as the absolute value of a tensor for ductile materials such as metals (with then $\boldsymbol{\epsilon}^\pi = \boldsymbol{\epsilon}^\mathrm{p}$), or

$$\dot{D}_{ij} = \left(\frac{\overline{Y}}{S}\right)^s \langle\dot{\boldsymbol{\epsilon}}^\pi\rangle^+_{ij} , \qquad (1.171)$$

with $\langle.\rangle^+$ as the positive part of a tensor for quasi-brittle materials such as concrete for (J. Mazars, Y. Berthaud and S. Ramtani 1990):

$$\begin{aligned}\boldsymbol{D} &= \begin{bmatrix} D & 0 & 0 \\ 0 & 0 & 0 \\ 0 & 0 & 0 \end{bmatrix} \quad \text{in tension} \\ \boldsymbol{D} &= \begin{bmatrix} 0 & 0 & 0 \\ 0 & D & 0 \\ 0 & 0 & D \end{bmatrix} \quad \text{in compression}\end{aligned} \qquad (1.172)$$

when the loading direction is $\vec{e}_1$.

1.4.5.3 Application to Metals, Concrete and Elastomers

To conclude, this section (Sect. 1.4.5) can be considered a synthetic presentation of constitutive models for metals, elastomers, concrete:

- For metals, $w_1 = 0$, $Q(q) = R(r)$ is the isotropic hardening, $\boldsymbol{X}$ is the kinematic hardening, $\|.\|$ is the von Mises norm $(.)_{\mathrm{eq}}$, $\boldsymbol{\epsilon}^\pi$ is the plastic strain $\boldsymbol{\epsilon}^\mathrm{p}$, and the generalized damage law (1.167) recovers $\dot{D} = (Y/S)^s \dot{p}$ where π equals the accumulated plastic strain p.
- For filled elastomers, and anticipating the choices made in Sect. 4.4.3 for the model of hyperelasticity with internal friction coupled with damage, w_1 is an hyperelasticity density such as Mooney or Hart–Smith densities, $R = 0$, and w_2 and F_X are quadratic functions. The variable $\boldsymbol{X}$ stands for the residual microstresses due to internal sliding with friction of the macromolecular chains on themselves and the black carbon filler particles. The internal inelastic strain $\boldsymbol{\epsilon}^\pi$ represents the average displacements incompatibilities due to friction on a microscale.
- For concrete, w_1, w_2, and F_X are quadratic, Q and $\boldsymbol{X}$ model the growth and the translation of the reversibility domain $f < 0$ in the $\boldsymbol{\sigma}^\pi$ space, and $\boldsymbol{\epsilon}^\pi$ represents the average slip discontinuities of the cracked media on a microscale.

1.5 Elasto-(Visco-)Plasticity Coupled with Damage

The whole set of constitutive equations is derived from the **dissipation potential** as explained in the general thermodynamics framework of Sect. 1.2.1. The different models differ by the choice of analytic forms of the functions f, F_X, and F_D for the potential

$$F = f + F_X + F_D . \tag{1.173}$$

1.5.1 Basic Equations without Damage Coupling ♂

Both plasticity and visco-plasticity introduce a **yield criterion** (f) by defining the elastic domain of the material by $f < 0$, and the plastic and visco-plastic states by

$$\begin{aligned} f = 0 , \quad \dot{f} = 0 & \quad \text{plasticity} \\ f = \sigma_v > 0 , & \quad \text{visco-plasticity} \end{aligned} \tag{1.174}$$

respectively. The term σ_v is the viscous stress given by the law of viscosity. More often, the von Mises criterion is used with isotropic hardening,

$$f = \sigma_{\text{eq}} - R - \sigma_y , \tag{1.175}$$

where σ_{eq} is the von Mises equivalent stress.

When isotropic and kinematic hardening are taken into consideration,

$$f = (\boldsymbol{\sigma} - \boldsymbol{X})_{\text{eq}} - R - \sigma_y , \tag{1.176}$$

with $(\boldsymbol{\sigma} - \boldsymbol{X})_{\text{eq}} = \sqrt{\frac{3}{2}(\sigma_{ij}^D - X_{ij}^D)(\sigma_{ij}^D - X_{ij}^D)}$.

The **isotropic hardening** R is related to the density of dislocations or flow arrests and it represents the growth in size of the yield surface. Exponential isotropic hardening is considered to ensure a quasi-saturation of the strain hardening when damage occurs, i.e.,

$$R = R(r) = R_\infty \left[1 - \exp(-br)\right] , \tag{1.177}$$

with R_∞ and b as temperature-dependent material parameters. The variable r is equal to the accumulated plastic strain p as long as there is no damage. A power law $R = K_p r^{1/M}$ or a linear law ($M = 1$) may also be used.

The **kinematic hardening** governed by the back stress $\boldsymbol{X}$ is related to the state of internal microstress concentration. It represents the translation of the yield surface as $\boldsymbol{X}$ defines the center of the current elastic domain in the stress space. Nonlinear kinematic hardening is modelled by the function potential F_X,

$$F_X = \frac{3\gamma}{4C} X_{ij} X_{ij} , \tag{1.178}$$

with C and γ temperature-dependent material parameters. As long as there is no damage, the classical evolution laws for the back stress are

$$X_{ij} = \frac{2}{3}C\epsilon_{ij}^{\mathrm{P}} \quad \text{for linear kinematic hardening}$$
$$\frac{\mathrm{d}}{\mathrm{d}t}\left(\frac{X_{ij}}{C}\right) = \frac{2}{3}\dot{\epsilon}_{ij}^{\mathrm{P}} - \frac{\gamma}{C}X_{ij}\dot{p} \quad \text{for nonlinear kinematic hardening.} \quad (1.179)$$

The last equation is the non-isothermal form of the Armstrong–Frederick law which is often used with $X_\infty = C/\gamma$ for easier identification, as in monotonic uniaxial loading $X = X_\infty[1 - \exp(-\gamma\epsilon_\mathrm{p})]$. The back stress $\boldsymbol{X}$ governed by the plastic strain is a deviatoric tensor ($X_{kk} = 0$ and $\boldsymbol{X} = \boldsymbol{X}^{\mathrm{D}}$). For Prager linear hardening, $X_{ij} = \frac{2}{3}C\epsilon_{ij}^{\mathrm{P}}$ can be recovered by setting $\gamma = 0$.

A strong effect of the temperature (T) is observed as there is a transition from visco-plastic time-dependent (viscous) behavior at high temperature, such as creep, to plastic time-independent behavior at lower temperature.

The constitutive equations can be derived from (1.25) as follows:

$$\begin{aligned}
\dot{\epsilon}_{ij}^{\mathrm{P}} &= \dot{\lambda}\frac{\partial F}{\partial \sigma_{ij}}, \\
\dot{r} &= -\dot{\lambda}\frac{\partial F}{\partial R}, \\
\dot{\alpha}_{ij} &= -\dot{\lambda}\frac{\partial F}{\partial X_{ij}}, \\
\dot{D} &= \dot{\lambda}\frac{\partial F}{\partial Y}, \quad \text{or} \quad \dot{D}_{ij} = \dot{\lambda}\frac{\partial F}{\partial Y_{ij}},
\end{aligned} \quad (1.180)$$

provided that

- *For plasticity*, the multiplier is determined by the consistency condition $f = 0$, $\dot{f} = 0$,

$$\dot{\lambda} = \frac{\dfrac{\partial f}{\partial \sigma_{ij}}\dot{\sigma}_{ij} + \dfrac{2}{3}\dfrac{\mathrm{d}C}{\mathrm{d}T}\alpha_{ij}\dfrac{\partial f}{\partial X_{ij}}\dot{T}}{\dfrac{\mathrm{d}R}{\mathrm{d}T}\dfrac{\partial f}{\partial R}\dfrac{\partial F}{\partial R} + \dfrac{2}{3}C\dfrac{\partial f}{\partial X_{ij}}\dfrac{\partial F}{\partial X_{ij}} - \dfrac{\partial f}{\partial D_{ij}}\dfrac{\partial F}{\partial D_{ij}}}. \quad (1.181)$$

- *For viscoplasticity*, the multiplier is a function of the accumulated plastic strain rate ($\dot{p} = \sqrt{\frac{2}{3}\dot{\epsilon}_{ij}^{\mathrm{P}}\dot{\epsilon}_{ij}^{\mathrm{P}}}$) given by a viscosity law. Different choices are possible:
  − Norton power law

$$\sigma_\mathrm{v} = K_N \dot{p}^{\frac{1}{N}} \quad \text{or} \quad \dot{p} = \left(\frac{f}{K_N}\right)^N, \quad (1.182)$$

1.5 Elasto-(Visco-)Plasticity Coupled with Damage

- Exponential law to ensure saturation at large plastic strain rate,

$$\sigma_v = K_\infty \left[1 - \exp\left(-\frac{\dot{p}}{n}\right)\right] \quad \text{or} \quad \dot{p} = \ln\left(1 - \frac{f}{K_\infty}\right)^{-n}, \quad (1.183)$$

where K_N, K_∞, N, and n are material parameters that depend on the temperature.

There are **many possibilities** in modelling plasticity and viscosity, depending on whether or not the yield stress, kinematic hardening, and isotropic hardening are equal to zero. The model must be chosen carefully because the material parameters for a given material have values that are specific to each model.

To be as clear as possible, the notations adopted in the book for the (visco-)plasticity parameters have an extra "0" in superscript or lowerscript when a zero yield stress is considered. They have an extra "y" in superscript or lowerscript when (visco-)plasticity is modelled with a non-zero yield stress but with one hardening only, either isotropic or kinematic. This is illustrated in the uniaxial case as:

- Yield function in monotonic tension $f = \sigma - X - R - \sigma_y$
  - If $\sigma_y = 0$ and $X = 0$, $R = K_p^0 \epsilon_p^{1/M_0}$
  - If $\sigma_y \neq 0$ and $X = 0$, $R = K_p^y \epsilon_p^{1/M_y}$ or $R = R_\infty^y (1 - \exp(-b_y \epsilon_p))$
  - If $\sigma_y \neq 0$ and $R = 0$, $X = C_y \epsilon_p$ or $X = X_\infty^y (1 - \exp(-\gamma_y \epsilon_p))$
  - If $\sigma_y \neq 0$, $X \neq 0$ and $R \neq 0$,
    $$\begin{cases} R = K_p \epsilon_p^{1/M} & \text{or } R = R_\infty (1 - \exp(-b\epsilon_p)), \\ X = C\epsilon_p & \text{or } X = X_\infty (1 - \exp(-\gamma \epsilon_p)), \end{cases}$$

- Viscosity law
  - If $\sigma_y = 0$, $X = 0$ and $R = 0$, $\dot{\epsilon}_p = \left(\dfrac{\sigma}{K_N^0}\right)^{N_0}$
  - If $\sigma_y \neq 0$, $X = 0$ and $R = 0$, $\dot{\epsilon}_p = \left(\dfrac{\sigma - \sigma_y}{K_N^y}\right)^{N_y}$
  - If $\sigma_y \neq 0$, $X \neq 0$ and $R \neq 0$, $\dot{\epsilon}_p = \left(\dfrac{\sigma - X - R - \sigma_y}{K_N}\right)^N$

  or $\dot{\epsilon}_p = \ln\left(1 - \dfrac{\sigma - X - R - \sigma_y}{K_\infty}\right)^{-n}$.

1.5.1.1 Fast Identification of (Visco-)Plasticity Material Parameters

An accurate value of the (visco-)plasticity material parameters is necessary for practical applications and may be obtained independently from the damage parameters if $p < p_D$. Altogether, there are 5 (+2) characteristic parameters of (visco-)plasticity to determine for each material and each temperature:

48   1 Background on Continuum Damage Mechanics

- Yield stress: σ_y.
- Isotropic hardening: R_∞ and b (or K_p and M).
- Kinematic hardening: C (linear) or X_∞ and γ (nonlinear).
- Viscosity: K_N and N or K_∞ and n.

1.5.1.1.1 Hardening parameters

For hardening, a direct and fast identification is possible from a tension test or tension-compression low cycles tests that are strain controlled, at increasing amplitude below the damage threshold, as R and X may be directly measured on the graph $\sigma(\epsilon_p)$.

- From a monotonic tensile curve it is only possible to identify one hardening which is either
    - An isotropic hardening by
        · A power law $R = \sigma - \sigma_y = K_p^y \epsilon_p^{1/M_y}$
        · An exponential law $R = \sigma - \sigma_y = R_\infty^y [1 - \exp(-b_y \epsilon_p)]$
    - Or a kinematic hardening by
        · A linear law $X = \sigma - \sigma_y = C_y \epsilon_p$
        · An exponential law $X = \sigma - \sigma_y = X_\infty^y [1 - \exp(-\gamma_y \epsilon_p)]$ with $X_\infty^y = R_\infty^y$, $\gamma_y = b_y$.

From the example of Fig. 1.12 for an Inconel alloy at room temperature with a yield stress depending upon the chosen hardening law:

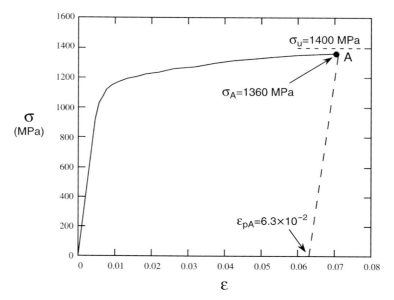

Fig. 1.12. Tension curve for identification of either isotropic or kinematic hardening parameters (Inconel alloy at room temperature, J. Dufailly 1987)

$\sigma_y = 920$ MPa, $K_p^y = 810$ MPa, $M_y = 4.5$ for the power R law

$\sigma_y = 1120$ MPa, $C_y = (\sigma_A - \sigma_y)/\epsilon_{pA} = 3800$ MPa for the linear X law

$\sigma_y = 1040$ MPa, $R_\infty^y = X_\infty^y = 315$ MPa, $b_y = \gamma_y = 55$ for the exp. laws

- From a tensile curve with repeated unloadings and compressions until the yield stresses in compression are reached, we can identify both isotropic and kinematic hardenings acting together as the back stress X is the ordinate of the middle of the elasticity straight line. The test should be performed at very low strain rate if the material exhibits viscosity.

  From the example of Fig. 1.13 on the same material used for Fig. 1.12, a curve fitting procedure gives the following for isotropic and kinematic hardenings considered altogether:
  a) $K_p = 280$ MPa, $M = 5.5$ for the isotropic hardening power law,
     or $R_\infty = 165$ MPa, $b = 80$ for the isotropic hardening exponential law
  b) $C = 7140$ MPa and $\gamma = 0$ for linear kinematic hardening,
     or $X_\infty = 450$ MPa, $\gamma = 60$ (and $C = \gamma X_\infty = 27000$ MPa) for nonlinear kinematic hardening

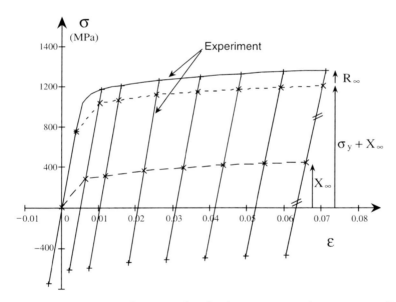

Fig. 1.13. Tension test with repeated unloadings up to a plastic strain offset of 10^{-5} for the identification of both isotropic and kinematic hardenings parameters (Inconel alloy at room temperature, J. Dufailly 1987)

- From the stabilized cycles of tension-compression experiments performed at different plastic strain ranges $\Delta\epsilon_p$, it is possible to identify the nonlinear kinematic hardening parameters C, γ, $X_\infty = C/\gamma$.

The integration of the nonlinear law (1.179) over each half cycle leads to

$$\frac{\Delta\sigma}{2} - k = X_\infty \tanh\left(\gamma\frac{\Delta\epsilon_p}{2}\right), \tag{1.184}$$

where $\Delta\sigma$ is the stress range of the considered cycle and $2k$ is the size of the elastic domain measured on each cycle unloading for each strain range ($k = \sigma_y + R$ if there is no viscosity, $k = \sigma_y + R + \sigma_v$ if there is). We can identify X_∞ as the asymptote of the curve $\frac{\Delta\sigma}{2} - k$ vs $\Delta\epsilon_p$ and then determine γ using a curve fitting procedure.

1.5.1.1.2 Viscosity parameters

The more accurate way to determine viscosity parameters is to perform relaxation tests after saturating both isotropic and kinematic hardenings. The saturation values R_∞ and X_∞ can be found from a very slow tensile test:

$$\sigma = \sigma_y + R_\infty + X_\infty + \sigma_v, \tag{1.185}$$

where σ_v is the viscous stress.

At constant strain, $\epsilon = \epsilon_e + \epsilon_p$, the strain rate $\dot\epsilon = 0$, and

$$\dot\epsilon_p = -\frac{\dot\sigma}{E} \tag{1.186}$$

coupled with the viscosity law leads to a nonlinear differential equation whose solution is the $\sigma(t)$ curve. The graph of $\sigma_v = \sigma - \sigma_y - R_\infty - X_\infty$ as a function of $\dot\epsilon_p = -\dot\sigma/E$ allows then for the identification of either

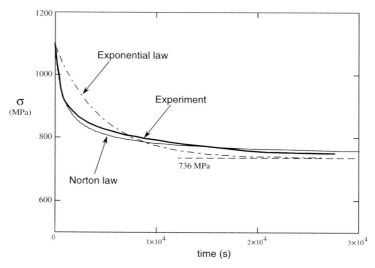

Fig. 1.14. Relaxation test for identification of viscosity parameters of an Inconel alloy at $T = 627°C$ (J. Dufailly 1987)

1.5 Elasto-(Visco-)Plasticity Coupled with Damage

- The Norton law $\dot{\epsilon}_p = \left(\dfrac{\sigma_v}{K_N}\right)^N$
- The exponential law $\sigma_v = K_\infty \left[1 - \exp\left(-\dfrac{\dot{\epsilon}_p}{n}\right)\right]$

From the example of Fig. 1.14 with $E = 160000$ MPa, a curve fitting procedure gives $K_N = 75000$ MPa·s$^{1/N}$, $N = 2.4$; $K_\infty = 10^7$ MPa, and $n = 1.4\,10^{-2}\mathrm{s}^{-1}$.

For better accuracy, the results of this fast identification may be taken as the starting solution of a numerical procedure of optimization if other tests are available (see Sect. 2.4).

1.5.1.2 Concerning the Choice of a Damage Model

Different possibilities arise corresponding to the physical effects really needed for the application in mind, the quantitative information available in the experimental data base and different levels of mathematical and numerical complexity. The following advice may be considered as guidelines:

- For **proportional loading** and more generally when simplicity is the most important issue, the isotropic damage models are preferred. When analytical solutions are needed, this choice is necessary and compatible with the consideration of the microdefects closure effect.
- For **numerical computations**, the use of the anisotropic damage models is encouraged as they do not introduce additional material parameters, except maybe the coefficient η which is equal to 3 in most cases. The difficulty related to the consideration of induced anisotropy is mostly numerical but Sect. 2.2 is devoted to the subject.
- For **non-proportional loading** (difficult to handle in closed form), choose the anisotropic damage models as they reproduce in a more realistic way the accumulation of the damages due to loadings applied in different directions.
- For **fatigue applications**, we advise using the models that include the microdefects closure effect as they lead to an evolution of the damage different in tension than in compression and then to a mean stress effect.

Last, choose a proper **damage threshold criterion** (see Sect. 1.4.1):

- For **monotonic applications** use the criterion $p = \epsilon_{pD}$,
- For **fatigue applications** use the criterion $p = p_D$. It is possible to consider $p_D = constant$ as long as no overloading occurs and then usually $p_D \gg \epsilon_{pD}$. The preferred way to calculate p_D for each loading amplitude is using

$$p_D = \epsilon_{pD} \left(\dfrac{\sigma_u - \sigma_f^\infty}{\dfrac{\sigma_{max} + |\sigma_{min}|}{2} - \sigma_f^\infty}\right)^m. \tag{1.187}$$

52   1 Background on Continuum Damage Mechanics

- For **numerical analysis** with a complex history of loading, use the stored energy criterion max $w_s = w_D$. It is then necessary to consider both nonlinear isotropic and kinematic hardenings with the thermodynamics correction of Sect. 1.4.1. The material parameters for the damage initiation are either w_D, A, m or ϵ_{pD}, A, m, with A and w_D given by (1.114) and (1.116). If the fatigue limit σ_f^∞ and the plasticity parameters are known, use the approximate formula (1.153) to calculate A.

1.5.2 Coupling with Isotropic Damage^{♂,♂♂}

The damage is represented by the scalar D. Its law of evolution is (see Sect. 1.4)

$$\dot{D} = \left(\frac{Y}{S}\right)^s \dot{p}, \quad \text{if} \quad r > p_D \quad \text{or} \quad \max w_s > w_D, \tag{1.188}$$

where the damage threshold is written in terms of stored energy or in an equivalent manner of accumulated plastic strain (see Sect. 1.4.1).

The full sets of constitutive equations for elasto-(visco-)plasticity coupled with isotropic damage with nonlinear isotropic and kinematic hardening are given in the following tables, where the case of linear kinematic hardening is recovered by setting $\gamma = 0$.

1.5.2.1 Isotropic Damage without Microdefects Closure Effect[♂] (Table 1.3)

The coupling of damage with elasticity as well as plasticity is made through the use of the effective stress $\tilde{\sigma}$ instead of σ in the elasticity law,

$$\epsilon_{ij}^e = \frac{1+\nu}{E}\tilde{\sigma}_{ij} - \frac{\nu}{E}\tilde{\sigma}_{kk}\delta_{ij}, \quad \tilde{\sigma}_{ij} = \frac{\sigma_{ij}}{1-D}, \tag{1.189}$$

and in the von Mises criterion,

$$f = (\tilde{\boldsymbol{\sigma}} - \boldsymbol{X})_{eq} - R - \sigma_y. \tag{1.190}$$

The plasticity remains incompressible and $\boldsymbol{\epsilon}^p$, $\boldsymbol{X}$, and $\boldsymbol{\alpha}$ are deviatoric tensors.

1.5 Elasto-(Visco-)Plasticity Coupled with Damage

Table 1.3. Elasto-(visco-)plasticity coupled with **isotropic damage**

Strain partition	$\epsilon_{ij} = \epsilon_{ij}^{\text{e}} + \epsilon_{ij}^{\text{p}}$
Thermo-elasticity	$\epsilon_{ij}^{\text{e}} = \dfrac{1+\nu}{E}\tilde{\sigma}_{ij} - \dfrac{\nu}{E}\tilde{\sigma}_{kk}\delta_{ij} + \alpha(T - T_{\text{ref}})\delta_{ij}$
(Visco-)plasticity	$\dot{\epsilon}_{ij}^{\text{p}} = \dfrac{3}{2}\dfrac{(\tilde{\sigma}_{ij}^{\text{D}} - X_{ij})}{(\tilde{\boldsymbol{\sigma}} - \boldsymbol{X})_{\text{eq}}}\dfrac{\dot{r}}{1-D}$ $\dot{p} = \dfrac{\dot{r}}{1-D}$ $R = R_{\infty}(1 - \exp(-br))$ $\dfrac{\text{d}}{\text{d}t}\left(\dfrac{X_{ij}}{\gamma X_{\infty}}\right) = \dfrac{2}{3}(1-D)\dot{\epsilon}_{ij}^{\text{p}} - \dfrac{X_{ij}}{X_{\infty}}\dot{r}$
Damage	$\dot{D} = \left(\dfrac{Y}{S}\right)^{s}\dot{p}$ if $r > p_{\text{D}}$ or $\max w_{\text{s}} > w_{\text{D}}$ up to D_{c} $Y = \dfrac{1}{2}E_{ijkl}\epsilon_{ij}^{\text{e}}\epsilon_{kl}^{\text{e}} = \dfrac{\tilde{\sigma}_{\text{eq}}^{2}R_{\nu}}{2E}$ $R_{\nu} = \dfrac{2}{3}(1+\nu) + 3(1-2\nu)\left(\dfrac{\sigma_{\text{H}}}{\sigma_{\text{eq}}}\right)^{2}$
Plastic multiplier	$\dot{r} = \dot{\lambda}$ given by $f = 0$ and $\dot{f} = 0$
Visco-plastic multiplier	$\dot{p} = \left(\dfrac{f}{K_N}\right)^{N}$ Norton law $\dot{p} = \ln\left(1 - \dfrac{f}{K_{\infty}}\right)^{-n}$ exponential law

1.5.2.2 Isotropic Damage with Microdefects Closure Effect♂♂ (Table 1.4)

The coupling of damage with elasticity is made through the use of the effective stress,

$$\tilde{\sigma}_{ij} = \frac{\langle \boldsymbol{\sigma} \rangle_{ij}^+}{1-D} + \frac{\langle \boldsymbol{\sigma} \rangle_{ij}^-}{1-hD} \\ + \frac{\nu}{1-2\nu} \left(\frac{\delta_{kl} \langle \boldsymbol{\sigma} \rangle_{kl}^+ - \langle \sigma_{kk} \rangle}{1-D} + \frac{\delta_{kl} \langle \boldsymbol{\sigma} \rangle_{kl}^- + \langle -\sigma_{kk} \rangle}{1-hD} \right) \delta_{ij}, \tag{1.191}$$

defined in Sect. 1.2.4 in the elasticity law, but not in the von Mises criterion in which $\boldsymbol{\sigma}/(1-D)$ is used. The justification is in the mechanism of plasticity itself controlled by slips and produced by shear stresses in the same manner regardless of their signs. Furthermore, as mentioned in Sect. 1.3.4, damage measurements mainly in compression by microhardness give the same values as those obtained by elasticity change in tension:

$$f = \left(\frac{\boldsymbol{\sigma}}{1-D} - \boldsymbol{X} \right)_{eq} - R - \sigma_y. \tag{1.192}$$

The plasticity remains incompressible and $\boldsymbol{\epsilon}^p$, $\boldsymbol{X}$, and $\boldsymbol{\alpha}$ are deviatoric tensors.

1.5 Elasto-(Visco-)Plasticity Coupled with Damage

Table 1.4. Elasto-(visco-)plasticity coupled with **isotropic damage** and with **microdefects closure effect**

Strain partition	$\epsilon_{ij} = \epsilon_{ij}^{\mathrm{e}} + \epsilon_{ij}^{\mathrm{p}}$
Thermo-elasticity	$\epsilon_{ij}^{\mathrm{e}} = \dfrac{1+\nu}{E}\tilde{\sigma}_{ij} - \dfrac{\nu}{E}\tilde{\sigma}_{kk}\delta_{ij} + \alpha(T - T_{\mathrm{ref}})\delta_{ij}$
(Visco-)plasticity	$\dot{\epsilon}_{ij}^{\mathrm{p}} = \dfrac{3}{2}\dfrac{\dfrac{\sigma_{ij}^{\mathrm{D}}}{1-D} - X_{ij}}{\left(\dfrac{\sigma}{1-D} - \boldsymbol{X}\right)_{\mathrm{eq}}}\dfrac{\dot{r}}{1-D}$ $\dot{p} = \dfrac{\dot{r}}{1-D}$ $R = R_\infty(1 - \exp(-br))$ $\dfrac{\mathrm{d}}{\mathrm{d}t}\left(\dfrac{X_{ij}}{\gamma X_\infty}\right) = \dfrac{2}{3}(1-D)\dot{\epsilon}_{ij}^{\mathrm{p}} - \dfrac{X_{ij}}{X_\infty}\dot{r}$
Damage	$\dot{D} = \left(\dfrac{Y}{S}\right)^s \dot{p}$ if $r > p_{\mathrm{D}}$ or $\max w_{\mathrm{s}} > w_{\mathrm{D}}$ up to D_{c} $Y = \dfrac{1+\nu}{2E}\left[\dfrac{\langle\boldsymbol{\sigma}\rangle_{ij}^{+}\langle\boldsymbol{\sigma}\rangle_{ij}^{+}}{(1-D)^2} + h\dfrac{\langle\boldsymbol{\sigma}\rangle_{ij}^{-}\langle\boldsymbol{\sigma}\rangle_{ij}^{-}}{(1-hD)^2}\right]$ $-\dfrac{\nu}{2E}\left[\dfrac{\langle\sigma_{kk}\rangle^2}{(1-D)^2} + h\dfrac{\langle-\sigma_{kk}\rangle^2}{(1-hD)^2}\right]$
Plastic multiplier	$\dot{r} = \dot{\lambda}$ given by $f = 0$ and $\dot{f} = 0$
Visco-plastic multiplier	$\dot{p} = \left(\dfrac{f}{K_N}\right)^N$ Norton law $\dot{p} = \ln\left(1 - \dfrac{f}{K_\infty}\right)^{-n}$ exponential law

1.5.3 Coupling with Anisotropic Damage[♂♂,♂♂♂]

The damage is represented by the second order tensor $\boldsymbol{D}$. Its evolution is governed by the law (see Sect. 1.4)

$$\dot{D}_{ij} = \left(\frac{\overline{Y}}{S}\right)^s |\dot{\epsilon}^{\mathrm{p}}|_{ij} \quad \text{if} \quad r > p_{\mathrm{D}} \quad \text{or} \quad \max w_{\mathrm{s}} > w_{\mathrm{D}}, \tag{1.193}$$

where the principal directions of the damage rate coincide with the absolute value (in terms of principal values) of the plastic strain rate. The damage threshold is written in terms of stored energy or in an equivalent manner of accumulated plastic strain (see Sect. 1.4.1).

The full sets of constitutive equations for elasto-(visco-)plasticity coupled with anisotropic damage with nonlinear isotropic and kinematic hardening are given in the following tables, where the case of linear kinematic hardening is recovered by setting $\gamma = 0$.

1.5.3.1 Anisotropic Damage without Microdefects Closure Effect[♂♂] (Table 1.5)

The coupling of damage with elasticity is made through the use of the effective stress,

$$\tilde{\sigma}_{ij} = (H_{ik}\sigma_{kl}^{\mathrm{D}}H_{lj})^{\mathrm{D}} + \frac{\sigma_{\mathrm{H}}}{1 - \eta D_{\mathrm{H}}}\delta_{ij}, \tag{1.194}$$

in the elasticity law as well as in the von Mises criterion,

$$f = (\tilde{\boldsymbol{\sigma}} - \boldsymbol{X})_{\mathrm{eq}} - R - \sigma_{\mathrm{y}}. \tag{1.195}$$

The plasticity remains incompressible and $\boldsymbol{\epsilon}^{\mathrm{p}}$, $\boldsymbol{X}$, and $\boldsymbol{\alpha}$ are deviatoric tensors.

Writing the following set of equations in terms of effective stress makes it fully similar to the set obtained in the case of isotropic damage.

1.5 Elasto-(Visco-)Plasticity Coupled with Damage

Table 1.5. Elasto-(visco-)plasticity coupled with **anisotropic damage**

Strain partition	$\epsilon_{ij} = \epsilon_{ij}^{\text{e}} + \epsilon_{ij}^{\text{p}}$
Thermo-elasticity	$\epsilon_{ij}^{\text{e}} = \dfrac{1+\nu}{E}\tilde{\sigma}_{ij} - \dfrac{\nu}{E}\tilde{\sigma}_{kk}\delta_{ij} + \alpha(T - T_{\text{ref}})\delta_{ij}$
(Visco-)plasticity	$\dot{\epsilon}_{ij}^{\text{p}} = [H_{ik}\dot{e}_{kl}^{\text{p}} H_{lj}]^{\text{D}}$ with $\dot{e}_{ij}^{\text{p}} = \dfrac{3}{2}\dfrac{\tilde{\sigma}_{ij}^{\text{D}} - X_{ij}}{(\tilde{\boldsymbol{\sigma}} - \boldsymbol{X})_{\text{eq}}}\dot{r}$ $\dot{p} = \dfrac{\left[\boldsymbol{H}(\tilde{\boldsymbol{\sigma}}^{\text{D}} - \boldsymbol{X})\boldsymbol{H}\right]_{\text{eq}}}{(\tilde{\boldsymbol{\sigma}} - \boldsymbol{X})_{\text{eq}}}\dot{r}$ $R = R_\infty\left(1 - \exp(-br)\right)$ $\dfrac{\text{d}}{\text{d}t}\left(\dfrac{X_{ij}}{\gamma X_\infty}\right) = \dfrac{2}{3}\dot{e}_{ij}^{\text{p}} - \dfrac{X_{ij}}{X_\infty}\dot{r}$
Damage	$\dot{D}_{ij} = \left(\dfrac{\overline{Y}}{S}\right)^s |\dot{\boldsymbol{\epsilon}}^{\text{p}}|_{ij}\quad \begin{array}{l}\text{if } r > p_{\text{D}} \text{ or } \max w_{\text{s}} > w_{\text{D}}\\ \text{up to } D_{\text{c}}\end{array}$ $\overline{Y} = \dfrac{1}{2}E_{ijkl}\epsilon_{ij}^{\text{e}}\epsilon_{kl}^{\text{e}} = \dfrac{\tilde{\sigma}_{\text{eq}}^2 \tilde{R}_\nu}{2E}$ $\tilde{R}_\nu = \dfrac{2}{3}(1+\nu) + 3(1-2\nu)\left(\dfrac{\tilde{\sigma}_{\text{H}}}{\tilde{\sigma}_{\text{eq}}}\right)^2$ $\tilde{\sigma}_{\text{eq}} = \left(\boldsymbol{H}\boldsymbol{\sigma}^{\text{D}}\boldsymbol{H}\right)_{\text{eq}},\ \tilde{\sigma}_{\text{H}} = \dfrac{\sigma_{\text{H}}}{1 - \eta D_{\text{H}}}$
Plastic multiplier	$\dot{r} = \dot{\lambda}$ given by $f = 0$ and $\dot{f} = 0$
Visco-plastic multiplier	$\dot{p} = \left(\dfrac{f}{K_N}\right)^N\quad$ Norton law $\dot{p} = \ln\left(1 - \dfrac{f}{K_\infty}\right)^{-n}\quad$ exponential law

1.5.3.2 Anisotropic Damage with Microdefects Closure Effect♢♢♢ (Table 1.6)

The coupling of damage with elasticity is made through the use of the effective stress,

$$\tilde{\sigma}_{ij} = (H^{\mathrm{p}}_{ik}\sigma^{\mathrm{D}}_{+kl}H^{\mathrm{p}}_{lj})^{\mathrm{D}} + (H^{\mathrm{n}}_{ik}\sigma^{\mathrm{D}}_{-kl}H^{\mathrm{n}}_{lj})^{\mathrm{D}} + \left[\frac{\langle\sigma_{\mathrm{H}}\rangle}{1-\eta D_{\mathrm{H}}} - \frac{\langle-\sigma_{\mathrm{H}}\rangle}{1-\eta h_{\mathrm{a}} D_{\mathrm{H}}}\right]\delta_{ij}, \tag{1.196}$$

defined in Sect. 1.2.4 in the elasticity law, where

$$H^{\mathrm{p}}_{ij} = (\mathbf{1}-\mathbf{D})^{-1/2}_{ij} \quad \text{and} \quad H^{\mathrm{n}}_{ij} = (\mathbf{1}-h_{\mathrm{a}}\mathbf{D})^{-1/2}_{ij}. \tag{1.197}$$

The full $(\boldsymbol{H}^{\mathrm{P}}\boldsymbol{\sigma}^{\mathrm{D}}\boldsymbol{H}^{\mathrm{P}})^{\mathrm{D}}$ is introduced in the yield criterion, not affected by the closure effect,

$$f = \left(\boldsymbol{H}^{\mathrm{P}}\boldsymbol{\sigma}^{\mathrm{D}}\boldsymbol{H}^{\mathrm{P}} - \boldsymbol{X}\right)_{\mathrm{eq}} - R - \sigma_{\mathrm{y}}. \tag{1.198}$$

The plasticity remains incompressible and $\boldsymbol{\epsilon}^{\mathrm{p}}$, $\boldsymbol{X}$, and $\boldsymbol{\alpha}$ are deviatoric tensors.

1.5 Elasto-(Visco-)Plasticity Coupled with Damage

Table 1.6. Elasto-(visco-)plasticity coupled with **anisotropic damage** and with **microdefects closure effect**

Strain partition	$\epsilon_{ij} = \epsilon_{ij}^{\mathrm{e}} + \epsilon_{ij}^{\mathrm{p}}$
Thermo-elasticity	$\epsilon_{ij}^{\mathrm{e}} = \dfrac{1+\nu}{E}\tilde{\sigma}_{ij} - \dfrac{\nu}{E}\tilde{\sigma}_{kk}\delta_{ij} + \alpha(T - T_{\mathrm{ref}})\delta_{ij}$
(Visco-)plasticity	$\dot{\epsilon}_{ij}^{\mathrm{p}} = \left[H_{ik}^{\mathrm{p}} \dot{e}_{kl}^{\mathrm{p}} H_{lj}^{\mathrm{p}}\right]^{\mathrm{D}}$ with $\dot{e}_{ij}^{\mathrm{p}} = \dfrac{3}{2}\dfrac{(H_{ik}^{\mathrm{p}}\sigma_{kl}^{\mathrm{D}} H_{lj})^{\mathrm{D}} - X_{ij}}{(\boldsymbol{H}^{\mathrm{P}}\boldsymbol{\sigma}^{\mathrm{D}}\boldsymbol{H}^{\mathrm{P}} - \boldsymbol{X})_{\mathrm{eq}}} \dot{r}$ $\dot{p} = \dfrac{\left[\boldsymbol{H}^{\mathrm{P}}\left((\boldsymbol{H}^{\mathrm{P}}\boldsymbol{\sigma}^{\mathrm{D}}\boldsymbol{H}^{\mathrm{P}})^{\mathrm{D}} - \boldsymbol{X}\right)\boldsymbol{H}^{\mathrm{P}}\right]_{\mathrm{eq}}}{(\boldsymbol{H}^{\mathrm{P}}\boldsymbol{\sigma}^{\mathrm{D}}\boldsymbol{H}^{\mathrm{P}} - \boldsymbol{X})_{\mathrm{eq}}}\dot{r}$ $R = R_\infty\left(1 - \exp(-br)\right)$ $\dfrac{\mathrm{d}}{\mathrm{d}t}\left(\dfrac{X_{ij}}{\gamma X_\infty}\right) = \dfrac{2}{3}\dot{e}_{ij}^{\mathrm{p}} - \dfrac{X_{ij}}{X_\infty}\dot{r}$
Damage	$\dot{D}_{ij} = \left(\dfrac{\overline{Y}}{S}\right)^s |\dot{\boldsymbol{\epsilon}}^{\mathrm{p}}|_{ij}$ if $r > p_{\mathrm{D}}$ or $\max w_{\mathrm{s}} > w_{\mathrm{D}}$ up to D_{c} $\overline{Y} = \dfrac{1+\nu}{2E}\mathrm{tr}\left[\left(\boldsymbol{H}^{\mathrm{P}}\boldsymbol{\sigma}_+^{\mathrm{D}}\boldsymbol{H}^{\mathrm{P}}\right)^2 + h_{\mathrm{a}}\left(\boldsymbol{H}^{\mathrm{n}}\boldsymbol{\sigma}_-^{\mathrm{D}}\boldsymbol{H}^{\mathrm{n}}\right)^2\right]$ $+ \dfrac{3(1-2\nu)}{2E}\left[\dfrac{\langle\sigma_{\mathrm{H}}\rangle^2}{(1-\eta D_{\mathrm{H}})^2} + h_{\mathrm{a}}\dfrac{\langle-\sigma_{\mathrm{H}}\rangle^2}{(1-\eta h_{\mathrm{a}} D_{\mathrm{H}})^2}\right]$
Plastic multiplier	$\dot{r} = \dot{\lambda}$ given by $f = 0$ and $\dot{f} = 0$
Visco-plastic multiplier	$\dot{p} = \left(\dfrac{f}{K_N}\right)^N$ Norton law $\dot{p} = \ln\left(1 - \dfrac{f}{K_\infty}\right)^{-n}$ exponential law

1.5.4 Non-Isothermal Behavior♂♂

The constitutive equations in the preceding tables are written for anisothermal cases and all the material parameters are functions of the temperature. If an incremental numerical procedure is used, the state laws must be properly differentiated with respect to the time. This is particularly important for the law of thermo-elasticity, in which appears an extra term when the temperature varies.

$$\dot{\epsilon}_{ij}^e = \frac{1+\nu}{E}\dot{\tilde{\sigma}}_{ij} - \frac{\nu}{E}\dot{\tilde{\sigma}}_{kk}\delta_{ij} + \alpha\dot{T} \\ + \left[\frac{\mathrm{d}}{\mathrm{d}T}\left(\frac{1+\nu}{E}\right)\tilde{\sigma}_{ij} - \frac{\mathrm{d}}{\mathrm{d}T}\left(\frac{\nu}{E}\right)\tilde{\sigma}_{kk}\delta_{ij} + \frac{\mathrm{d}\alpha}{\mathrm{d}T}(T - T_{\mathrm{ref}})\delta_{ij}\right]\dot{T}. \quad (1.199)$$

There are no such temperature derivatives in the laws of (visco-)plasticity and damage as they are derived from the dissipation potential that is written in terms of powers.

Note that the evolution law for nonlinear kinematic hardening is simpler for anisothermal loading if it is written in terms of $\boldsymbol{\alpha}$, as with the state law $X_{ij} = \frac{2}{3}C\alpha_{ij}$. We can then obtain

$$\dot{\alpha}_{ij} = (1-D)\dot{\epsilon}_{ij}^{\mathrm{P}} - \gamma\alpha_{ij}\dot{r} \quad \text{or} \quad \dot{\alpha}_{ij} = \dot{e}_{ij}^{\mathrm{P}} - \gamma\alpha_{ij}\dot{r}, \quad (1.200)$$

with $\dot{e}_{ij}^{\mathrm{P}} = \dot{\epsilon}_{ij}^{\mathrm{P}}$ and $\dot{r} = \dot{p}$ as long as there is no damage. This avoids taking into account the supplementary thermal term due to the derivative of $C = \gamma X_\infty$ with respect to T in the Armstrong–Frederick law, as

$$\frac{\mathrm{d}}{\mathrm{d}t}\left(\frac{X_{ij}}{\gamma X_\infty}\right) = \frac{2}{3}\dot{e}_{ij}^{\mathrm{P}} - \frac{X_{ij}}{X_\infty}\dot{r} \quad (1.201)$$

becomes

$$\dot{X}_{ij} = \frac{2}{3}C\dot{e}_{ij}^{\mathrm{P}} - \gamma X_{ij}\dot{r} + \frac{X_{ij}}{C}\frac{\mathrm{d}C}{\mathrm{d}T}\dot{T} \quad (1.202)$$

Note also that in plasticity, the expression (1.181) for the multiplier $\dot{\lambda}$ introduces a $\dot{T}$ term.

1.5.5 Two-Scale Model for Damage at Microscale♂
(J. Lemaitre and I. Doghri 1988, R. Desmorat and J. Lemaitre 2000)

The damage threshold criterion $\max w_s = w_D$ or $p = p_D$ states that damage occurs for a given amount of plasticity. In the same spirit, the damage evolution and its anisotropy are both governed by the plastic strains. The damage model previously described applies then only when plastic strains are sensible on a mesoscale, i.e. mostly for ductile failure, low cycle fatigue, and creep (Chaps. 3, 4, 5). For brittle failures with no permanent strains, other damage models are necessary (see Chap. 7). For quasi-brittle failures or high cycle fatigue, for which the loading is below the conventional yield stress, plasticity (or dissipation) and damage occur but at a scale much smaller than the RVE scale. A multiscale approach is thus necessary (see also Sects. 6.2.3 and 7.2.4).

Two scales are defined in accordance with the growth of microscale damage in a mesoscopic RVE:

- The **mesoscale** or the classical scale of continuum mechanics.
- The **microscale** or the scale of microdefects present inside the RVE but whose effects on the mechanical properties are not measurable, except for monotonic and fatigue supture properties. From the mechanical point of view of the homogenization procedures, such defects are considered altogether as a "weak" inclusion in a meso-RVE, the matrix (see Fig. 1.15).

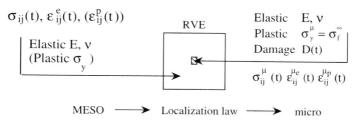

Fig. 1.15. Microelement embedded in an elastic (plastic) RVE

1.5.5.1 Constitutive Equations

The overall elasto-(visco-)plasticity material behavior is the same on the two different scales but as the inclusion is "weak," it will be subjected to damage, with the matrix remaining undamaged.

On a mesoscale, the full set of elasto-(visco-)plasticity constitutive equations with a yield stress of σ_y and with both isotropic and kinematic hardening may be considered. For quasi-brittle failure or high cycle fatigue applications (with a state of stress under the yield stress) only elasticity is considered.

On a microscale, the behavior is modeled by elasto-(visco-)plasticity coupled with damage. The **weakness** of the inclusion is related to its yield stress σ_y^μ taken equal to the **asymptotic fatigue limit** σ_f^∞ of the material, below which no plasticity exists and then no damage can occur. The yield criterion is written as

$$f^\mu = \left(\frac{\sigma^\mu}{1-D^\mu} - X^\mu\right)_{eq} - \sigma_f^\infty, \qquad (1.203)$$

where $\boldsymbol{X}^\mu$ is the microscale back stress. To simplify, only linear kinematic hardening is assumed on the microscale but with the same plastic modulus C_y as the modulus measured on the mesoscale. The microscale damage is assumed to be isotropic and its evolution law is written in terms of microstress, micro-energy density release rate, and micro-accumulated plastic strain, i.e., $\dot{D}^\mu = (Y^\mu/S)^s \dot{p}^\mu$. As the mean stress effect needs to be represented in fatigue, the micro-energy density rate Y^μ takes into account the different behaviors in tension and in compression through the parameter h. Finally, the plastic multiplier, $\dot{\lambda} = \dot{p}^\mu(1 - D^\mu)$, is calculated from

- The consistency condition $f^\mu = 0$ and $\dot{f}^\mu = 0$ for micro-plasticity or
- The Norton law $\dot{p}^\mu = \left(\dfrac{f^\mu}{K_N}\right)^N$ for micro-viscoplasticity.

The full set of equations is given in Table 1.7. The variables in the inclusion (stress tensor $\boldsymbol{\sigma}^\mu$, total strain tensor $\boldsymbol{\epsilon}^\mu$, elastic strain tensor $\boldsymbol{\epsilon}^{\mu e}$ and plastic strain tensor $\boldsymbol{\epsilon}^{\mu p}$, damage D^μ ...) have a "μ"-superscript to denote microscale when the variables at the mesoscale are simply represented by $\boldsymbol{\sigma}$, $\boldsymbol{\epsilon}$, $\boldsymbol{\epsilon}^e$, $\boldsymbol{\epsilon}^p$ ... with no superscripts. Note that damage being considered only on the microscale, the "μ"-superscript is omitted by setting $D^\mu \equiv D$.

1.5.5.2 Scale Transition

The scale transition for inner defects is made by use of the Eshelby–Kröner localization law: the total and plastic strains at microscale level are related to the strains at the mesolevel as in

$$\epsilon_{ij}^\mu = \epsilon_{ij} + \beta(\epsilon_{ij}^{\mu p} - \epsilon_{ij}^p), \qquad (1.204)$$

where β is given by an Eshelby analysis of a spherical inclusion:

$$\beta = \frac{2}{15}\frac{4 - 5\nu}{1 - \nu}. \qquad (1.205)$$

For surface defects, due to free edge conditions the localization law must be changed (see Sect. 6.4.6).

1.5.5.3 Use of the Two-Scale Model

The mesostresses ($\boldsymbol{\sigma}(t)$) and strains ($\boldsymbol{\epsilon}(t)$, ($\boldsymbol{\epsilon}^p(t)$)) are determined from a classical structure calculation (made in elasticity for quasi-brittle and high cycle fatigue applications). Their history are the inputs for the post-processing calculation of the strains, stresses, hardening, and damage at microscale. Their values are obtained from the time integration of the constitutive equations at microlevel coupled altogether with the Eshelby–Kröner law of localization.

When the damage (D) reaches the critical value (D_c) there is initiation of a crack in the inclusion which corresponds to a mesocrack in the RVE if

1.5 Elasto-(Visco-)Plasticity Coupled with Damage

Table 1.7. Two-scale damage model equations

Mesoscale σ $\epsilon, \epsilon^{\text{e}}, \epsilon^{\text{p}}$ $\boldsymbol{X}, R$	General case Quasi-brittle failure High cycle fatigue	Elasto-(visco-)plasticity constitutive equations Elasticity $\epsilon_{ij} = \dfrac{1+\nu}{E}\sigma_{ij} - \dfrac{\nu}{E}\sigma_{kk}\delta_{ij}$ with $\sigma_{\text{f}}^{\infty} < \sigma_{\text{eq}} < \sigma_{\text{y}}$
Scale transition	Eshelby–Kröner localization law	$\epsilon_{ij}^{\mu} = \epsilon_{ij} + \beta(\epsilon_{ij}^{\mu\text{p}} - \epsilon_{ij}^{\text{p}})$
Microscale σ^{μ} $\epsilon^{\mu}, \epsilon^{\mu\text{e}}, \epsilon^{\mu\text{p}}$	(Visco-)plasticity coupled with isotropic damage	$\epsilon_{ij}^{\mu\text{e}} = \dfrac{1+\nu}{E}\dfrac{\sigma_{ij}^{\mu}}{1-D} - \dfrac{\nu}{E}\dfrac{\sigma_{kk}^{\mu}}{1-D}\delta_{ij}$ $f^{\mu} = (\tilde{\boldsymbol{\sigma}}^{\mu} - \boldsymbol{X}^{\mu})_{\text{eq}} - \sigma_{\text{f}}^{\infty}$ with $\tilde{\sigma}_{ij}^{\mu} = \dfrac{\sigma_{ij}^{\mu}}{1-D}$ $\dot{\epsilon}_{ij}^{\mu\text{p}} = \dfrac{3}{2}\dfrac{\tilde{\sigma}_{ij}^{\mu\text{D}} - X_{ij}^{\mu}}{(\tilde{\boldsymbol{\sigma}}^{\mu} - \boldsymbol{X}^{\mu})_{\text{eq}}}\dfrac{\dot{\lambda}}{1-D}$
$\boldsymbol{X}^{\mu}$		$\dot{X}_{ij}^{\mu} = \dfrac{2}{3}C_{\text{y}}\dot{\epsilon}_{ij}^{\mu\text{p}}(1-D)$
D		$\dot{D} = \left(\dfrac{Y^{\mu}}{S}\right)^{s}\dot{p}^{\mu}$ if $p^{\mu} > p_{\text{D}}$ and up to D_{c} with $p_{\text{D}} = \epsilon_{\text{pD}}\left[\dfrac{\sigma_{\text{u}} - \sigma_{\text{f}}^{\infty}}{\dfrac{\sigma_{\text{eq}}^{\mu\,\text{max}} + \sigma_{\text{eq}}^{\mu\,\text{min}}}{2} - \sigma_{\text{f}}^{\infty}}\right]^{m}$ $Y^{\mu} = \dfrac{1+\nu}{2E}\left[\dfrac{\langle\boldsymbol{\sigma}^{\mu}\rangle_{ij}^{+}\langle\boldsymbol{\sigma}^{\mu}\rangle_{ij}^{+}}{(1-D)^{2}} + h\dfrac{\langle\boldsymbol{\sigma}^{\mu}\rangle_{ij}^{-}\langle\boldsymbol{\sigma}^{\mu}\rangle_{ij}^{-}}{(1-hD)^{2}}\right]$ $-\dfrac{\nu}{2E}\left[\dfrac{\langle\sigma_{kk}^{\mu}\rangle^{2}}{(1-D)^{2}} + h\dfrac{\langle-\sigma_{kk}^{\mu}\rangle^{2}}{(1-hD)^{2}}\right]$

its size is comparable to the value of the mesocrack initiated (δ_{0}) calculated using (1.252) in Sect. 1.6.3.

All the material parameters at the mesoscale are identified with large plastic strain. Nevertheless, we advise adjusting s and m on a test where fracture occurs in quasi-brittle conditions.

1.5.5.4 More Precise Identification of the Asymptotic Fatigue Limit
(S. Calloch and C. Doudard 2003)

In the two-scale damage model the asymptotic fatigue limit (σ_f^∞) plays an important role as the yield stress at the microlevel. A definition as the asymptote of the Wöhler curve is mathematically appealing but difficult to practice because to reach points close to this limit, we may need experiments of months, years, centuries and more duration! Fortunately, there is an other way by evaluating the dissipation through an elevation of temperature in cyclic tests (M.P. Luong 1998, A. Galtier 2002).

Damage (and fatigue in particular) is always related to plastic strain, at microlevel for high cycle fatigue. In cyclic tests, the corresponding dissipation $\sigma_{ij}\dot{\epsilon}_{ij}^P$ induces an elevation of temperature which can be measured if the tests are performed at a frequency large enough to compensate the heat lost by conduction. The ideal would be adiabatic tests. Then, if several successive increasing levels of periodic stresses are applied on one specimen and the stationary temperature T is recorded, σ_f^∞ corresponds to the level for which an elevation of temperature is just detected. Of course this value depends on the accuracy of the temperature measurement but, with thermocouples and some elementary precautions, an offset of 0.05°C may be detected, thus providing a good accuracy for σ_f^∞.

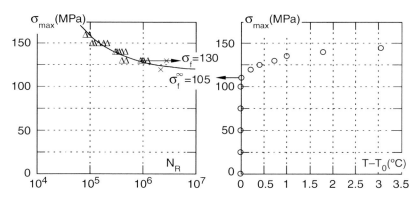

Fig. 1.16. Wöhler curve of a dual phase steel and asymptotic fatigue limit detected by temperature elevation (S. Calloch and C. Doudard 2003)

In the following example of dual-phase steel, the successive stress levels at 10 Hz are 25, 50, 75, 100, 105, 120, 125, 130, 135, 140, and 145 MPa for about 1000 cycles each.

The graph of the elevation of temperature $T-T_0$ as a function of the stress level in Fig. 1.16 allows us to consider the asymptotic fatigue limit $\sigma_f^\infty = 105$ MPa when the engineering fatigue limit is $\sigma_f = 130$ MPa at $N_R = 10^6$ cycles.

1.6 Localization and Mesocrack Initiation

The aim of this section is to give an answer to difficult questions concerning a mesocrack initiation: what is a crack initiation from a continuum mechanics point of view? What is the size and the direction of the crack initiated? A theoretical difficulty is encountered as continuum mechanics defines its scale of application as the scale of the Representative Volume Element, in which the mechanical fields are homogeneous and continuous, whereas the free surface creation of a crack is a geometrical, non-homogeneous process.

1.6.1 Critical Damage Criterion♂

The simplest and most practical solution to model a crack initiation is to use the critical damage criterion which states that a mesocrack is initiated when the damage reaches a critical value D_c.

- For isotropic damage, this occurs when $D = D_c$ (see Sect. 1.4.1).
- For anisotropic damage with principal damage components D_I, $I = 1, 2, 3$, this takes place when the maximum principal damage reaches D_c, i.e., when $\max D_I = D_c$ as already stated in Sect. 1.4.3.

The term D_c is considered as a material constant, which is not so easy to measure. For most materials, $0.2 \leq D_c \leq 0.5$.

1.6.2 Strain Damage Localization Criterion♂♂♂

One may consider the mesocrack initiation as the ultimate state of the strain localization process which occurs in softening materials. Due to severe loading, localization bands occur in which strain and damage are strongly localized and in which microcracks develop and then degenerate into a mesocrack (R. Billardon, I. Doghri, A. Benallal and G. Geymonat 1989).

This is a realistic criterion for a mesocrack initiation in its physical definition but it has two main drawbacks:

1. Such a criterion needs tough calculations and is difficult to use in finite element computations.
2. It is strongly dependent on the damage evolution law as well as the plastic or visco-plastic behavior of the materials, such as anisotropy, ratcheting, and non-proportional loading effects.

Nevertheless, at least for the uniaxial tensile case, it allows for the theoretical calculation of the value of the critical damage D_c without any need for experiments.

Two main approaches are available for the study of the instabilities related to the phenomenon of localization: the bifurcation approach which applies to time-independent constitutive laws; the perturbation approach which applies to both time-independent and time-dependent constitutive laws. The consideration of viscosity effects makes the solution of the mechanical problem unique in general but instabilities may still occur.

1.6.2.1 Bifurcation Approach of Localization
(R. Hill 1962, J.R. Rice 1973, J.R. Rice and J.W. Rudnicki 1975)

The bifurcation approach consists of the analysis of the loss of uniqueness in the rate mechanical problem. It is is written in terms of velocity field $\vec{v}$, strain rate $\dot{\epsilon}$ and stress rate $\dot{\sigma}$. In the quasi-static case,

$$\begin{aligned}
&\dot{\sigma}_{ij,j} + \dot{f}_i = 0 && \text{in the structure,} \\
&\dot{\epsilon}_{ij} = \tfrac{1}{2}\left(v_{i,j} + v_{j,i}\right) && \text{in the structure,} \\
&\dot{\sigma}_{ij} N_j = \dot{F}_i^{\text{given}} && \text{on the given load boundary,} \\
&v_i = \dot{U}_i^{\text{given}} && \text{on the given displacement boundary, and} \\
&\dot{\sigma}_{ij} = L_{ijkl}\dot{\epsilon}_{ij} && \text{in the structure,}
\end{aligned} \qquad (1.206)$$

with $\vec{f}$ as the body force, $\vec{U}^{\text{given}}$ as the applied displacements, $\vec{F}^{\text{given}}$ as the applied loads, and $\vec{N}$ as the outward normal of the boundary. The last equation is the rate form of the time-independent constitutive laws. Its introduces the fourth order tensor tangent operator $\underline{\boldsymbol{L}}$ which will be introduced later for elasto-plasticity coupled with damage (1.216).

Let us consider a homogeneous infinite body with a known homogenous solution $\vec{v}^{(1)}$ and let us assume that at a given loading time, a second possible solution of the rate problem $\vec{v}^{(2)}$ exists. Two bifurcation modes may be exhibited:

1. *A harmonic eigenmode* (j is the pure imaginary number $j = \sqrt{-1}$),

$$[v_i] = v_i^{(1)} - v_i^{(2)} = g_i \exp\left(j\xi n_k x_k\right) \qquad (1.207)$$

   which, if reported within the equilibrium, leads exactly to the loss of ellipticity of the equilibrium equations,

$$\xi^2 \left(\vec{n}\underline{\boldsymbol{L}}\vec{n}\right)_{ij} g_j = 0 \quad \text{and} \quad \det\left(\vec{n}\underline{\boldsymbol{L}}\vec{n}\right) = 0, \qquad (1.208)$$

   where ξ is an arbitrary wave number of mode $\vec{g}$ and $\left(\vec{n}\underline{\boldsymbol{L}}\vec{n}\right)$ is the acoustic tensor related to the localization plane with a normal $\vec{n}$: $\left(\vec{n}\underline{\boldsymbol{L}}\vec{n}\right)_{ij} = n_k L_{ikjl} n_l$.

2. *A bifurcation mode involving discontinuous velocity gradient*. If the subscripts $^{(1)}$ and $^{(2)}$ stand now for the fields in each side of the discontinuity surface (or planar band) of normal $\vec{n}$, the continuity of the normal stress

and Maxwell compatibility equations read

$$\dot{\sigma}_{ij}^{(1)} n_j = \dot{\sigma}_{ij}^{(2)} n_j \quad \text{and} \quad v_{i,j}^{(1)} - v_{i,j}^{(2)} = g_i n_j, \tag{1.209}$$

with $\dot{\sigma}_{ij}^{(1)} = L_{ijkl}^{(1)} v_{k,l}^{(1)}$ and $\dot{\sigma}_{ij}^{(2)} = L_{ijkl}^{(2)} v_{k,l}^{(2)}$ due to the usual symmetry $L_{ijkl} = L_{ijlk}$. One ends up with

$$\left(\vec{n} \underline{\boldsymbol{L}}^{(1)} \vec{n} \right)_{ij} g_j + n_j \left(L_{ijkl}^{(1)} - L_{ijkl}^{(2)} \right) v_{k,l}^{(2)} = 0. \tag{1.210}$$

The first case of continuous bifurcation corresponds to a material still plastifying (and damaging) on both sides of the discontinuity surface: $L_{ijkl}^{(1)} = L_{ijkl}^{(2)} = L_{ijkl}$ given by (1.216) and

$$\det \left(\vec{n} \underline{\boldsymbol{L}} \vec{n} \right) = 0. \tag{1.211}$$

The vector $\vec{g}$ is then the eigenvector associated with a vanishing eigenvalue of the acoustic tensor. If $\vec{g}$ is orthogonal to $\vec{n}$, one gets a shear band. If $\vec{g}$ is parallel to $\vec{n}$, an opening mode results instead.

The second case of discontinuous bifurcation corresponds to a body still loaded with $L_{ijkl}^{(1)} = L_{ijkl}$ at one side of the planar band but unloaded in elasticity at the other side ($L_{ijkl}^{(2)} = \tilde{E}_{ijkl}$). This leads to the condition $\det \left(\vec{n} \underline{\boldsymbol{L}} \vec{n} \right) < 0$.

The condition $\det \left(\vec{n} \underline{\boldsymbol{L}} \vec{n} \right) \leq 0$ is a **necessary** and **sufficient** condition for strain localization to occur in homogeneous, non-viscous bodies.

1.6.2.2 Critical Damage from the Bifurcation Analysis

The tangent stiffness operators may be derived from the full sets of elastoplasticity coupled with the damage constitutive equations of Sect. 1.5. No kinematic hardening and no unilateral effect are considered here in order to keep the calculation simple! The constitutive equations are written in a concise way, as in

$$\dot{\epsilon}_{ij}^{\text{P}} = a_{ij} \dot{p} \quad \text{with} \quad a_{ij} = \frac{3}{2} \frac{(H_{ik} \tilde{\sigma}_{kl}^{\text{D}} H_{lj})^{\text{D}}}{(\boldsymbol{H} \, \tilde{\boldsymbol{\sigma}}^{\text{D}} \boldsymbol{H})_{\text{eq}}}, \tag{1.212}$$

$$\dot{r} = \dot{p} \Delta \quad \text{with} \quad \Delta = \frac{\tilde{\sigma}_{\text{eq}}}{\left(\boldsymbol{H} \, \tilde{\boldsymbol{\sigma}}^{\text{D}} \boldsymbol{H} \right)_{\text{eq}}}, \tag{1.213}$$

$$\dot{D}_{ij} = d_{ij} \dot{p} \quad \text{with} \quad d_{ij} = \begin{cases} \left(\dfrac{Y}{S} \right)^s \delta_{ij}, & \text{isotropic damage,} \\ \left(\dfrac{Y}{S} \right)^s |\boldsymbol{a}|_{ij}, & \text{anisotropic damage,} \end{cases} \tag{1.214}$$

and where $H_{ij} = (\mathbf{1} - \mathbf{D})_{ij}^{-1/2}$ for anisotropic damage and $H_{ij} = \delta_{ij}/\sqrt{1-D}$ for isotropic damage.

The accumulated plastic strain rate is determined by the consistency condition $\dot{f} = 0$:

$$\dot{p} = \frac{b_{ij}\dot{\tilde{\sigma}}_{ij}}{R'(r)\Delta}, \quad \text{with} \quad b_{ij} = \frac{3}{2}\frac{\tilde{\sigma}_{ij}^{\mathrm{D}}}{\tilde{\sigma}_{\mathrm{eq}}}. \tag{1.215}$$

For both isotropic and anisotropic damage models,

$$L_{ijkl} = \tilde{E}_{ijkl} - \frac{2G}{R'\Delta + 2G\, a_{mn}b_{mn}}\left[\tilde{E}_{ijpq}a_{pq}b_{kl} + M_{ijpq}^{-1}\frac{\partial \tilde{\sigma}_{pq}}{\partial D_{rs}}d_{rs}b_{kl}\right], \tag{1.216}$$

with $\tilde{E}_{ijkl} = M_{ijpq}^{-1}E_{pqkl}$ as the elastic effective tensor and M_{ijkl} is given by (1.15). Readers can refer to Sect. 2.2.4 for details concerning the calculation of the tangent operator.

For isotropic damage, the strain localization condition, $\det(\vec{n}\mathbf{L}\vec{n}) = 0$, written for uniaxial tension leads to the expression of a critical damage which depends on the normal of the planar localization band,

$$D_{\mathrm{c}}(\vec{n}) \approx 1 - \kappa(\theta)\left(\frac{\sigma_{\mathrm{u}}^2}{2ES}\right)^s \frac{\sigma_{\mathrm{u}}}{G}, \tag{1.217}$$

with $\vec{n} = [\cos\theta, \sin\theta, 0]^{\mathrm{T}}$ and

$$\kappa(\theta) = \frac{2 - 4\nu + (5 - 4\nu)\sin^2\theta\cos^2\theta}{3 - 6\sin^2\theta\cos^2\theta + (1 - 2\nu)(2\sin^2\theta - \cos^2\theta)}. \tag{1.218}$$

This allows for the determination of the normal to the localization plane: for $\nu = 0.3$, $\theta = 42°$. In fact, the ratio of the ultimate stress to the shear modulus (σ_{u}/G) remains very small for any material such that κ is of the order of magnitude of unity and D_{c} in tension is found to be close to 1.

For anisotropic damage in pure tension, the result "a critical damage almost equal to 1" is also obtained but in terms of hydrostatic effective damage: strain localization occurs when $\eta D_{\mathrm{Hc}} \approx 1$ or $D_{\mathrm{Hc}} \approx 1/\eta$, i.e., in tension when

$$\boxed{D_{1\mathrm{c}} = 2D_{2\mathrm{c}} = 2D_{3\mathrm{c}} \approx \frac{3}{2\eta}.} \tag{1.219}$$

With $\eta = 3$, this gives $D_{1\mathrm{c}} \approx 0.5$. This value is in accordance with the critical damages often measured in experiments with poor accuracy due to the difficulty in defining the initiation of a mesocrack in practice! The stress at strain localization has a non-vanishing value,

$$\sigma = \frac{3\sigma_{\mathrm{u}}}{\dfrac{2}{1-D_1} + \dfrac{1}{1-D_2}}, \tag{1.220}$$

which is $\sigma_{\mathrm{R}} = \sigma_{\mathrm{u}}/2$ for $\eta = 3$. In the absence of any measurements, $D_{\mathrm{c}} = 0.5$ is a good candidate for all applications.

1.6.2.3 Perturbation Analysis – Stability
(A. Molinari 1985, A. Benallal 1988)

To check the possibility of bifurcation toward solutions that are different from the homogeneous one, a way to proceed is to slightly perturb the reference solution ($\vec{v}^0$). If the evolution of the perturbed solution ($\vec{v} = \vec{v}^0 + \delta\vec{v}$) remains close to $\vec{v}^0$, one expects stability. If not, instability occurs.

From a mathematical point of view, this practical approach applied to mechanical problems is fully similar to the study of the stability of a set of first order differential equations written as

$$\frac{\partial U}{\partial t} = \boldsymbol{G}(U, \mu_{\text{load}}), \qquad (1.221)$$

where $\boldsymbol{G}$ is in general a nonlinear differential operator, μ_{load} is a parameter (for example, representing the loading intensity), and U is the vector solution (for example, made of the velocity and all the necessary variables to describe the material behavior).

Different perturbation and stability analyses are possible:

- Study of the decay (or lack of decay) in the linearized perturbation δU:

$$\frac{\partial}{\partial t}(U^0 + \delta U) = \boldsymbol{G}^0(U, \mu_{\text{load}}) + \frac{\partial \boldsymbol{G}}{\partial U}(U^0)\delta U \qquad (1.222)$$

or

$$\frac{\partial}{\partial t}(\delta U) = \boldsymbol{A}\delta U \qquad \boldsymbol{A} = \frac{\partial \boldsymbol{G}}{\partial U}(U^0). \qquad (1.223)$$

If one of the eigenvalues (λ_A) of the operator $\boldsymbol{A}$ has a positive real part, the perturbation δU will grow.

- Study of the decay (or lack of decay) in a relative perturbation defined as

$$Z = \frac{\delta U}{\|U^0\|}, \qquad (1.224)$$

and

$$\frac{\partial Z}{\partial t} = \left[\frac{\partial \boldsymbol{G}}{\partial U}(U^0) - \frac{1}{\|U^0\|}\frac{\partial}{\partial t}\|U^0\|\mathbf{1}\right]Z, \qquad (1.225)$$

with $\mathbf{1}$ as the identity operator. This set of differential equations leads to a different value for λ_A from that in the previous case. If the perturbations are sought in the form $\delta U = U^\star \exp(\lambda_A t)$ and $\delta Z = Z^\star \exp(\lambda_{\text{rel}} t)$, λ_{rel} compared to λ_A yields

$$\lambda_{\text{rel}} = \lambda_A - \frac{1}{\|U^0\|}\frac{\partial}{\partial t}\|U^0\| \qquad (1.226)$$

and the "relative" instability occurs when

$$\operatorname{Re}(\lambda_A) \geq \frac{1}{\|U^0\|}\frac{\partial}{\partial t}\|U^0\| \qquad (1.227)$$

instead of $\operatorname{Re}(\lambda_A) \geq 0$.

The perturbation and stability analysis applies to rather complex problems such as elasto-plasticity or elasto-visco-plasticity coupled (or not coupled) with damage, quasi-static or dynamic loadings, temperature and heat conduction, or any other diffusive phenomenon.

1.6.2.4 General Thermo-Mechanical Case
(A. Benallal and V. Cano 1996)

Let us consider the general case of structures made of generalized standard materials submitted to dynamic loadings. The thermodynamics description of such material behaviors is given in Sect. 1.2. It introduces the Helmholtz free energy potential, $\rho\psi$, written in terms of strains $\boldsymbol{\epsilon}$ and internal variables V_K (the damage D or $\boldsymbol{D}$ is one of those), a yield criterion f, and a dissipation potential F.

The equilibrium equations are represented by

$$\sigma_{ij,j} + f_i = \rho \frac{\partial^2 u_i}{\partial t^2}, \qquad (1.228)$$

with $\vec{u}$ as the displacement, $\vec{f}$ as the body force, and ρ as the density. The compatibility equations are of the form

$$\epsilon_{ij} = \frac{1}{2}\left(u_{i,j} + u_{j,i}\right). \qquad (1.229)$$

The energy conservation principle (first principle of thermodynamics) reads as

$$-q_{i,i} + \omega = \rho C_\epsilon \dot{T} - A_K \dot{V}_K + T\left[\frac{\partial A_K}{\partial T}\dot{V}_K - \frac{\partial \sigma_{ij}}{\partial T}\dot{\epsilon}_{ij}\right], \qquad (1.230)$$

with ω as the heat source, $C_\epsilon = -T\left.\dfrac{\partial^2\psi}{\partial T^2}\right|_\epsilon$ as the heat capacity, and A_K as the associated variables. The constitutive equations are the

$$\begin{aligned}
\text{elasticity law} && \sigma_{ij} &= \rho\frac{\partial \psi}{\partial \epsilon_{ij}}, \\
\text{state laws} && A_K &= -\rho\frac{\partial \psi}{\partial V_K}, \\
\text{evolution laws} && \dot{V}_K &= -\dot{\lambda}\frac{\partial F}{\partial A_K}, \\
\text{conduction law} && q_i &= -k\,T_{,i},
\end{aligned} \qquad (1.231)$$

where k is the conductivity and $\dot{\lambda} = \dot{r}$ is the Lagrange multiplier in addition to the boundary and initial conditions.

In order to derive the differential operator $\boldsymbol{G}$ for the general thermo-mechanical problem, let us set

1.6 Localization and Mesocrack Initiation

$$U = \begin{bmatrix} u_i \\ v_i \\ T \\ \sigma_{ij} \\ V_K \end{bmatrix} \quad \text{with} \quad v_i = \frac{\partial u_i}{\partial t}, \quad (1.232)$$

with a set of equations to be solved:

$$\dot{u}_i = v_i$$

$$\dot{v}_i = \frac{1}{\rho}(\sigma_{ij,j} + f_i)$$

$$\dot{T} = \frac{k}{\rho C_\epsilon} T_{,kk} + \frac{T}{C_\epsilon} \frac{\partial^2 \psi}{\partial \epsilon_{ij} \partial T} v_{i,j} - \frac{1}{C_\epsilon} \left[\frac{\partial \psi}{\partial V_K} - T \frac{\partial^2 \psi}{\partial V_K \partial T} \right] \dot{V}_K + \frac{\omega}{\rho C_\epsilon} \quad (1.233)$$

$$\dot{\sigma}_{ij} = \rho \frac{\partial^2 \psi}{\partial \epsilon_{ij} \partial T} \dot{T} + \rho \frac{\partial^2 \psi}{\partial \epsilon_{ij} \partial \epsilon_{kl}} v_{k,l} + \rho \frac{\partial^2 \psi}{\partial \epsilon_{ij} \partial V_K} \dot{V}_K$$

$$\dot{V}_K = -\dot{r} \frac{\partial F}{\partial A_K}$$

For an infinite homogeneous body we take the Fourier transform of this set of equations or, in an equivalent manner, seek the perturbed solution of the form

$$U(\mathbf{x}, t) = U^0(t) + \hat{U}(t) \exp(\mathrm{j} \xi n_k x_k) \quad \text{and} \quad \hat{U} = \begin{bmatrix} \hat{u}_i \\ \hat{v}_i \\ \hat{T} \\ \hat{\sigma}_{ij} \\ \hat{V}_K \end{bmatrix}, \quad (1.234)$$

where ξ acts as a wave number. We then linearize the previous differential equations, identify the differential operator $\mathbf{A}$ and solve for the solution as

$$\hat{U} = U^\star \exp(\lambda_A t), \quad (1.235)$$

where λ_A is the stability parameter to be determined ($\mathrm{Re}(\lambda_A) < 0$ implies stability). We will then obtain a set of linear equations and after some substitutions, we can formally express them as

$$\begin{aligned} \lambda_A u_\mathrm{p}^\star &= v_\mathrm{p}^\star, \\ \rho \lambda_A^2 u_\mathrm{p}^\star &= \mathrm{j}\xi \sigma_{pq} n_q, \\ V_K^\star &= \mathrm{j}\xi C_{Kpq}^\lambda n_q u_\mathrm{p}^\star + R_K^\lambda T^\star, \\ \sigma_{pq}^\star &= \mathrm{j}\xi L_{pqrs}^\lambda n_s u_r^\star + M_{pq}^\lambda T^\star, \\ \left(\rho C_\epsilon + \frac{k\xi^2}{\lambda_A} + \rho C_{V_K}^\lambda \right) T^\star &= \mathrm{j}\xi P_{pq}^\lambda n_q u_\mathrm{p}^\star, \end{aligned} \quad (1.236)$$

where the 4 tensors L^λ_{ijkl}, C^λ_{Kij}, M^λ_{ij}, and P^λ_{ij}, the vector R^λ_K, and the scalar $\rho C^\lambda_{V_K}$ introduced depend on the stability parameter λ_A but not the wave number ξ. They are temperature-dependent and are affected by viscosity and inertia.

- The fourth order tensor L^λ_{ijkl} generalizes the expression for the tangent operator L_{ijkl} to viscous materials loaded in dynamics at constant temperature as the rate law

$$\dot{\sigma}_{pq} = L_{pqrs}\dot{\epsilon}_{rs} = L_{pqrs}\dot{u}_{r,s}. \quad (1.237)$$

Coupled with the consideration of the perturbed fields,

$$\sigma_{pq} = \sigma^0_{pq} + \sigma^\star_{pq}\exp(\lambda_A t)\exp(\mathrm{j}\xi n_k x_k),$$
$$u_r = u^0_r + u^\star_r \exp(\lambda_A t)\exp(\mathrm{j}\xi n_k x_k) \quad (1.238)$$

lead to

$$\sigma^\star_{pq} = \mathrm{j}\xi L_{pqrs} n_s u^\star_r. \quad (1.239)$$

Note that L^λ_{ijkl} is not a material property as it depends on λ_A.
- The tensor M^λ_{ij} stands for the thermal effect on stresses and $M^\lambda_{ij}T^\star$ represents the perturbation of the thermal stress,
- The tensor C^λ_{Kij} and R^λ_K represent the contributions in terms of internal variables due to the mechanical and the thermal loadings, respectively.
- The heat capacity contribution ($C^\lambda_{V_K}$) is due to the consideration of internal variables to model the material behavior and P^λ_{ij} models the heat source due to the material deformation.

Finally, we obtain a set of two equations of the two unknowns $\vec{u}^\star$ and $T^\star$

$$\begin{bmatrix} \rho\lambda_A^{\star 2}\delta_{ik} + \xi^2 n_\mathrm{p} L^\lambda_{ipkq} n_q & -\mathrm{j}\xi M^\lambda_{ip} n_\mathrm{p} \\ -\frac{\mathrm{j}\xi}{\rho} P^\lambda_{kq} n_q & 1 + \frac{k\xi^2}{\rho C_\epsilon \lambda_A} + \frac{C^\lambda_{V_K}}{C_\epsilon} \end{bmatrix} \begin{bmatrix} u^\star_k \\ T^\star \end{bmatrix} = 0 \quad (1.240)$$

and the parameter λ_A is a solution of the polynomial equations,

$$\det(\vec{n}\boldsymbol{\mathcal{L}}\vec{n}) = 0, \quad (1.241)$$

$$\mathcal{L}_{ijkl} = \rho\lambda_A^2 \delta_{ik}\delta_{jl} + \xi^2\left[L^\lambda_{ijkl} + \frac{M^\lambda_{ij}P^\lambda_{kl}}{\rho C_\epsilon + \frac{k\xi^2}{\lambda_A} + \rho C^\lambda_{V_K}}\right]. \quad (1.242)$$

When $\mathrm{Re}(\lambda_A) \geq 0$ (or $\mathrm{Re}(\lambda_A) \geq \frac{1}{\|U^0\|}\frac{\partial}{\partial t}\|U^0\|$) the perturbation (or the relative perturbation) starts to grow.

The adiabatic, isothermal, and static cases are limiting cases of this perturbation analysis. They are solved by mathematically setting:

- $k = 0$ for the adiabatic case
- $\rho C_\epsilon \to \infty$ for the isothermal case

- $\rho = 0$ for the quasi-static case except in the heat equation in which ρC_ϵ is kept constant

For elasto-visco-plastic materials (without damage), the parameter λ_A is bound and the question of when the localization process takes place has not been clearly answered. For constitutive equations coupled with damage, the strain damage localization criterion $\lambda_A = \lambda_c = \infty$ coincides with $\det(\vec{n}\tilde{\boldsymbol{E}}\vec{n}) = 0$ (inertia effects neglected) and therefore with $\eta D_\mathrm{H} \approx 1$ as well. One ends up with $\eta = 3$ to the 3D strain-damage localization criterion

$$\boxed{\operatorname{tr}\boldsymbol{D} \approx 1} \qquad (1.243)$$

which corresponds in tension to

$$D_{1c} = 2D_{2c} = 2D_{3c} \approx \frac{1}{2}. \qquad (1.244)$$

1.6.2.5 Effects of Viscosity, Temperature, and Inertia

Qualitatively,

- Viscosity has a regularizing effect.
- Heat conduction may have a destabilizing effect. The smaller the wave number is, the larger is the destabilization.
- Dynamics has a stabilizing effect which is much more important for the large wave numbers than for the lower ones. Inertia effect may usually be neglected.

1.6.3 Size and Orientation of the Crack Initiated♂

Based on the principles of Continuum Damage Mechanics, the size of the mesocrack is the size on which the constitutive equations operate, i.e., the size of the Representative Volume Element: infinitely small for a mathematician but something concrete for an engineer!

If the question asked is whether or not a crack is initiated, the answer is given by information on the loading conditions for which the damage reaches its critical value.

If the damage analysis must be followed by a fracture analysis based on an existing initial finite crack, its size and orientation has to be determined. One way is to consider that the mesocrack initiated by **Continuum Damage Mechanics** involving a **dissipated energy** ϕ_Dp can also be described by a process of **Fracture Mechanics** involving the same amount of **dissipated energy** $\phi_\mathrm{F} = \phi_\mathrm{Dp}$ (J. Mazars 1984). This bridges the gap between damage mechanics and fracture mechanics at least in the sense of energy. Since δ_0 is the linear size of the mesocrack, also the size of the RVE for this purpose, the energy ϕ_{Dp} involved in the process of initiation by damage of a mesocrack is composed of two energies:

- The energy dissipated in the damage process itself, including isotropic damage,

$$\phi_D = \delta_0^3 \int_0^{D_c} Y dD. \tag{1.245}$$

Assuming saturated hardening and a triaxiality function of $R_\nu = 1$ as the crack is often initiated at the surface where the state of stress is also often uniaxial, we have

$$\phi_D \approx \delta_0^3 \frac{\sigma_u^2}{2E} D_c. \tag{1.246}$$

- The energy dissipated in the associated plasticity process is

$$\phi_P = \delta_0^3 \int_0^{\text{fracture}} \sigma_{ij} d\epsilon_{ij}^P. \tag{1.247}$$

Assuming either proportional loading or uniaxial loading leads to

$$\phi_P = \delta_0^3 \int_0^{p_R} \sigma_{eq}(p) dp, \tag{1.248}$$

where p_R is the accumulated plastic strain at rupture (for which $\sigma_{eq} \approx \sigma_u$ and $p_R = \epsilon_{pR}$ in pure tension). Then

$$\phi_{Dp} = \phi_D + \phi_P \approx \delta_0^3 \left(\frac{\sigma_u^2}{2E} D_c + \sigma_u \epsilon_{pR} \right). \tag{1.249}$$

The first term is preponderant for brittle materials as the second term is dominant for ductile materials.

On the same RVE, the energy ϕ_F is evaluated through the concepts of classical fracture mechanics dealing with a crack area $A_0 = \delta_0^2$ and a strain energy release rate G:

$$\phi_F = \int_0^{A_0} G dA. \tag{1.250}$$

For simplicity, let us consider the upper bound on ϕ_F given by the maximum value G_c of G, the material toughness. This approximation is close to reality for brittle materials and is a step function of the R-curve for ductile materials. Then,

$$\phi_F \approx G_c \delta_0^2. \tag{1.251}$$

Writing the energy balance $\phi_{Dp} = \phi_F$ gives the length δ_0 which matches Continuum Damage Mechanics and fracture mechanics:

$$\boxed{\delta_0 \approx \frac{G_c}{\frac{\sigma_u^2}{2E} D_c + \sigma_u \epsilon_{pR}}.} \tag{1.252}$$

1.6 Localization and Mesocrack Initiation

This formula gives values generally obtained by metallurgy or physics considerations, as shown below:

- $\delta_0 \approx 0.1$ to 0.2 mm for light alloys
- $\delta_0 \approx 0.2$ to 0.5 mm for steels
- $\delta_0 \approx 0.1$ to 1 mm for ceramics
- $\delta_0 \approx 0.5$ to 1 mm for polymers
- $\delta_0 \approx 10$ to 100 mm for concretes

Concerning the orientation of the plane in which lies the initiated mesocrack, its normal $\vec{n}$ is the result of the localization analysis of Sect. 1.6.2,

$$\boxed{\det(\vec{n}\boldsymbol{L}\vec{n}) = 0\,.} \quad (1.253)$$

If no localization analysis has been performed, the anisotropic damage evolution gives the orientation of the mesocrack as the plane with normal $\vec{n}$ (in which lies the maximum value of the damage vector $\boldsymbol{D}\vec{n}$ which corresponds to the maximum principal value of the damage (see Sect. 1.4.3)):

$$\boxed{\vec{n} = \vec{n}(\max D_I)\,.} \quad (1.254)$$

If only an isotropic damage analysis has been performed, we use the orientation of the maximum principal stress,

$$\boxed{\vec{n} = \vec{n}(\max \sigma_I)\,.} \quad (1.255)$$

To conclude, Table 1.8 gives the possible choices in modelling the damage threshold, critical damage, size, and orientation of the mesocrack initiated.

Table 1.8. Conditions for initiation of damage and mesocrack

Damage threshold	Monotonic loading	$p = \epsilon_{\text{pD}}$	$\epsilon_{\text{pD}} = \epsilon_{\text{p}}(\sigma = \sigma_{\text{u}})$
	Cyclic loading	$p = p_{\text{D}}$	$p_{\text{D}} = \epsilon_{\text{pD}} \left[\dfrac{\sigma_{\text{u}} - \sigma_{\text{f}}^{\infty}}{\dfrac{\sigma_{\text{eq}}^{\max} + \sigma_{\text{eq}}^{\min}}{2} - \sigma_{\text{f}}^{\infty}} \right]^m$
	Numerical prediction	$\max w_{\text{s}} = w_{\text{D}}$	w_{D} given by (1.116)
Critical damage	Isotropic damage	$D = D_{\text{c}}$	$D_{\text{c}} = 1 - \dfrac{\sigma_{\text{R}}}{\sigma_{\text{u}}}$
	Anisotropic damage	$\max D_I = D_{\text{c}}$	$D_{\text{c}} = 0.5$
	Numerical prediction	$\det(\vec{n}\underline{\boldsymbol{L}}\vec{n}) = 0$	$\eta D_{\text{H}} \approx 1$
Direction of the crack plane	Isotropic damage	Max principal $\boldsymbol{\sigma}$	$\vec{n}$
	Anisotropic damage	Max principal $\boldsymbol{D}$	$\vec{n}$
	Numerical prediction	Localization	$\vec{n}$
Size of the mesocrack	Checking initiation	RVE	$A_{0\ (mm)^2} \approx \begin{vmatrix} (0.2)^2 & \text{metals} \\ (1)^2 & \text{polymers} \\ (100)^2 & \text{concrete} \end{vmatrix}$
	Checking propagation	Equivalence with Fracture Mechanics	$A_0 = \dfrac{G_{\text{c}}^2}{\left(\dfrac{\sigma_{\text{u}}^2}{2E} D_{\text{c}} + \sigma_{\text{u}}\epsilon_{\text{pR}}\right)^2}$

2

Numerical Analysis of Damage

Damage is essentially a nonlinear phenomenon often coupled with (visco-)plasticity, also a nonlinear phenomenon. Therefore we can not expect simple closed-form solutions of mesocrack initiation problems except for rough approximations of simple cases.

For early design of mechanical components, the coupling of the strain behavior with the damage may be neglected and a **post-processing** of damage evolution is possible after a classical structure analysis (D. Hayhurst and F.A. Leckie 1974). This approach is the **uncoupled analysis** based on a reference (visco-)plastic computation. For localized (visco-)plasticity and damage, this reference computation can even be purely elastic with a local energetic correction (such as the Neuber method) to estimate the plastic strain. Fast!

For accurate engineering applications and when the coupling between strains and damage is strong, (visco-)plasticity, damage, and possible cracks distributed over the whole structure deem a **fully coupled analysis** necessary. The constitutive equations need to be implemented within a finite element computer code and the numerical analysis encounters the classical difficulties of convergence of linearized schemes. It needs special algorithms, much care, and large computer times (A. Benallal, R. Billardon and I. Doghri 1988).

When the damage is localized on the mesoscale, there is the possibility to use the two-scale damage model of Chap. 1 in which damage occurs on a microscale only (J. Lemaitre and I. Doghri 1992). This **locally coupled analysis** mainly applies to high cycle fatigue (fatigue in the elastic range), eventually with initial plastic strain and damage. The analysis can be performed by post-processing an elastic computation for "elastic" fatigue or a (visco-)plastic computation if the process of creation of non-trivial initial conditions has been modelled.

Another area that requires numerical analysis is the **precise identification of the material parameters**. Even if a fast, rough identification is possible using simple methods, an adjustment or updating of the parameters from more complex tests or similar studies needs numerous iterations and robust optimization techniques based on sensitivity analyses. A table at the end of the chapter gives values of damage parameters for several materials.

In order to make the expressions simpler, intrinsic notations are used in this chapter when describing the schemes for the numerical implementation of the constitutive equations.

2.1 Uncoupled Analysis

Structure calculations performed with nonlinear material behavior become very costly in time very quickly. Their use is therefore difficult to perform for designs in which the geometry of the structure and the choice of the materials are the results of an optimization process.

A fast way to evaluate the damage is by post-processing classical calculations of elasticity or elasto-(visco-)plasticity in order to perform a posteriori time integration of the damage evolution law of Sect. 1.4 (see Fig. 2.1):

$$D = \int_{t_\mathrm{D}}^{t} \left(\frac{Y(t)}{S}\right)^s \dot{p}(t)\mathrm{d}t \quad \text{or} \quad \boldsymbol{D} = \int_{t_\mathrm{D}}^{t} \left(\frac{\overline{Y}(t)}{S}\right)^s |\dot{\boldsymbol{\epsilon}}^\mathrm{P}(t)|\mathrm{d}t, \qquad (2.1)$$

where t_D is the time taken for damage initiation, corresponding to the damage threshold $p_\mathrm{D} = \int_0^{t_\mathrm{D}} \dot{p}(t)\mathrm{d}t$ (see Sect. 1.4.1).

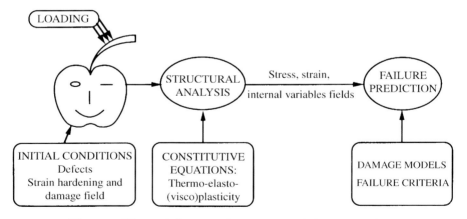

Fig. 2.1. Uncoupled approach to structural damage analysis

The time histories of the plastic strain, the energy density release rate Y for isotropic damage, and the effective strain energy $\overline{Y}$ for anisotropic damage are the necessary inputs for the damage post-processing. They may come directly from an elasto-(visco-)plastic computation or indirectly from an elastic computation followed by an initial post-processing of (visco-)plasticity.

Note that the material parameters depend on the temperature which may vary during the loading. In that case, their time dependencies have to be taken into account when integrating the damage law.

The procedure for the uncoupled analysis of damage is then:

- Perform a structure calculation to obtain the histories of Y or $\overline{Y}$ and ϵ^{p}
- Evaluate D or $\boldsymbol{D}$ by means of the time integration of Eq. (2.1)
- Estimate the time or the load corresponding to the critical damage D_{c}
- Define a safety margin on the most appropriate parameter: the plastic strain p, the number of cycles N, the time t, or the damage equivalent stress $\sigma^\star$, through a safety factor Saf (larger than unity) in order to ensure

$$p \le \frac{p_{\mathrm{R}}}{Saf} \quad \text{or} \quad N \le \frac{N_{\mathrm{R}}}{Saf} \quad \text{or} \quad t \le \frac{t_{\mathrm{R}}}{Saf} \quad \text{or} \quad \sigma^\star \le \frac{\sigma^\star_{\mathrm{R}}}{Saf} \qquad (2.2)$$

as a safe non-failure design criterion.

Most of the time, neglecting the coupling of the damage with the strain behavior yields an upper bound on these parameters. The integration of the damage rate equation (2.1) concerns any kind of loading: monotonic, cyclic, multi-level, random ... Examples are given in the next chapters.

2.1.1 Uniaxial Loading[ठ,ठ̈ठ̈]

Let us first consider the 1D case of tension or tension-compression with a state of stress

$$\boldsymbol{\sigma} = \begin{bmatrix} \sigma & 0 & 0 \\ 0 & 0 & 0 \\ 0 & 0 & 0 \end{bmatrix}, \qquad \sigma = \sigma(t) \qquad (2.3)$$

and a state of plastic strain

$$\boldsymbol{\epsilon}^{\mathrm{p}} = \begin{bmatrix} \epsilon_{\mathrm{p}} & 0 & 0 \\ 0 & -\frac{\epsilon_{\mathrm{p}}}{2} & 0 \\ 0 & 0 & -\frac{\epsilon_{\mathrm{p}}}{2} \end{bmatrix}, \qquad \epsilon_{\mathrm{p}} = \epsilon_{\mathrm{p}}(t). \qquad (2.4)$$

The accumulated plastic strain is then

$$p = p(t) = \int_0^t |\dot{\epsilon}_{\mathrm{p}}(t)| \mathrm{d}t, \qquad (2.5)$$

the von Mises stress is $\sigma_{\mathrm{eq}} = |\sigma|$, and the stress triaxiality has the value $1/3$.

The damage calculated by means of the isotropic model is

$$D = \int_{t_{\mathrm{D}}}^t \left(\frac{Y(t)}{S}\right)^s |\dot{\epsilon}_{\mathrm{p}}(t)| \mathrm{d}t. \qquad (2.6)$$

For the anisotropic model, the same expression is obtained on D_1 (the component in the loading direction) as for tension such that

$$\boldsymbol{D} = D_1 \begin{bmatrix} 1 & 0 & 0 \\ 0 & \frac{1}{2} & 0 \\ 0 & 0 & \frac{1}{2} \end{bmatrix} \quad (2.7)$$

and then

$$D_1 = \int_{t_D}^{t} \left(\frac{\overline{Y}(t)}{S}\right)^s |\dot{\epsilon}_p| \mathrm{d}t . \quad (2.8)$$

The expressions for the functions $Y(t)$ or $\overline{Y}(t)$ depend on whether or not we consider the microdefects closure effect (quasi-unilateral conditions) through the value of the microdefects closure parameter h or h_a of Sect. 1.2.4.

2.1.1.1 Damage without Microdefects Closure Effect

For both isotropic and anisotropic damage models, the effective energy density release rate functions have the same expression if written in terms of the uniaxial elastic strain ϵ_e:

$$Y = \frac{1}{2} E \epsilon_e^2(t) \quad \text{or} \quad \overline{Y} = \frac{1}{2} E \epsilon_e^2(t) . \quad (2.9)$$

From a reference elasto-(visco-)plastic computation, the knowledge of the time history of the elastic strain allows for the determination of $Y(t)$ or $\overline{Y}(t)$ and the damage, as follows:

$$D = \int_{t_D}^{t} \left(\frac{E \epsilon_e^2(t)}{2S}\right)^s |\dot{\epsilon}_p(t)| \mathrm{d}t \quad \text{or} \quad D_1 = \int_{t_D}^{t} \left(\frac{E \epsilon_e^2(t)}{2S}\right)^s |\dot{\epsilon}_p(t)| \mathrm{d}t . \quad (2.10)$$

Due to the uncoupled analysis where the stress and the strain are calculated without the softening effect of damage, Y and $\overline{Y}$ written in terms of stresses are preferred as they give an upper bound on D. In order to keep the post-processing simple enough for design purposes, the preferable case of isotropic damage is considered for which

$$Y = \frac{\sigma^2(t)}{2E(1-D)^2} . \quad (2.11)$$

If the stress and plastic strain histories are known, we have

$$\int_0^D (1-D)^{2s} \mathrm{d}D = \int_{t_D}^{t} \left(\frac{\sigma^2(t)}{2ES}\right)^s |\dot{\epsilon}_p(t)| \mathrm{d}t \quad (2.12)$$

or

$$\boxed{D = 1 - \left[1 - (2s+1)\int_{t_D}^{t}\left(\frac{\sigma^2(t)}{2ES}\right)^s |\dot{\epsilon}_p(t)|\mathrm{d}t\right]^{\frac{1}{2s+1}}} . \quad (2.13)$$

The **anisotropic** case is similar, with an effective triaxiality function $\tilde{R}_\nu$ now that depends on the damage (Eq. (1.41) of Sect. 1.2.3). For uniaxial tension, it is a function of D_1 alone because $D_2 = D_3 = D_1/2$ and $D_H = 2D_1/3$, leading to

$$\tilde{R}_\nu(D_1) = \frac{2}{3}(1+\nu) + 3(1-2\nu)\left[\left(1 - \frac{2\eta}{3}D_1\right)\left(\frac{2}{1-D_1} + \frac{1}{1-\frac{D_1}{2}}\right)\right]^{-2} \quad (2.14)$$

and the evolution of the damage is obtained as an implicit function of D_1, which is easy to solve by use of mathematics software:

$$\boxed{\int_0^{D_1} \tilde{R}_\nu^{-s} dD_1 = \int_{t_D}^t \left(\frac{\sigma^2(t)}{2ES}\right)^s |\dot{\epsilon}_p(t)| dt.} \quad (2.15)$$

2.1.1.2 Damage with Microdefects Closure Effect⚛

Considering the quasi-unilateral conditions is important when dealing with compressive loading and cyclic or fatigue applications as they lead to a damage rate that is much smaller in compression than in tension. The representation of the mean stress effect in fatigue then follows.

Microdefects closure material parameters h or h_a are introduced within the expressions for the Y and $\overline{Y}$ functions. This makes the models more difficult for closed-form solutions. We advise using the approximation $\overline{Y}(\sigma, \boldsymbol{D}, h_a) \approx Y(\sigma, D = 0, h = 0.2)$ if damage anisotropy needs to be taken into account. For the resolution with the exact $\overline{Y}$ function for anisotropic damage with microdefects closure effect, please refer to Sect. 2.1.2 on proportional loading.

For the uniaxial tension-compression case, one has

$$Y = \frac{1}{2E}\left[\frac{\langle\sigma\rangle^2}{(1-D)^2} + h\frac{\langle-\sigma\rangle^2}{(1-hD)^2}\right], \quad (2.16)$$

which may be written in terms of elastic strain

$$Y = \frac{E}{2}\left[\langle\epsilon_e\rangle^2 + h\langle-\epsilon_e\rangle^2\right]. \quad (2.17)$$

Then,

$$\boxed{D = \int_{t_D}^t \left(\frac{Y}{S}\right)^s |\dot{\epsilon}_p| dt \quad \text{with} \quad Y = \begin{cases} Y^+ = \dfrac{\sigma^2(t)}{2E(1-D)^2} & \text{in tension,} \\ Y^- = h\dfrac{\sigma^2(t)}{2E(1-hD)^2} & \text{in compression.} \end{cases}}$$
$$(2.18)$$

2.1.2 Proportional Loading$^{\circ,\circ\circ\circ}$

Uniaxial loading is a particular case of proportional loading for which the stress tensor at a given point M of a structure remains proportional to a **time-independent** tensor $\boldsymbol{\Sigma}$ such that

$$\sigma_{ij}(M,t) = \sigma_\Sigma(t)\, \Sigma_{ij}(M)\,. \tag{2.19}$$

The main sufficient condition for a structure to be considered as proportionally loaded is when there is only one applied load or when all the loads vary proportionally to **one parameter**.

With the normalization $\Sigma_{\text{eq}} = \sqrt{\frac{3}{2}\Sigma_{ij}^{\text{D}}\Sigma_{ij}^{\text{D}}} = 1$, the scalar σ_Σ is the signed von Mises stress and $|\sigma_\Sigma| = \sigma_{\text{eq}}$. In tension-compression,

$$\boldsymbol{\Sigma} = \begin{bmatrix} 1 & 0 & 0 \\ 0 & 0 & 0 \\ 0 & 0 & 0 \end{bmatrix}, \quad \boldsymbol{\Sigma}^{\text{D}} = \begin{bmatrix} \frac{2}{3} & 0 & 0 \\ 0 & -\frac{1}{3} & 0 \\ 0 & 0 & -\frac{1}{3} \end{bmatrix}, \quad \text{and}\quad \sigma_\Sigma = \sigma\,. \tag{2.20}$$

For elasto-(visco-)plasticity that is coupled (or not coupled) with isotropic damage, the plastic strain tensor remains proportional to the deviatoric part of $\boldsymbol{\Sigma}$, as in

$$\epsilon_{ij}^{\text{P}}(M,t) = \frac{3}{2}\epsilon_{\text{p}\Sigma}(t)\,\Sigma_{ij}^{\text{D}}(M)\quad\text{and}\quad \Sigma_{ij}^{\text{D}} = \Sigma_{ij} - \frac{1}{3}\Sigma_{kk}\delta_{ij}\,, \tag{2.21}$$

introducing the scalar signed equivalent plastic strain $\epsilon_{\text{p}\Sigma}$ such as $|\epsilon_{\text{p}\Sigma}| = \sqrt{\frac{2}{3}\epsilon_{ij}^{\text{P}}\epsilon_{ij}^{\text{P}}}$. Due to the 3/2 term in (2.21), the accumulated plastic strain has the same expression as for the uniaxial tensile case:

$$\dot{p} = |\dot{\epsilon}_{\text{p}\Sigma}|\quad\text{or}\quad p = p(t) = \int_0^t |\dot{\epsilon}_{\text{p}\Sigma}|\mathrm{d}t\,. \tag{2.22}$$

The **anisotropic** damage case is more complex and the previous equation $\epsilon_{ij}^{\text{P}} = \frac{3}{2}\epsilon_{\text{p}\Sigma}\Sigma_{ij}^{\text{D}}$ of proportional loading in terms of plastic strains may be seen as an approximation for the general anisotropic 3D case but it strictly applies for tension-compression and for shear.

2.1.2.1 Kinematic Hardening as an Additional Isotropic Hardening in Monotonic Proportional Loading$^{\circ}$

For elasto-(visco-)plasticity with isotropic and kinematic hardenings R and $\boldsymbol{X}$, one has $\sigma_{ij}^{\text{D}} = \sigma_\Sigma \Sigma_{ij}^{\text{D}}$, $\epsilon_{ij}^{\text{P}} = \frac{3}{2}\epsilon_{\text{p}\Sigma}\Sigma_{ij}^{\text{D}}$ and

$$X_{ij} = X(t)\,\Sigma_{ij}^{\text{D}}\,,\qquad X_{\text{eq}} = \sqrt{\frac{3}{2}X_{ij}X_{ij}} = |X|\,. \tag{2.23}$$

The yield criterion is

$$f = (\boldsymbol{\sigma}^D - \boldsymbol{X})_{eq} - R - \sigma_y = \left((\sigma_\Sigma - X)\boldsymbol{\Sigma}^D\right)_{eq} - R - \sigma_y, \quad (2.24)$$

with

$$\left((\sigma_\Sigma - X)\boldsymbol{\Sigma}^D\right)_{eq} = \sqrt{\frac{3}{2}(\sigma_\Sigma - X)^2 \Sigma_{ij}^D \Sigma_{ij}^D} = |\sigma_\Sigma - X|\, \Sigma_{eq} = |\sigma_\Sigma - X|. \quad (2.25)$$

In cases of increasing **monotonic proportional loadings** ($\sigma_\Sigma > 0$, $X > 0$, $X < \sigma_\Sigma$) one ends up with

$$\boxed{f = |\sigma_\Sigma - X| - R - \sigma_y = \sigma_{eq} - (X_{eq} + R) - \sigma_y,} \quad (2.26)$$

which shows that kinematic hardening may be considered in these cases as an **additional isotropic hardening**. This remark is consistent with the consideration of one hardening only (either kinematic or isotropic) for computations of structures undergoing monotonic loadings.

2.1.2.2 Damage Post-processing Without Microdefects Closure Effect ⟲

For both isotropic and anisotropic damage, the effective elastic energy density functions have the same expression if written in terms of elastic strain $\boldsymbol{\epsilon}^e$:

$$Y = \frac{1}{2} E_{ijkl} \epsilon_{ij}^e \epsilon_{kl}^e \quad \text{or} \quad \overline{Y} = \frac{1}{2} E_{ijkl} \epsilon_{ij}^e \epsilon_{kl}^e. \quad (2.27)$$

As for the uniaxial case, information on the time history of the elastic strains allows for the determination of $Y(t)$ or $\overline{Y}(t)$ and the damage.

Expressions written in terms of stresses are preferred. For isotropic damage,

$$Y = \frac{\sigma_{eq}^2 R_\nu}{2E(1-D)^2} = \frac{\sigma_\Sigma^2 R_\nu}{2E(1-D)^2}, \quad (2.28)$$

where the triaxiality function R_ν only depends on the tensor $\boldsymbol{\Sigma}$ and has a constant value:

$$R_\nu = \frac{2}{3}(1+\nu) + \frac{1}{3}(1-2\nu)(\Sigma_{kk})^2 = const. \quad (2.29)$$

The possibilities for the time integration of the damage evolution law are

$$D = \int_{t_D}^{t} \left(\frac{E_{ijkl}\epsilon_{ij}^e \epsilon_{kl}^e}{2S}\right)^s |\dot{\epsilon}_{p\Sigma}| dt. \quad (2.30)$$

If the elastic and plastic strains are known or (the most conservative choice), then

$$\int_0^D (1-D)^{2s} \mathrm{d}D = R_\nu^s \int_{t_\mathrm{D}}^t \left(\frac{\sigma_\mathrm{eq}^2}{2ES}\right)^s |\dot\epsilon_{\mathrm{p}\Sigma}| \mathrm{d}t$$

$$\Leftrightarrow \boxed{D = 1 - \left[1 - (2s+1)R_\nu^s \int_{t_\mathrm{D}}^t \left(\frac{\sigma_\mathrm{eq}^2(t)}{2ES}\right)^s |\dot\epsilon_{\mathrm{p}\Sigma}(t)| \mathrm{d}t\right]^{\frac{1}{2s+1}}} \quad (2.31)$$

If the von Mises stress $\sigma_\mathrm{eq} = |\sigma_\Sigma|$, the stress triaxiality $T_\mathrm{X} = \dfrac{\sigma_\mathrm{H}}{\sigma_\mathrm{eq}} = \Sigma_\mathrm{H} = \dfrac{1}{3}\Sigma_{kk}$ and the accumulated plastic strain rate $\dot p = |\dot\epsilon_{\mathrm{p}\Sigma}|$ are known, or

$$D = R_\nu^s \int_{t_\mathrm{D}}^t \left(\frac{\sigma_\mathrm{eq}^2}{2ES}\right)^s |\dot\epsilon_{\mathrm{p}\Sigma}| \mathrm{d}t \quad (2.32)$$

as an approximation of the last formula.

The case of **anisotropic damage** is similar but is made more difficult for closed-form solutions because of the dependency of the triaxiality function on the damages D_1, D_2, and D_3. Nevertheless, it may always be worked out by using

$$\boxed{D_{ij} = \frac{3}{2}|\boldsymbol{\Sigma}^\mathrm{D}|_{ij} \int_{t_\mathrm{D}}^t \left(\frac{\overline Y}{S}\right)^s |\dot\epsilon_{\mathrm{p}\Sigma}| \mathrm{d}t\,,} \quad (2.33)$$

where $\overline Y = \dfrac{1}{2}E_{ijkl}\epsilon^\mathrm{e}_{ij}\epsilon^\mathrm{e}_{kl}$ and $|\boldsymbol{\Sigma}^\mathrm{D}|$ denotes the absolute value of the deviatoric tensor $\boldsymbol{\Sigma}^\mathrm{D}$ in terms of principal components.

2.1.2.3 Damage Post-processing with Microdefects Closure Effect♂♂♂

For the isotropic damage model, the only change in the damage law is the expression of Y, which now introduces the microdefects closure parameter h (1.49):

$$Y = \frac{1+\nu}{2E}\left[\frac{\langle\boldsymbol\sigma\rangle^+_{ij}\langle\boldsymbol\sigma\rangle^+_{ij}}{(1-D)^2} + h\frac{\langle\boldsymbol\sigma\rangle^-_{ij}\langle\boldsymbol\sigma\rangle^-_{ij}}{(1-hD)^2}\right] - \frac{\nu}{2E}\left[\frac{\langle\sigma_{kk}\rangle^2}{(1-D)^2} + h\frac{\langle-\sigma_{kk}\rangle^2}{(1-hD)^2}\right], \quad (2.34)$$

with $h \approx 0.2$ for metals.

By considering the definition $\sigma_{ij} = \sigma_\Sigma \Sigma_{ij}$ of proportional loading and by neglecting the effects of the damage on Y, one gets

$$Y \approx \frac{\sigma_\Sigma^2(t)R_{\nu h}}{2E}, \quad (2.35)$$

if $\sigma_\Sigma > 0$, $R_{\nu h} = R_{\nu h}^+ = (1+\nu)\left[\langle\boldsymbol{\Sigma}\rangle_{ij}^+ \langle\boldsymbol{\Sigma}\rangle_{ij}^+ + h\langle\boldsymbol{\Sigma}\rangle_{ij}^- \langle\boldsymbol{\Sigma}\rangle_{ij}^-\right]$
$$ -\nu\left[\langle\Sigma_{kk}\rangle^2 + h\langle-\Sigma_{kk}\rangle^2\right] = const, \quad (2.36)$$

if $\sigma_\Sigma < 0$, $R_{\nu h} = R_{\nu h}^- = (1+\nu)\left[\langle\boldsymbol{\Sigma}\rangle_{ij}^- \langle\boldsymbol{\Sigma}\rangle_{ij}^- + h\langle\boldsymbol{\Sigma}\rangle_{ij}^+ \langle\boldsymbol{\Sigma}\rangle_{ij}^+\right]$
$$ -\nu\left[\langle-\Sigma_{kk}\rangle^2 + h\langle\Sigma_{kk}\rangle^2\right] = const. \quad (2.37)$$

For the anisotropic damage model the microdefects closure parameter h_a is introduced. It is different from h, with $h_a \approx 0$ in most cases (1.60). The effective elastic energy density is then

$$\overline{Y} = \frac{1+\nu}{2E}\mathrm{tr}\left(\boldsymbol{H}^\mathrm{P}\boldsymbol{\sigma}_+^\mathrm{D}\boldsymbol{H}^\mathrm{P}\right)^2 + \frac{3(1-2\nu)}{2E}\frac{\langle\sigma_\mathrm{H}\rangle^2}{(1-\eta D_\mathrm{H})^2} \quad (2.38)$$

so that

$$R_{\nu h}^+ = (1+\nu)\langle\boldsymbol{\Sigma}^\mathrm{D}\rangle_{ij}^+ \langle\boldsymbol{\Sigma}^\mathrm{D}\rangle_{ij}^+ + \frac{1}{3}(1-2\nu)\langle\Sigma_{kk}\rangle^2 = const \quad (2.39)$$

$$R_{\nu h}^- = (1+\nu)\langle\boldsymbol{\Sigma}^\mathrm{D}\rangle_{ij}^- \langle\boldsymbol{\Sigma}^\mathrm{D}\rangle_{ij}^- + \frac{1}{3}(1-2\nu)\langle-\Sigma_{kk}\rangle^2 = const. \quad (2.40)$$

2.1.3 Post-processing a (Visco-)Plastic Computation

A classical elasto-(visco-)plastic computation without any damage gives the time histories of the von Mises stress σ_eq, the hydrostatic stress σ_H, the stress triaxiality $T_\mathrm{X} = \sigma_\mathrm{H}/\sigma_\mathrm{eq}$, the elastic energy density $\frac{1}{2}E_{ijkl}\epsilon^\mathrm{e}_{ij}\epsilon^\mathrm{e}_{kl}$, the accumulated plastic strain rate $\dot{p}$, and the stored energy density w_s. These quantities are used directly as inputs for the time integration of the damage evolution law, providing an estimation of the damage as a function of the time, the accumulated plastic strain, or the number of cycles for fatigue applications.

To make the post-processing procedure fast, the time integration of the damage law may be performed only at the most loaded point location where the damage equivalent stress $\sigma^\star = \sigma_\mathrm{eq} R_\nu^{1/2}$ has the largest value in the whole structure. Such a criterion takes the stress triaxiality effect into account through the function R_ν. In that case the hypothesis of a proportional loading may often be made, allowing for the use of the analytical expressions of Sects. 2.1.1 and 2.1.2.

The post-processing procedure may also be automatically applied at each Gauss point of the structure. Then the most loaded point simply is the point where the damage is the largest. The advantage is that one obtains maps of the damage field, while the drawback is of course a longer computer time.

If the loading is periodic, it is only necessary to run the elasto-(visco-)plastic computation up to the stabilized cycle (10 to 50) as the information collected for the time integration of the damage law will be sufficient. For large numbers of cycles, a "jump" in cycles procedure may be applied to avoid the calculation of too many cycles (see Sect. 2.1.5).

2.1.4 Post-processing an Elastic Computation ⚥

Often, the plasticity is localized so that the stresses and the strains of the entire structure may be first calculated in elasticity. This has many advantages including reduced computational time cost and the possibility to use the main linearity property to obtain the stress and strain fields for many other applied loadings. The results depend only on the Young's modulus and Poisson ratio.

The (visco-)plastic quantities p, w_s and the quantities strongly dependent on the (visco-)plastic behavior σ_{eq}, σ_H, Y, or $\overline{Y}$ need to be estimated from an initial post-processing procedure as those quantities are required for the time integration of the damage law. Examples of these caculations include the Neuber method or the Strain Energy Density (SED) method fully described in Sects. 3.2.4, 4.2.4 and 5.2.4. These methods are based on a local energy equivalence at the stress concentration points between an elastic and an elasto-plastic computation of the same structure with the same loading history. Applied as post-processors of an elastic computation, the Neuber and SED methods allow for the determination of the von Mises stress and the accumulated plastic strain where plasticity is localized (small scale yielding).

The damage post-processing is then carried out as described in Sects. 2.1.1 or 2.1.2, yielding closed-form expressions for the mesocrack initiation conditions with an explicit dependence on plasticity and damage material parameters.

The knowledge of the stress triaxiality T_X is of first importance as it partly governs the damage growth through the triaxiality function:

$$R_\nu = \frac{2}{3}(1+\nu) + 3(1-2\nu)T_X^2, \quad \text{where} \quad T_X = \frac{\sigma_H}{\sigma_{eq}}. \qquad (2.41)$$

It is not given by the Neuber or SED methods. Nevertheless some analytical results may be obtained for plane deformation problems:

- For a **plane stress** state, the points located along free edges are exposed to pure tension (or compression). The value of the triaxiality ratio is simply $1/3$.
- For a **plane strain** state in **elasticity**, the triaxiality ratio evaluated at points located along free edges only depends on Poisson ratio,

$$T_X = \frac{1+\nu}{3\sqrt{1-\nu+\nu^2}}, \qquad (2.42)$$

where T_X depends on neither the loading type nor its intensity. For $\nu = 0$, $T_X = 1/3$ for some composite materials. For $\nu = 1/3$, $T_X = 0.5$. For $\nu \approx 0.5$, $T_X = 1/\sqrt{3} \approx 0.58$ for rubber-like materials.
- For a **plane strain** state in **(visco-)plasticity**, the closed-form expression of the stress triaxiality may be derived analytically for points located along free edges.

In plasticity with linear hardening the yield function is $f = \sigma_{\text{eq}} - C_{\text{y}} p - \sigma_{\text{y}}$, where C_{y} is the plastic "tangent" modulus and $C_{\text{y}} \approx (\sigma_{\text{u}} - \sigma_{\text{y}})/\epsilon_{\text{pu}}$. The boundary conditions (free edges of normal $\vec{e}_1$: $\sigma_{i1} = 0$), the plane strain condition $\epsilon_{i3} = 0$, and the elasto-plastic behavior considered altogether lead to

$$\dot{\epsilon}_{33} = \frac{\dot{\sigma}_{33} - \nu \dot{\sigma}_{22}}{E} + \frac{(2\sigma_{33} - \sigma_{22})\dot{p}}{2\sigma_{\text{eq}}} = 0. \quad (2.43)$$

With the notations,

$$\chi = \frac{C_{\text{y}}(1-2\nu)}{2(C_{\text{y}} + E)}, \quad u_0 = \frac{1}{\sqrt{1-\nu+\nu^2}}, \quad \mathcal{W} = \frac{2\sqrt{3}\chi}{3+4\chi^2}\left(1 - \frac{2\chi}{1-2\nu}\right),$$

and $\omega = \dfrac{2\chi}{1-2\nu} + \dfrac{2\chi^2}{3+4\chi^2}\left(1 - \dfrac{2\chi}{1-2\nu}\right),$

$$(2.44)$$

and without any assumption about the loading, the closed-form expression for the $T_{\text{X}}(p)$ law is then governed by the parametric representation

$$T_{\text{X}}(u) = \frac{u}{2} - \frac{1}{6}\sqrt{4 - 3u^2},$$

$$p(u) = \frac{\sigma_{\text{y}}}{C_{\text{y}}}\left(\frac{\sqrt{1 - \frac{3u_0^2}{4}} - \chi \cdot u_0}{\sqrt{1 - \frac{3u^2}{4}} - \chi \cdot u}\right)^{\omega} \cdot \exp\left\{\mathcal{W} \cdot \left[\arcsin\frac{\sqrt{3}u}{2}\right]_{u_0}^{u}\right\} - \frac{\sigma_{\text{y}}}{C_{\text{y}}},$$

$$(2.45)$$

which may be seen as a law $T_{\text{X}} = T_{\text{X}}(p)$ or as $T_{\text{X}} = T_{\text{X}}(\sigma_{\text{eq}})$: the triaxiality ratio on a free edge depends on the von Mises stress only.

For Poisson ratios larger than 0.3, there is a slight difference between T_{X} evaluated in elasticity and T_{X} evaluated in plasticity: the triaxiality remains between 0.5 and 0.58 but its variation shows that the plane strain loadings are not truly proportional. As soon as elastic strains are negligible, T_{X} reaches a saturation value $T_{\text{X sat}}$. Table 2.1 gives the values of the triaxiality at saturation as a function of the ratio C_{y}/E for different Poisson ratios. For metals, it is $T_{\text{X sat}} = 0.58$.

Table 2.1. Triaxiality in plane strain condition

C_{y}/E	10^{-6}	0.1	0.25	0.5
$T_{\text{X sat}}(\nu = 0)$	0.58	0.56	0.54	0.50
$T_{\text{X sat}}(\nu = 0.3)$	0.58	0.57	0.56	0.55
$T_{\text{X sat}}(\nu = 0.5)$	0.58	0.58	0.58	0.58

If the hardening law is not linear, one has to fit a linear law in the plastic strain range under consideration and apply the parametric expression (2.45) as well.

Consideration of the viscosity effects introduces the viscous stress σ_v function of the loading rate. For strain rates at which the viscosity effects are important, the yield stress has to be replaced by $\sigma_y + \sigma_v$ but the saturation value is independent of the loading rate.

To conclude, $T_X = 0.58$ at a free edge point is a good value to consider in order to evaluate R_ν and to achieve plane strain Neuber or SED method.

- For **3D cases**, the hypothesis of a stress triaxiality in plasticity equal to the (known) stress triaxiality in elasticity can be made. This may be seen as a weak form of proportional loading.

2.1.5 Jump-in-Cycles Procedure in Fatigue♂
(R. Billardon and I. Doghri 1988)

For loadings that are periodic or periodic by blocks in fatigue, the step by step computation in time becomes prohibitive when the number of cycles becomes large (10^4, 10^6, 10^8, ...). Therefore a simplified method which allows the program to "jump" full blocks of $\overline{\Delta N}$ cycles is necessary. The time integration is performed over one cycle once in a while and the computation time may be reduced by a factor of almost $\overline{\Delta N}$.

This "jump"-in-cycles procedure is as follows:

1. Before any damage growth, i.e., up to $w_s = w_D$ or $p = p_D$, run the computation until reaching a stabilized cycle N_s and let $\left.\dfrac{\delta p}{\delta N}\right|_{N_s}$ be the accumulated plastic strain increment over this single cycle. Assume then that during the next $\overline{\Delta N}$ cycles, p remains linear versus the number of cycles N and calculate the number of cycles to be advanced by $\overline{\Delta N}$ cycles as

$$\overline{\Delta N} = \frac{\overline{\Delta p}}{\left.\dfrac{\delta p}{\delta N}\right|_{N_s}}, \qquad (2.46)$$

where $\overline{\Delta p}$ is a given value which determines the accuracy of the procedure and $\overline{\Delta p} \approx p_D/50$ is a good compromise between accuracy and time cost. The accumulated plastic strain is updated as

$$p(N_s + \overline{\Delta N}) = p(N_s) + \overline{\Delta p} \qquad (2.47)$$

and is equal to $r(N_s + \overline{\Delta N})$ as long as there is no damage. The stored energy density is updated as

$$w_s(N_s + \overline{\Delta N}) = w_s(N_s) + \int_{p(N_s)}^{p(N_s+\overline{\Delta N})} \frac{A}{m} R(p) p^{\frac{1}{m}-1} dp. \qquad (2.48)$$

The stresses, strains, plastic strains, kinematic hardening at the end of the cycle N_s, the accumulated plastic strain $p(N_s + \overline{\Delta N})$, and the stored

energy density $w_s(N_s + \overline{\Delta N})$ are then the initial values for the computation of the first time increment of the cycle $(N_s + \overline{\Delta N} + 1)$.
Repeat the jumps until damage occurs.

2. Once damage occurs, first run the computation at constant damage until a new stabilized cycle N_s is reached. Then

- Calculate the accumulated plastic strain increment $\left.\dfrac{\delta p}{\delta N}\right|_{N_s}$.
- Calculate the damage increment $\left.\dfrac{\delta D}{\delta N}\right|_{N_s}$ or $\left.\dfrac{\delta D_H}{\delta N}\right|_{N_s}$ over this cycle.
- Assume again that during the next $\overline{\Delta N}$ cycles, p and D (or D_H) remain linear versus N and calculate the number of cycles to be advanced as

$$\overline{\Delta N} = \min\left(\frac{\overline{\Delta p}}{\left.\dfrac{\delta p}{\delta N}\right|_{N_s}}, \frac{\overline{\Delta D}}{\left.\dfrac{\delta D}{\delta N}\right|_{N_s}}\right), \qquad (2.49)$$

where $\overline{\Delta D}$ is a given value which determines the accuracy on the damage. Here again $\overline{\Delta D} \approx D_c/50$ is a good compromise between accuracy and time cost and take $\overline{\Delta p} = \left(\dfrac{S}{Y_{\max}}\right)^s \overline{\Delta D}$ with $Y_{\max}$ as the maximum value of functions Y or $\overline{Y}$ over the cycle N_s.

The accumulated plastic strain and the damage are finally updated as

$$p(N_s + \overline{\Delta N}) = p(N_s) + \left.\frac{\delta p}{\delta N}\right|_{N_s} \overline{\Delta N},$$

$$r(N_s + \overline{\Delta N}) = r(N_s) + \left.\frac{\delta r}{\delta N}\right|_{N_s} \overline{\Delta N},$$

$$D(N_s + \overline{\Delta N}) = D(N_s) + \left.\frac{\delta D}{\delta N}\right|_{N_s} \overline{\Delta N} \quad \text{for isotropic damage, and}$$

$$D_{ij}(N_s + \overline{\Delta N}) = D_{ij}(N_s)\left[1 + \frac{1}{D_H(N_s)}\left.\frac{\delta D_H}{\delta N}\right|_{N_s} \overline{\Delta N}\right] \quad \begin{array}{l}\text{for anisotropic} \\ \text{damage,}\end{array}$$

(2.50)

and the stored energy density as

$$w_s(N_s + \overline{\Delta N}) = w_s(N_s) + \int_{r(N_s)}^{r(N_s+\overline{\Delta N})} \frac{A}{m} R(r) r^{\frac{1}{m}-1} dp. \qquad (2.51)$$

The stresses, strains, plastic strains, kinematic hardening at the end of the cycle N_s, the accumulated plastic strain $p(N_s + \overline{\Delta N})$, the stored energy density $w_s(N_s + \overline{\Delta N})$, and the damage $D(N_s + \overline{\Delta N})$ or $\boldsymbol{D}(N_s + \overline{\Delta N})$ are then the initial values for the computation of the first time increment of the cycle $(N_s + \overline{\Delta N} + 1)$.

2.2 Fully-Coupled Analysis

For complex loadings, such as non-proportional or anisothermal, or for severe loadings leading to large scale yielding, the stress and damage states cannot be determined from an uncoupled analysis. Stress redistribution and stress triaxiality changes enhanced by the damage may be a major accelerating factor towards structural failure.

It is then necessary to solve the continuum mechanics equilibrium equations in a fully coupled manner with the elasto-(visco-)plasticity coupled with damage laws (see Fig. 2.2). This is in general a difficult task as the problem is strongly nonlinear and the computation of strain localization gives rise to non-unique, mesh-dependent solutions.

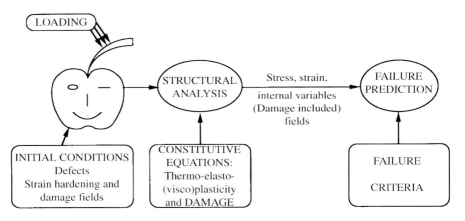

Fig. 2.2. Fully coupled approach to structural damage analysis

Based on both time and space discretizations, the **finite element analysis** (FEA) is commonly used to solve complex engineering structure problems. The FEA is briefly described here for nonlinear material behavior. The main part of this section concerns the description of the **implicit numerical schemes** used for the implementation of the damage models of Chap. 1 as subroutines of commercial finite element computer codes.

Temperature-dependent loadings and behaviors are considered. As the influence of the mechanical behavior on heat transfer remains small in many situations (be careful with contact problems and high speed dynamic effects ...), one assumes that the history of the temperature field is determined from an initial thermal computation. It affects the mechanical problem through the temperature dependency of the material parameters and the thermal strain $\epsilon^{\text{th}} = \alpha(T - T_{\text{ref}})\mathbf{1}$. The schemes described remain the same whether or not the viscosity effect is taken into account. They allow us to represent the continuous transition plasticity/visco-plasticity due to temperature variation.

One considers here the quasistatic case. The acceleration quantities and the inertia effect are not taken into account.

2.2.1 Nonlinear Material Behavior FEA♂♂
(J.J. Moreau 1974, Q.S. Nguyen 1977, J.C. Simo and R.L. Taylor 1986)

The finite element analysis uses a mesh of the considered structure in accordance with both the computation option (2D, axisymmetric, 3D) and the degree of approximation for the elementary displacement field (linear or quadratic shape functions in general).

The approximations for the displacement $\vec{u}(M)$ on each element V_e have the general expression

$$\vec{u} \approx \{u\} = \{u(M)\} = [N]\{U^e\}, \qquad (2.52)$$

where shape functions $N_{ij}(M)$ are introduced. The vector $\{U^e\}$ represents the nodal unknowns (displacements) of an element; it is part of the nodal displacements vector $\{U\}$ for the whole structure. For 2D problems with n-node elements, $\{U^e\}$ has $2n$ components and $[N]$ is a $2 \times (2n)$ matrix. For 3D problems with n-node elements, $\{U^e\}$ has $3n$ components and $[N]$ is a $3 \times (3n)$ matrix.

The Voigt vectorial representation for the strain tensor $\{\epsilon\}$ and the stress tensor $\{\sigma\}$ may be considered:

$$\{\epsilon\} = \begin{bmatrix} \epsilon_{11} \\ \epsilon_{22} \\ \epsilon_{33} \\ \sqrt{2}\epsilon_{23} \\ \sqrt{2}\epsilon_{31} \\ \sqrt{2}\epsilon_{12} \end{bmatrix}, \quad \{\sigma\} = \begin{bmatrix} \sigma_{11} \\ \sigma_{22} \\ \sigma_{33} \\ \sqrt{2}\sigma_{23} \\ \sqrt{2}\sigma_{31} \\ \sqrt{2}\sigma_{12} \end{bmatrix}. \qquad (2.53)$$

The strains are calculated from information on the displacements by use of classical differential operators $[D]$ and $[B]$, such that

$$\{\epsilon\} = [D]\{u\} = [B]\{U^e\} \quad \text{and} \quad [B] = [D][N]. \qquad (2.54)$$

The **virtual work principle** allows us to derive the global matricial form for the equilibrium equations,

$$\{R_{\text{GE}}\} \equiv \sum_{\text{all elements}} \int_{V_e} [B]^{\text{T}} \{\sigma\} \, \text{d}V - \{F\} = 0, \qquad (2.55)$$

where the vector $\{F\}$ accounts for the applied loading. This last equation also defines the global FE residual $\{R_{\text{GE}}\}$ which vanishes for the equilibrated solution.

The local nonlinear constitutive equations are often linearized and solved by use of iterative methods (Newton–Raphson or Quasi-Newton methods ... or others). The consistent tangent operator $\boldsymbol{L}^c$ and its matricial representation $[L^c]$ are then defined as

$$\boldsymbol{L}^c = \frac{\partial \Delta \boldsymbol{\sigma}}{\partial \Delta \boldsymbol{\epsilon}} \quad \text{or} \quad [L^c] = \frac{\partial \{\Delta \epsilon\}}{\partial \{\Delta \sigma\}}, \qquad (2.56)$$

where $\{\Delta \sigma\}$ is the stress increment obtained for the strain increment $\{\Delta \epsilon\}$. $[L^c]$ will be used to build a FE tangent matrix (see next section), ensuring good convergence of the iterative scheme for the resolution of the global equilibrium (2.55).

To sum up, the full mechanical problem is made of a global equation concerning the whole structure and two local equations:

- The global equilibrium $\{R_{\text{GE}}\} = 0$,
- The compatibility equation $\{\epsilon\} = [B]\{U^e\}$,
- The constitutive equations considered at each Gauss (or integration) point.

Of course the boundary conditions and the initial conditions are included as well.

Due to the nonlinearity in material behavior and the loading dependency of the tangent operator, it is difficult to ensure both global equilibrium and local behavior simultaneously. An **iterative** process takes place which consists of each instant t_{n+1} in iterations made of a global equilibrium resolution, followed by local integrations (see Fig. 2.3):

- The global FE resolution of the structure equilibrium is made with tangent operators $[L^c(M)]$. The initial value for $[L^c]$ is the effective elasticity matrix $[\tilde{E}]$. The solution obtained called "**elastic predictor**" is a first estimate $\{U_{n+1}^1\}$ for the nodal displacements vector at time t_{n+1} and for the strains $\{\epsilon_{n+1}^1\}$.
- The local time integrations of the nonlinear behavior equations called "**plastic corrector**" are performed in order to satisfy the state and the evolution laws ("local" means "performed at each Gauss point"). Practically, this integration is made through the use of subroutines
    - whose inputs are the first estimate for the strains $\{\epsilon_{n+1}^1\}$ and the values of the state and internal variables at time t_n
  and
    - whose outputs are the updated estimates for the state and internal variables at time t_{n+1}, the updated value of the consistent tangent operator (if possible), or any good approximation.

A global convergence criterion is checked at the end of the local time integration of the constitutive equations, such as the absolute convergence, written in terms of residuals $\|\{R_{\text{GE}}\}\| < admitted\ error$, or the relative convergence,

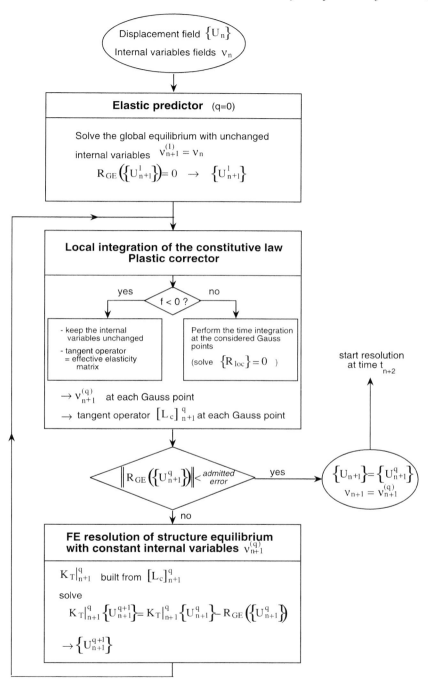

Fig. 2.3. FEA algorithm for nonlinear constitutive equations

written in terms of relative residuals $\frac{\|\{\Delta R_{\mathrm{GE}}\}\|}{\|\{F\}\|} <$ *admitted error*. When satisfied, the estimate for $\{U_{n+1}^q\}$, for the state and internal variables, is kept as the FE solution at time t_{n+1}. If the convergence is not easily reached, the time step is reduced and the whole process is restarted with a smaller t_{n+1}.

The computation of the structure response at time t_{n+2} is made following the same procedure. The relative criterion is usually preferred because dimensionless but it is inapplicable when unloadings with $\|\{F\}\| = 0$ occur.

2.2.2 FE Resolution of the Global Equilibrium

The global equilibrium equations at time t_{n+1} are

$$R_{\mathrm{GE}}(\{U_{n+1}\}) = 0, \qquad (2.57)$$

with $\{U_{n+1}\}$ as the FE solution (to be determined) of the mechanical problem at time t_{n+1}. If this nonlinear equation is solved iteratively by use of the Newton method, solutions $\{U_{n+1}^{q+1}\}$ are calculated at each q-th global iteration such as

$$R_{\mathrm{GE}}(\{U_{n+1}^q\}) + K_{\mathrm{T}}|_{n+1}^q \cdot \left(\{U_{n+1}^{q+1}\} - \{U_{n+1}^q\}\right) = 0. \qquad (2.58)$$

The previous equation introduces the tangent stiffness matrix K_{T} built from the knowledge at each point of the consistent tangent matrix $[L^c]$:

$$K_{\mathrm{T}}|_{n+1}^q = \left[\frac{\partial R_{\mathrm{GE}}}{\partial \{U\}}\right]_{n+1}^q = \sum_{\text{all elements}} \int_{V_e} [B]^{\mathrm{T}} [L^c]_{n+1}^q [B] \, \mathrm{d}V_e. \qquad (2.59)$$

Despite its expected rapid convergence, the Newton–Raphson process solving the global equilibrium equations can be expensive and inconvenient because its main disadvantage for non-associated plasticity coupled with damage models is that $[L^c]$ and K_{T} are not symmetric, so specific solvers are required.

In fact, any approximations for $[L^c]$ and K_{T} may be used as these matrices only affect the convergence rate of the iterative process (Fig. 2.4). Modified Newton methods use, for instance, the tangent matrix calculated at the first iteration. It is even possible to use the elasticity matrix $[E]$ or the effective elasticity matrix $[\tilde{E}]$ as a constant iterative matrix over a time increment. Quasi-Newton methods use a secant approximation K_{S} of K_{T}.

Last, note that the arc-length methods (G.A. Wempner 1971, E. Riks 1972, 1979, M.A. Crisfield 1980) can converge when classical displacement control methods fail. Nevertheless, the capture of instabilities such as severe strain localization may demand specific enhanced arc-length algorithms (M.A. Crisfield 1996).

In any case, most robust algorithms are available in the commercial FE computer codes. Do not hesitate to switch from one method to another when convergence difficulties are encountered!

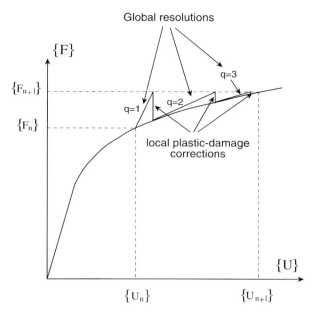

Fig. 2.4. Iterative process for nonlinear FEA

To initiate the process, the resolution starts with a (thermo-)elastic computation (the **elastic predictor**) with $[L^c]^1_{t=t_{n+1}} = [\tilde{E}]_{t=t_n}$ (the effective elasticity matrix) locally and gives $\{U^1_{n+1}\}$ as the initial first estimate. The corresponding first estimate for the strain at time t_{n+1} needed for the local time integrations is

$$\{\epsilon^1_{n+1}\} = \{\epsilon_n\} + \{\Delta\epsilon\} \quad \text{and} \quad \{\Delta\epsilon\} = [B]\left(\{U^1_{n+1}\} - \{U_n\}\right). \quad (2.60)$$

After each q-th global resolution, a q-th local time integration (the **plastic corrector**) gives the value of the tangent operator, $[L^c]^q_{n+1}$, a new first estimate for the FE solution, $\{U_{n+1}\} = \{U^{q+1}_{n+1}\}$, and $\{\epsilon_{n+1}\} = \{\epsilon^{q+1}_{n+1}\}$ is set with

$$\{\epsilon^{q+1}_{n+1}\} = \{\epsilon_n\} + \{\Delta\epsilon^q\} \quad \text{and} \quad \{\Delta\epsilon^q\} = [B]\left(\{U^{q+1}_{n+1}\} - \{U_n\}\right). \quad (2.61)$$

Remark on dynamics – *Taking dynamic effects into consideration does not change the numerical resolution of the mechanical problem by too much. From a theoretical point of view a matrix K_{dyn} (built with the acceleration terms of dynamics equilibrium not explicited stated above) has to be added to the previous tangent matrix K_{T}. As for the damage models of Chap. 1, the density remains constant and K_{dyn} does not depend on the damage behavior. From a practical point of view, the expression for K_{dyn} depends on the numerical scheme chosen for the time discretization of the acceleration quantities and explicit algorithms may be needed for* the local integration of the constitutive laws. Such a work needs much care and is not described here.

2.2.3 Local Integration Subroutines♂♂

After each resolution of the global structure equilibrium, the value of the stress $\{\sigma_{n+1}\}$, the thermodynamics internal variables, the accumulated plastic strain, the stored energy, and the value of $[L^c]$ are updated. This is made by performing the local time integration of the constitutive laws of Sect. 1.5 at the structure's Gauss points for which $f > 0$ (a negative yield criterion $f < 0$ corresponds to an elastic state and does not need any correction).

The inputs for the local integration subroutine are the values at current time t_{n+1} of the strain tensor ϵ_{n+1} (represented by the previous $\{\epsilon_{n+1}\}$), the thermodynamics variables σ_n, ϵ_n, ϵ_n^e, ϵ_n^p, r_n, R_n, α_n, X_n, D_n, or D_n at time t_n, as well as the values of the accumulated plastic strain p_n and the stored energy w_{sn}. The calculation of p is not strictly necessary as it is replaced by the variable r, but is useful for comparisons with measures and fitting of experimental data. Recall that, as long as there is no damage, the equality $r = p$ stands and the damage threshold $p = p_D$ is equivalent to $r = p_D$. The outputs are the same variables as above but calculated at time t_{n+1} and the updated consistent tangent operator (previous matrix $[L^c]$ or any good approximation for $[L^c]$).

If the damage model does not exist in the chosen FE code, it has to be implemented as a subroutine (such as ABAQUS UMAT FORTRAN subroutines) written in the adequate language of programming. The corresponding implicit algorithms are fully described in Sects. 2.2.4 and 2.2.5.

Due to the strong nonlinearity of the behavior and the possible occurrence of instabilities, implicit schemes are preferred. During the same computation, they allow switching between plasticity models at low temperature to viscoplasticity models at higher temperature (and reciprocally) as the Norton parameters vary with T.

For high-temperature, visco-plastic damage computations, explicit schemes (very sensitive to the time step) may be used with shorter computation times if the value of D_c is not too large (≤ 0.3). The algorithms for the time integration of the constitutive equations are simpler: the values at time t_{n+1} of the accumulated plastic strain and the state variables are calculated from the explicit discretization of a nonlinear system of first order differential equations (use of Runge–Kutta schemes, for example). This makes the computation of structures undergoing visco-plasticity/plasticity transition more difficult. Plasticity is then numerically recovered by setting $N \to \infty$ and the numerical convergence is difficult.

Finally, do not use explicit schemes for states of damage close to the occurrence of strain localization as instabilities are expected.

2.2.4 Single Implicit Algorithm for Damage Models[66]
(R. Desmorat and J. Besson 2001)

The damage models of Tables 1.3 and 1.5, Sect. 1.5, use the effective stress for damage coupling with both elasticity and plasticity. This feature allows for the use of the existing schemes developed for elasto-(visco-) plasticity (A. Benallal, R. Billardon and I. Doghri 1988).

The algorithm described here applies to anisothermal computations, the temperature maps being determined for each time increment by a previous heat transfer computation. In order to make the presentation as well as the programming synthetic, both isotropic and anisotropic damage variables are represented by the second order tensors $\boldsymbol{D}$ and $\boldsymbol{H} = (\boldsymbol{1} - \boldsymbol{D})^{-\frac{1}{2}}$. Isotropy represents the limiting case where $\boldsymbol{D} = D\,\boldsymbol{1}$, $\boldsymbol{H} = 1/\sqrt{1-D}$, and $\eta = 1$.

The evolution laws for the internal variables are derived from the normality rule applied to the potential $F = f + F_X + F_D$, where f is the yield criterion, F_X stands for the nonlinearity of the kinematic hardening and F_D stands for the damage evolution. Remember that

$$f = (\tilde{\boldsymbol{\sigma}} - \boldsymbol{X})_{\text{eq}} - R - \sigma_y, \quad \text{and} \quad F_X = \frac{3\gamma}{4C}\boldsymbol{X} : \boldsymbol{X}, \qquad (2.62)$$

with $(.)_{\text{eq}} = \sqrt{\frac{3}{2}(.)^{\text{D}} : (.)^{\text{D}}}$ the von Mises norm. We will define the normals as

$$\boldsymbol{n}^X = \frac{\partial F}{\partial \tilde{\boldsymbol{\sigma}}} = -\frac{\partial f}{\partial \boldsymbol{X}} = \frac{3}{2}\frac{\tilde{\boldsymbol{\sigma}}^{\text{D}} - \boldsymbol{X}}{(\tilde{\boldsymbol{\sigma}}^{\text{D}} - \boldsymbol{X})_{\text{eq}}},$$

$$\boldsymbol{n} = \frac{\partial F}{\partial \boldsymbol{\sigma}} = \left(\boldsymbol{H}\boldsymbol{n}^X \boldsymbol{H}\right)^{\text{D}}, \qquad (2.63)$$

$$\boldsymbol{m} = -\frac{\partial F}{\partial \boldsymbol{X}} = \boldsymbol{n}^X - \gamma \boldsymbol{\alpha}.$$

The evolution laws for the plastic strain and the kinematic hardening then become

$$\dot{\boldsymbol{\epsilon}}^{\text{p}} = \dot{\lambda}\frac{\partial F}{\partial \boldsymbol{\sigma}} = \dot{r}\boldsymbol{n} \quad \text{and} \quad \dot{\boldsymbol{\alpha}} = -\dot{\lambda}\frac{\partial F}{\partial \boldsymbol{X}} = \dot{r}\boldsymbol{m} \qquad (2.64)$$

and the accumulated plastic strain rate is

$$\dot{p} = \sqrt{\frac{2}{3}\boldsymbol{n}:\boldsymbol{n}}\;\dot{r}, \qquad (2.65)$$

which reads $\dot{p} = \dot{r}/(1-D)$ for isotropic damage.

With the state laws

$$\tilde{\boldsymbol{\sigma}} = \underline{\underline{\boldsymbol{E}}} : [\boldsymbol{\epsilon}^{\text{e}} - \alpha(T - T_{\text{ref}})\boldsymbol{1}] \quad \text{and} \quad \boldsymbol{X} = \frac{2}{3}C\boldsymbol{\alpha} \qquad (2.66)$$

and with the definition of $\boldsymbol{H}$, the normals $\boldsymbol{n}^X$, $\boldsymbol{n}$, and $\boldsymbol{m}$ may be expressed at given T as functions of $\boldsymbol{\epsilon}^e$, $\boldsymbol{\alpha}$, and $\boldsymbol{D}$ only:

$$\boldsymbol{n}^X = \boldsymbol{n}^X(\boldsymbol{\epsilon}^e, \boldsymbol{\alpha}), \qquad \boldsymbol{n} = \boldsymbol{n}(\boldsymbol{\epsilon}^e, \boldsymbol{\alpha}, \boldsymbol{D}), \qquad \boldsymbol{m} = \boldsymbol{m}(\boldsymbol{\epsilon}^e, \boldsymbol{\alpha}). \qquad (2.67)$$

A set of 4 nonlinear equations of the 4 independent variables $\mathcal{W} = \{\boldsymbol{\epsilon}^e, \boldsymbol{\alpha}, r, \boldsymbol{D}\}$ then has to be solved in a coupled manner, as follows:

$$\begin{aligned} &\dot{\boldsymbol{\epsilon}} - \dot{\boldsymbol{\epsilon}}^e - \dot{r}\boldsymbol{n} = 0, \\ &\dot{\boldsymbol{\alpha}} - \dot{r}\boldsymbol{m} = 0, \\ &f - \sigma_v = (\tilde{\boldsymbol{\sigma}} - \boldsymbol{X})_{\text{eq}} - R - \sigma_y - \sigma_v = 0, \\ &\begin{cases} \dot{\boldsymbol{D}} - \left(\dfrac{Y}{S}\right)^s \dot{p}\,\mathbf{1} = 0 & \text{isotropic damage,} \\ \dot{\boldsymbol{D}} - \left(\dfrac{\overline{Y}}{S}\right)^s |\boldsymbol{n}|\dot{r} = 0 & \text{anisotropic damage,} \end{cases} \end{aligned} \qquad (2.68)$$

where $|\boldsymbol{n}|$ means the absolute value of $\boldsymbol{n}$ in terms of principal components and where

- $R = R(r)$
- $\sigma_v = \begin{cases} K_N \dot{p}^{\frac{1}{N}} = K_N \left(\sqrt{\frac{2}{3}\boldsymbol{n}:\boldsymbol{n}}\,\dot{r}\right)^{\frac{1}{N}} & \text{Norton law} \\ K_\infty \left[1 - \exp\left(-\frac{1}{n}\sqrt{\frac{2}{3}\boldsymbol{n}:\boldsymbol{n}}\,\dot{r}\right)\right] & \text{exponential law} \end{cases}$
- $Y(\boldsymbol{\epsilon}^e) = \overline{Y}(\boldsymbol{\epsilon}^e) = \dfrac{1}{2}\boldsymbol{E} : \boldsymbol{\epsilon}^e : \boldsymbol{\epsilon}^e$

For the sake of relative simplicity, the dilatancy terms are dropped out next. Keeping them does not yield any particular difficulties.

2.2.4.1 Discretization by the θ-method

The previous set of nonlinear equations is discretized in time by considering the resolution variables at the intermediate time, $t_{n+\theta} = t_n + \theta \Delta t$, with θ as the numerical parameter of the so-called θ-method.

- The value $0 \le \theta \le 1$ is used for the 3 differential equations.
- In plasticity, it is of first importance to ensure the validity of the yield criterion at each computation time t_k. The value $\theta = 1$ (Euler fully implicit scheme) is set for the nonlinear equation $f_{n+1} = 0$ and the incremental form of the plasticity loading/unloading conditions is $\Delta r \ge 0$, $f_{n+1} \le 0$, and $\Delta r f_{n+1} = 0$.
  Any value $0 < \theta \le 1$ can be used in visco-plasticity.

2.2 Fully-Coupled Analysis

In plasticity coupled with damage, the local residual is then defined as

$\{R_{\text{loc}}\} = \{R_{\epsilon^e}, R_\alpha, R_r, , R_D\}^T,$

$$\{R_{\text{loc}}\} = \begin{cases} R_{\epsilon^e} = \Delta\epsilon^e - \Delta\epsilon + \Delta r\, \boldsymbol{n}_{n+\theta}\,, \\ R_\alpha = \Delta\boldsymbol{\alpha} - \Delta r\, \boldsymbol{m}_{n+\theta}\,, \\ R_r = f_{n+1} = (\tilde{\boldsymbol{\sigma}}_{n+1} - \boldsymbol{X}_{n+1})_{\text{eq}} - R(r_{n+1}) - \sigma_y\,, \\ R_D = \begin{cases} \Delta D - \left(\dfrac{Y(\epsilon^e_{n+\theta})}{S}\right)^s \dfrac{\Delta r}{1 - D_{n+\theta}}\,\mathbf{1}\,, & \text{for isotropic damage,} \\[6pt] \Delta \boldsymbol{D} - \left(\dfrac{\overline{Y}(\epsilon^e_{n+\theta})}{S}\right)^s |\boldsymbol{n}_{n+\theta}|\,\Delta r\,, & \text{for anisotropic damage,} \end{cases} \end{cases}$$
(2.69)

where for any of the 4 variables ϵ^e, $\boldsymbol{\alpha}$, r, and $\boldsymbol{D}$, the subscript $n+\theta$ means "variable at time $t_n + \theta\Delta t$": $\mathcal{W}_{n+\theta} = \mathcal{W}_n + \theta \Delta \mathcal{W}$, i.e.,

$$\begin{aligned} \epsilon^e_{n+\theta} &= \epsilon^e_n + \theta \Delta \epsilon^e\,, \\ \boldsymbol{\alpha}_{n+\theta} &= \boldsymbol{\alpha}_n + \theta \Delta \boldsymbol{\alpha}\,, \\ r_{n+\theta} &= r_n + \theta \Delta r\,, \\ \boldsymbol{D}_{n+\theta} &= \boldsymbol{D}_n + \theta \Delta \boldsymbol{D}\,, \end{aligned}$$
(2.70)

and for any function g of those variables,

$$g_{n+\theta} = g\left(\epsilon^e_{n+\theta}, \boldsymbol{\alpha}_{n+\theta}, r_{n+\theta}, \boldsymbol{D}_{n+\theta}\right). \tag{2.71}$$

In visco-plasticity coupled with damage, the residual R_r is changed to R_r^{v},

$$R_r^{\text{v}} = \Delta r - \dfrac{\Delta t}{\sqrt{\tfrac{2}{3} \boldsymbol{n}_{n+\theta} : \boldsymbol{n}_{n+\theta}}} \left\langle \dfrac{f_{n+\theta}}{K_N} \right\rangle^N. \tag{2.72}$$

Performing the time integration of the constitutive law consists of reaching the local convergence $\|\{R_{\text{loc}}\}\| < \textit{admitted error}$ at a given strain $\{\epsilon\} = \{\epsilon_{n+1}\}$ or $\epsilon = \epsilon_{n+1}$ by correcting the variables ϵ^e, $\boldsymbol{\alpha}$, r and $\boldsymbol{D}$. The Newton iterative scheme may be used again and one has then to solve

$$\{R_{\text{loc}\,n+1}^{q'}\} + \left[\dfrac{\partial \{R_{\text{loc}}\}}{\partial \Delta \mathcal{W}}\right]^{q'}_{n+1} \cdot \left(\mathcal{W}^{q'+1}_{n+1} - \mathcal{W}^{q'}_{n+1}\right) = 0\,, \tag{2.73}$$

where the expression for the jacobian matrix $[\partial\{R\}/\partial \Delta \mathcal{W}]^{q'}_{n+1}$ (or any good approximation) of the partial derivative of each discretized equation with respect to each variable increment $\Delta \mathcal{W} = \{\Delta \epsilon^e, \Delta \boldsymbol{\alpha}, \Delta r, \Delta \boldsymbol{D}\}$ is needed for convergence reasons. The convergence of the Newton scheme for the local integration is not guaranteed in general. For the damage models under consideration, the convergence is enhanced by using Amarthe derivatives derived in next subsection and 3D computations have proven the method efficient.

At convergence, $\Delta \boldsymbol{\epsilon}^\text{e} = \Delta \boldsymbol{\epsilon}^{\text{e}\,q'+1}$, $\Delta \boldsymbol{\alpha} = \Delta \boldsymbol{\alpha}^{q'+1}$, $\Delta r = \Delta r^{q'+1}$, and $\Delta \boldsymbol{D} = \Delta \boldsymbol{D}^{q'+1}$ is set and

$$\begin{aligned}
\boldsymbol{\epsilon}^\text{e}_{n+1} &= \boldsymbol{\epsilon}^\text{e}_n + \Delta \boldsymbol{\epsilon}^\text{e}\,, \\
\boldsymbol{\alpha}_{n+1} &= \boldsymbol{\alpha}_n + \Delta \boldsymbol{\alpha}\,, \\
r_{n+1} &= r_n + \Delta r\,, \\
\boldsymbol{D}_{n+1} &= \boldsymbol{D}_n + \Delta \boldsymbol{D}\,.
\end{aligned} \quad (2.74)$$

2.2.4.2 Closed-form expression of the Jacobian matrix

The derivatives of $\{R\}$ taken with respect to $\Delta \mathcal{W}$ define the Jacobian matrix as $[\text{Jac}] = \dfrac{\partial \{R\}}{\partial \Delta \mathcal{W}}$. For any function g one has

$$\frac{\partial g}{\partial \Delta \mathcal{W}} = \frac{\partial g}{\partial \mathcal{W}} : \frac{\partial \mathcal{W}}{\partial \Delta \mathcal{W}} = \theta \frac{\partial g}{\partial \mathcal{W}} \quad (2.75)$$

and all the necessary $\dfrac{\partial}{\partial \Delta \mathcal{W}}$ terms are given thereafter.

First, let us consider $\boldsymbol{\epsilon}^\text{e}$, $\boldsymbol{\alpha}$, r, and $\boldsymbol{D}$ as independent variables such that

$$\begin{aligned}
\frac{\partial \boldsymbol{n}^\text{X}}{\partial \tilde{\boldsymbol{\sigma}}} &= -\frac{\partial \boldsymbol{n}^\text{X}}{\partial \boldsymbol{X}} = \frac{1}{(\tilde{\boldsymbol{\sigma}} - \boldsymbol{X})_\text{eq}} \left[\frac{3}{2} \boldsymbol{I} - \frac{1}{2} \boldsymbol{1} \otimes \boldsymbol{1} - \boldsymbol{n}^\text{X} \otimes \boldsymbol{n}^\text{X} \right], \\
\frac{\partial \boldsymbol{n}^\text{X}}{\partial r} &= 0\,, \quad \text{and} \\
\frac{\partial \boldsymbol{n}^\text{X}}{\partial \boldsymbol{D}} &= 0\,,
\end{aligned} \quad (2.76)$$

where $(\tilde{\boldsymbol{\sigma}} - \boldsymbol{X})_\text{eq} = \left(\boldsymbol{E} : \boldsymbol{\epsilon}^\text{e} - \dfrac{2}{3} C \boldsymbol{\alpha} \right)_\text{eq}$.

To calculate the derivatives of $\boldsymbol{n}$, a fourth order tensor $\underline{\boldsymbol{Q}}$ is defined such that $\boldsymbol{n} = (\boldsymbol{H} \boldsymbol{n}^\text{X} \boldsymbol{H})^\text{D} = \underline{\boldsymbol{Q}} : \boldsymbol{n}^\text{X}$ and

$$\underline{\boldsymbol{Q}} = \boldsymbol{H} \underline{\otimes} \boldsymbol{H} - \frac{1}{3} \boldsymbol{1} \otimes \boldsymbol{H}^2\,, \quad (2.77)$$

where the special tensorial product $\boldsymbol{H} \underline{\otimes} \boldsymbol{H}$ applied on the symmetric second order tensor $\boldsymbol{H}$ stands for $(\boldsymbol{H} \underline{\otimes} \boldsymbol{H})_{ijkl} = H_{ik} H_{jl}$, leading to

$$\begin{aligned}
\frac{\partial \boldsymbol{n}}{\partial \tilde{\boldsymbol{\sigma}}} &= -\frac{\partial \boldsymbol{n}}{\partial \boldsymbol{X}} = \underline{\boldsymbol{Q}} : \frac{\partial \boldsymbol{n}^\text{X}}{\partial \tilde{\boldsymbol{\sigma}}} \\
\frac{\partial \boldsymbol{n}}{\partial r} &= 0\,.
\end{aligned} \quad (2.78)$$

Table 2.2. Jacobian terms for isotropic and anisotropic damage models

Derivatives of $\boldsymbol{R}_{\epsilon^e}$ with respect to $\Delta \mathcal{W}$

$$\frac{\partial \boldsymbol{R}_{\epsilon^e}}{\partial \Delta \boldsymbol{\epsilon}^e} = \underline{\boldsymbol{I}} + \theta \Delta r \, \underline{\boldsymbol{Q}}_{n+\theta} : \left.\frac{\partial \boldsymbol{n}^{\mathrm{X}}}{\partial \tilde{\boldsymbol{\sigma}}}\right|_{n+\theta} : \underline{\boldsymbol{E}} \qquad \frac{\partial \boldsymbol{R}_{\epsilon^e}}{\partial \Delta r} = \boldsymbol{n}_{n+\theta}$$

$$\frac{\partial \boldsymbol{R}_{\epsilon^e}}{\partial \Delta \boldsymbol{\alpha}} = \frac{2}{3} C \theta \Delta r \, \underline{\boldsymbol{Q}}_{n+\theta} : \left.\frac{\partial \boldsymbol{n}^{\mathrm{X}}}{\partial \boldsymbol{X}}\right|_{n+\theta} \qquad \frac{\partial \boldsymbol{R}_{\epsilon^e}}{\partial \Delta D} = \theta \Delta r \left.\frac{\partial \boldsymbol{n}}{\partial D}\right|_{n+\theta}$$

Derivatives of $\boldsymbol{R}_{\alpha}$ with respect to $\Delta \mathcal{W}$

$$\frac{\partial \boldsymbol{R}_{\alpha}}{\partial \Delta \boldsymbol{\epsilon}^e} = -\theta \Delta r \left.\frac{\partial \boldsymbol{n}^{\mathrm{X}}}{\partial \tilde{\boldsymbol{\sigma}}}\right|_{n+\theta} : \underline{\boldsymbol{E}} \qquad \frac{\partial \boldsymbol{R}_{\alpha}}{\partial \Delta r} = -\boldsymbol{m}_{n+\theta}$$

$$\frac{\partial \boldsymbol{R}_{\alpha}}{\partial \Delta \boldsymbol{\alpha}} = (1 + \gamma \theta \Delta r) \underline{\boldsymbol{I}}$$
$$\qquad - \frac{2}{3} C \theta \Delta r \left.\frac{\partial \boldsymbol{n}^{\mathrm{X}}}{\partial \boldsymbol{X}}\right|_{n+\theta} \qquad \frac{\partial \boldsymbol{R}_{\alpha}}{\partial \Delta D} = -\theta \Delta r \left.\frac{\partial \boldsymbol{n}^{\mathrm{X}}}{\partial D}\right|_{n+\theta}$$

Derivatives of R_r with respect to $\Delta \mathcal{W}$ (take $\theta = 1$ for the plastic case)

$$\frac{\partial R_r}{\partial \Delta \boldsymbol{\epsilon}^e} = \theta \, \boldsymbol{n}^{\mathrm{X}}_{n+\theta} : \underline{\boldsymbol{E}} \qquad \frac{\partial R_r}{\partial \Delta r} = -\theta R'(r_{n+\theta})$$

$$\frac{\partial R_r}{\partial \Delta \boldsymbol{\alpha}} = -\frac{2}{3} C \theta \, \boldsymbol{n}^{\mathrm{X}}_{n+\theta} \qquad \frac{\partial R_r}{\partial \Delta D} = 0$$

Derivatives of $\boldsymbol{R}_{\mathrm{D}}$ with respect to $\Delta \mathcal{W}$

Isotropic damage	Anisotropic damage

$$\frac{\partial \boldsymbol{R}_{\mathrm{D}}}{\partial \Delta \boldsymbol{\epsilon}^e} = -\frac{s\theta Y_{n+\theta}^{s-1} \Delta r}{S^s (1 - D_{n+\theta})} \boldsymbol{1} \otimes \underline{\boldsymbol{E}} : \boldsymbol{\epsilon}_{n+\theta}^e \qquad \frac{\partial \boldsymbol{R}_{\mathrm{D}}}{\partial \Delta \boldsymbol{\epsilon}^e} \approx -\frac{s\theta \overline{Y}_{n+\theta}^{s-1} \Delta r}{S^s} |\boldsymbol{n}_{n+\theta}| \otimes \underline{\boldsymbol{E}} : \boldsymbol{\epsilon}_{n+\theta}^e$$

$$\frac{\partial \boldsymbol{R}_{\mathrm{D}}}{\partial \Delta \boldsymbol{\alpha}} = 0 \qquad \frac{\partial \boldsymbol{R}_{\mathrm{D}}}{\partial \Delta \boldsymbol{\alpha}} \approx 0$$

$$\frac{\partial \boldsymbol{R}_{\mathrm{D}}}{\partial \Delta r} = -\left(\frac{Y_{n+\theta}}{S}\right)^s \frac{1}{1 - D_{n+\theta}} \boldsymbol{1} \qquad \frac{\partial \boldsymbol{R}_{\mathrm{D}}}{\partial \Delta r} = -\left(\frac{\overline{Y}_{n+\theta}}{S}\right)^s |\boldsymbol{n}_{n+\theta}|$$

$$\frac{\partial \boldsymbol{R}_{\mathrm{D}}}{\partial \Delta D} = \underline{\boldsymbol{I}} - \left(\frac{Y_{n+\theta}}{S}\right)^s \qquad \frac{\partial \boldsymbol{R}_{\mathrm{D}}}{\partial \Delta D} \approx \underline{\boldsymbol{I}}$$
$$\qquad \times \frac{\theta \Delta r}{(1 - D_{n+\theta})^2} \frac{1}{3} \boldsymbol{1} \otimes \boldsymbol{1}$$

The differentiation of the property $H^2(1-D) = (1-D)H^2 = 1$ allows for the calculation of the derivatives of H^2 and H with respect to D:

$$\frac{\partial H^2}{\partial D} = H^2 \underline{\otimes} H^2,$$
$$\frac{\partial H}{\partial D} = \underline{A}^{-1} : \left(H^2 \underline{\otimes} H^2\right), \tag{2.79}$$

where $\underline{A}$ is the fourth order tensor $\underline{A} = H \otimes 1 + 1 \otimes H$ or its symmetric part already introduced in Chap. 1 (Eq. (1.38) of Sect. 1.2.3). The expression for $\frac{\partial n}{\partial D}$ is finally derived from the knowledge of

$$\frac{\partial(H\ n^X H)}{\partial H} = 1 \underline{\otimes} \left(H\ n^X\right) + \left(H\ n^X\right) \underline{\otimes} 1 \tag{2.80}$$

as

$$\frac{\partial n}{\partial D} = \frac{\partial(H\ n^X H)}{\partial H} : \frac{\partial H}{\partial D} - \frac{1}{3} 1 \otimes n^X : \left(H^2 \underline{\otimes} H^2\right). \tag{2.81}$$

Table 2.2 gives an abstract of all the Jacobian terms for plasticity coupled with damage. When the viscosity effects are taken into account, one needs the terms $\frac{\partial R_r^v}{\partial W}$ at time $t_{n+\theta}$. For Δr, we consider the Norton law (2.72) here so that

$$\frac{\partial R_r^v}{\partial \Delta r} = 1 - \frac{N\Delta t f_{n+\theta}^{N-1}}{K_N^N} \frac{\partial R_r}{\partial \Delta r}. \tag{2.82}$$

For $\Delta \mathcal{W}_i = \Delta \epsilon^e, \Delta \alpha, \Delta D$, we have

$$\begin{aligned}\frac{\partial R_r^v}{\partial \Delta \mathcal{W}_i} &= -\frac{N\Delta t f_{n+\theta}^{N-1}}{K_N^N \sqrt{\frac{2}{3} n_{n+\theta} : n_{n+\theta}}} \frac{\partial \Delta R_r}{\partial \Delta \mathcal{W}_i} \\ &\quad - \frac{\Delta t f_{n+\theta}^N}{K_N^N} \frac{\frac{2}{3}\theta\ n_{n+\theta}}{\sqrt{\frac{2}{3} n_{n+\theta} : n_{n+\theta}}} : \left.\frac{\partial n}{\partial \mathcal{W}_i}\right|_{n+\theta}. \end{aligned} \tag{2.83}$$

2.2.4.3 Updating the Thermodynamics Variables

Once ϵ_{n+1}^e, α_{n+1}, r_{n+1}, and D_{n+1} are known, the remaining thermodynamics variables are updated explicitly:

- Effective stress tensor $\tilde{\sigma}_{n+1} = \underline{E} : \epsilon_{n+1}^e$
- Plastic strain tensor $\epsilon_{n+1}^p = \epsilon_{n+1} - \epsilon_{n+1}^e$
- Strain hardening variables $R_{n+1} = R(r_{n+1})$, $X_{n+1} = \frac{2}{3} C \alpha_{n+1}$
- Accumulated plastic strain $p_{n+1} = p_n + \sqrt{\frac{2}{3} n_{n+\theta} : n_{n+\theta}}\ \Delta r$ equivalent to $p_{n+1} = p_n + \Delta r/(1 - D_{n+\theta})$ for isotropic damage

- Stored energy (Be careful for the integration: the correction term $z(r) = \frac{A}{m} r^{(1-m)/m}$ of Sect. 1.4.1 is singular in $r = 0$)

$$w_{s\,n+1} = AR(r_{n+1})r_{n+1}^{1/m} + \frac{1}{3} C\, \boldsymbol{\alpha}_{n+1} : \boldsymbol{\alpha}_{n+1} - \hat{w}_{n+1},$$

$$\hat{w}_{n+1} = \hat{w}_n + \frac{A}{2} \left[R'(r_n) r_n^{1/m} + R'(r_{n+1}) r_{n+1}^{1/m} \right] \Delta r,$$

with $R'(r) = bR_\infty \exp(-br)$ for exponential isotropic hardening.
- Damage $\boldsymbol{H}_{n+1} = (1 - D_{n+1})^{-\frac{1}{2}}$, $Y_{n+1} = \overline{Y}_{n+1} = \frac{1}{2} \boldsymbol{E} : \boldsymbol{\epsilon}^{\text{e}}_{n+1} : \boldsymbol{\epsilon}^{\text{e}}_{n+1}$
- Stress tensor $\boldsymbol{\sigma}_{n+1} = \boldsymbol{M}^{-1}_{n+1} : \tilde{\boldsymbol{\sigma}}_{n+1} = \boldsymbol{\sigma}^{\text{D}}_{n+1} + \tilde{\sigma}_{\text{H}\,n+1} \mathbf{1}$ (see Eq. (2.91)), also written as

$$\begin{cases} \boldsymbol{\sigma}^{\text{D}}_{n+1} = \boldsymbol{H}^{-1}_{n+1} \tilde{\boldsymbol{\sigma}}_{n+1} \boldsymbol{H}^{-1}_{n+1} - \dfrac{(1 - D_{n+1}) : \tilde{\boldsymbol{\sigma}}_{n+1}}{3(1 - D_{\text{H}\,n+1})} (1 - D_{n+1}), \\ \sigma_{\text{H}\,n+1} = (1 - \eta D_{\text{H}\,n+1}) \tilde{\sigma}_{\text{H}\,n+1}. \end{cases} \quad (2.84)$$

Note that for isotropic damage, the last equation simply recovers $\boldsymbol{\sigma}_{n+1} = (1 - D_{n+1})\, \tilde{\boldsymbol{\sigma}}_{n+1}$.

2.2.4.4 Consistent Tangent Operator
(J.C. Simo and R.L. Taylor 1985, J. Besson 1999)

One of the advantages of the implicit integration by Newton methods is the direct calculation of the consistent tangent operator $\underline{\boldsymbol{L}}^c$ or its matrix representation $[L^c]$.

Let us break down the residual $\{R_{\text{loc}}\}$ into 2 parts,

$$\{R_{\text{loc}}\} = \{R_{\text{i}}\} - \{R_{\text{e}}\} \quad \text{and} \quad \{R_{\text{e}}\} = \begin{bmatrix} \Delta\boldsymbol{\epsilon} \\ 0 \\ 0 \\ 0 \end{bmatrix} \quad (2.85)$$

where $\{R_{\text{i}}\}$ corresponds to the contribution due to the internal variables and $\{R_{\text{e}}\}$ to the contribution due to the applied loading. Once the convergence is reached, we perform a perturbation of (2.85), $\delta\{R_{\text{loc}}\} = \delta\{R_{\text{i}}\} - \delta\{R_{\text{e}}\} = \{0\}$. With the definition of the Jacobian matrix, we have

$$\delta\{R_{\text{i}}\} = [\text{Jac}]\, \delta\Delta W. \quad (2.86)$$

One has then

$$\begin{bmatrix} \delta\Delta\boldsymbol{\epsilon}^{\text{e}} \\ \delta\Delta\boldsymbol{\alpha} \\ \delta\Delta r \\ \delta\Delta D \end{bmatrix} = [\text{Jac}]^{-1} \begin{bmatrix} \delta\Delta\boldsymbol{\epsilon} \\ 0 \\ 0 \\ 0 \end{bmatrix}, \quad (2.87)$$

which shows that the first column of the inverse of the Jacobian matrix at convergence is the vector

$$\begin{bmatrix} [\text{Jac}]^{-1}_{\epsilon^e,\epsilon^e} \\ [\text{Jac}]^{-1}_{\alpha,\epsilon^e} \\ [\text{Jac}]^{-1}_{r,\epsilon^e} \\ [\text{Jac}]^{-1}_{D,\epsilon^e} \end{bmatrix} = \begin{bmatrix} \dfrac{\partial \Delta \epsilon^e}{\partial \Delta \epsilon} \\ \dfrac{\partial \Delta \alpha}{\partial \Delta \epsilon} \\ \dfrac{\partial \Delta r}{\partial \Delta \epsilon} \\ \dfrac{\partial \Delta D}{\partial \Delta \epsilon} \end{bmatrix} \quad (2.88)$$

whose value at time $t_{n+\theta}$ is calculated within Newton iterative resolution.

The tensorial definition of the effective stress $\tilde{\boldsymbol{\sigma}} = \underline{\boldsymbol{M}} : \boldsymbol{\sigma}$ with

$$\underline{\boldsymbol{M}} = \boldsymbol{H} \underline{\otimes} \boldsymbol{H} - \frac{1}{3}\left[\boldsymbol{H}^2 \otimes \boldsymbol{1} + \boldsymbol{1} \otimes \boldsymbol{H}^2\right] + \frac{1}{9}\left(\text{tr}\boldsymbol{H}^2\right)\boldsymbol{1} \otimes \boldsymbol{1} + \frac{1}{3(1-\eta D_\text{H})}\boldsymbol{1} \otimes \boldsymbol{1} \quad (2.89)$$

gives back

$$\tilde{\boldsymbol{\sigma}} = \left(\boldsymbol{H}\boldsymbol{\sigma}^\text{D}\boldsymbol{H}\right)^\text{D} + \frac{\sigma_\text{H}}{1-\eta D_\text{H}}\boldsymbol{1}, \quad (2.90)$$

and may be inverted in

$$\boldsymbol{\sigma} = \underline{\boldsymbol{M}}^{-1} : \tilde{\boldsymbol{\sigma}} = \boldsymbol{H}^{-1}\tilde{\boldsymbol{\sigma}}\boldsymbol{H}^{-1} - \frac{(\boldsymbol{1}-\boldsymbol{D}):\tilde{\boldsymbol{\sigma}}}{3(1-D_\text{H})}(\boldsymbol{1}-\boldsymbol{D}) + (1-\eta D_\text{H})\tilde{\sigma}_\text{H}\boldsymbol{1}, \quad (2.91)$$

with $\boldsymbol{H}^{-1} = (\boldsymbol{1}-\boldsymbol{D})^{1/2}$ and

$$\underline{\boldsymbol{M}}^{-1} = \boldsymbol{H}^{-1}\underline{\otimes}\boldsymbol{H}^{-1} - \frac{(\boldsymbol{1}-\boldsymbol{D}) \otimes (\boldsymbol{1}-\boldsymbol{D})}{3(1-D_\text{H})} + \frac{1}{3}(1-\eta D_\text{H})\boldsymbol{1} \otimes \boldsymbol{1}. \quad (2.92)$$

Then,

$$\delta\Delta\boldsymbol{\sigma} = \underline{\boldsymbol{M}}^{-1} : \delta\Delta\tilde{\boldsymbol{\sigma}} + \tilde{\boldsymbol{\sigma}} : \frac{\partial \underline{\boldsymbol{M}}^{-1}}{\partial \boldsymbol{D}} : \delta\Delta\boldsymbol{D} \quad (2.93)$$

and finally with

$$\begin{aligned} \delta\Delta\tilde{\boldsymbol{\sigma}} &= \underline{\boldsymbol{E}} : \delta\Delta\boldsymbol{\epsilon}^e = \underline{\boldsymbol{E}} : [\text{Jac}]^{-1}_{\epsilon^e,\epsilon^e} : \delta\Delta\boldsymbol{\epsilon} \\ \delta\Delta\boldsymbol{D} &= [\text{Jac}]^{-1}_{D,\epsilon^e} : \delta\Delta\boldsymbol{\epsilon}, \end{aligned} \quad (2.94)$$

the expression for the consistent tangent operator is:

$$\boxed{\underline{\boldsymbol{L}}^\text{c} = \underline{\boldsymbol{M}}^{-1} : \underline{\boldsymbol{E}} : [\text{Jac}]^{-1}_{\epsilon^e,\epsilon^e} + \tilde{\boldsymbol{\sigma}} : \frac{\partial \underline{\boldsymbol{M}}^{-1}}{\partial \boldsymbol{D}} : [\text{Jac}]^{-1}_{D,\epsilon^e}.} \quad (2.95)$$

2.2.5 Damage Models with Microdefects Closure Effect◊◊◊
(R. Desmorat and J. Besson 2001)

For the damage models with the microdefects closure effect (consideration of the material parameters h or h_a of Chap. 1), the effective stress defined in elasticity by $\tilde{\boldsymbol{\sigma}} = \underline{\underline{\boldsymbol{E}}} : [\boldsymbol{\epsilon}^e - \alpha(T - T_{\text{ref}})\mathbf{1}]$ is not used for the coupling with plasticity.

For the isotropic damage model, $\tilde{\boldsymbol{\sigma}}$ is given by Eq. (1.48) of Sect. 1.2.4 but now

$$f = (\boldsymbol{s} - \boldsymbol{X})_{\text{eq}} - R - \sigma_y \quad \text{with} \quad \boldsymbol{s} = \frac{\boldsymbol{\sigma}^D}{1 - D} \neq \tilde{\boldsymbol{\sigma}}^D, \qquad (2.96)$$

where $(.)_{\text{eq}} = \sqrt{\frac{3}{2}(.)^D : (.)^D}$ is the von Mises norm. For the anisotropic damage model $\tilde{\boldsymbol{\sigma}}$ is given by (1.57) and

$$f = (\boldsymbol{s} - \boldsymbol{X})_{\text{eq}} - R - \sigma_y \quad \text{with} \quad \boldsymbol{s} = (\mathbf{1} - \boldsymbol{D})^{-1/2} \boldsymbol{\sigma}^D (\mathbf{1} - \boldsymbol{D})^{-1/2} \neq \tilde{\boldsymbol{\sigma}}^D. \quad (2.97)$$

Again, isotropic and anisotropic damage may be represented by the single second order tensor $\boldsymbol{D}$, with isotropy being the limiting case $\boldsymbol{D} = D\mathbf{1}$.

As the yield criterion remains unchanged, the normals $\boldsymbol{n}^X$, $\boldsymbol{n}$ and $\boldsymbol{m}$ have the same expressions as for the case without the microdefects closure effect:

$$\boldsymbol{n}^X = \frac{3}{2} \frac{\boldsymbol{s}^D - \boldsymbol{X}}{(\boldsymbol{s} - \boldsymbol{X})_{\text{eq}}},$$
$$\boldsymbol{n} = \left[(\mathbf{1} - \boldsymbol{D})^{-1/2} \boldsymbol{n}^X (\mathbf{1} - \boldsymbol{D})^{-1/2}\right]^D, \qquad (2.98)$$
$$\boldsymbol{m} = \boldsymbol{n}^X - \gamma \boldsymbol{\alpha},$$

and can be considered as functions of $\boldsymbol{\alpha}$, $\boldsymbol{D}$ and $\boldsymbol{\sigma}$ only.

A set of 5 nonlinear equations function of the 5 independent variables $\{\boldsymbol{\epsilon}^e, \boldsymbol{\alpha}, r, \boldsymbol{D}, \boldsymbol{\sigma}\}$ has to be solved in a coupled manner:

$$\dot{\boldsymbol{\epsilon}} - \dot{\boldsymbol{\epsilon}}^e - \dot{r}\boldsymbol{n} = 0,$$
$$\dot{\boldsymbol{\alpha}} - \dot{r}\boldsymbol{m} = 0,$$
$$f - \sigma_v = (\boldsymbol{s} - \boldsymbol{X})_{\text{eq}} - R - \sigma_y - \sigma_v = 0,$$
$$\begin{cases} \dot{\boldsymbol{D}} - \left(\frac{Y}{S}\right)^s \dot{p}\mathbf{1} = 0 & \text{for isotropic damage,} \\ \dot{\boldsymbol{D}} - \left(\frac{\overline{Y}}{S}\right)^s |\boldsymbol{n}|\dot{r} = 0 & \text{for anisotropic damage,} \end{cases} \qquad (2.99)$$
$$\boldsymbol{\epsilon}^e - \underline{\underline{\boldsymbol{E}}}^{-1} : \tilde{\boldsymbol{\sigma}} - \alpha(T - T_{\text{ref}})\mathbf{1} = 0,$$

with

- $\tilde{\boldsymbol{\sigma}}$ as the effective stress such as the elasticity law. For isotropic damage,

$$\boldsymbol{\epsilon}^{\mathrm{e}} - \frac{1+\nu}{E}\left[\frac{\langle\boldsymbol{\sigma}\rangle^+}{1-D} + \frac{\langle\boldsymbol{\sigma}\rangle^-}{1-hD}\right] + \frac{\nu}{E}\left[\frac{\langle\mathrm{tr}\,\boldsymbol{\sigma}\rangle}{1-D} - \frac{\langle-\mathrm{tr}\,\boldsymbol{\sigma}\rangle}{1-hD}\right]\mathbf{1} - \alpha(T-T_{\mathrm{ref}})\mathbf{1} = 0 \tag{2.100}$$

and for anisotropic damage,

$$\begin{aligned}\boldsymbol{\epsilon}^{\mathrm{e}} &- \frac{1+\nu}{E}\left[(\boldsymbol{H}^{\mathrm{p}}\boldsymbol{\sigma}_+^{\mathrm{D}}\boldsymbol{H}^{\mathrm{p}})^{\mathrm{D}} + (\boldsymbol{H}^{\mathrm{n}}\boldsymbol{\sigma}_-^{\mathrm{D}}\boldsymbol{H}^{\mathrm{n}})^{\mathrm{D}}\right] \\ &- \frac{1-2\nu}{E}\left[\frac{\langle\sigma_{\mathrm{H}}\rangle}{1-\eta D_{\mathrm{H}}} - \frac{\langle-\sigma_{\mathrm{H}}\rangle}{1-\eta h_{\mathrm{a}}D_{\mathrm{H}}}\right]\mathbf{1} - \alpha(T-T_{\mathrm{ref}})\mathbf{1} = 0\,,\end{aligned} \tag{2.101}$$

with $\boldsymbol{H}^{\mathrm{p}} = (\mathbf{1}-\boldsymbol{D})^{-1/2}$ and $\boldsymbol{H}^{\mathrm{n}} = (\mathbf{1}-h_{\mathrm{a}}\boldsymbol{D})^{-1/2}$.

- $\boldsymbol{X} = \frac{2}{3}C\boldsymbol{\alpha}$, $R = R(r)$ for hardening.
- $\sigma_{\mathrm{v}} = K_N \dot{p}^{\frac{1}{N}}$ or $\sigma_{\mathrm{v}} = K_\infty\left[1 - \exp\left(-\frac{\dot{p}}{n}\right)\right]$ for the viscosity law.
- $Y = Y(\boldsymbol{D},\boldsymbol{\sigma})$, $\overline{Y} = \overline{Y}(\boldsymbol{D},\boldsymbol{\sigma})$. For isotropic damage,

$$\begin{aligned}Y &= \frac{1+\nu}{2E}\left[\frac{\langle\boldsymbol{\sigma}\rangle^+ : \langle\boldsymbol{\sigma}\rangle^+}{(1-D)^2} + h\frac{\langle\boldsymbol{\sigma}\rangle^- : \langle\boldsymbol{\sigma}\rangle^-}{(1-hD)^2}\right] \\ &- \frac{\nu}{2E}\left[\frac{\langle\mathrm{tr}\,\boldsymbol{\sigma}\rangle^2}{(1-D)^2} + h\frac{\langle-\mathrm{tr}\,\boldsymbol{\sigma}\rangle^2}{(1-hD)^2}\right]\end{aligned} \tag{2.102}$$

and for anisotropic damage,

$$\begin{aligned}\overline{Y} &= \frac{1+\nu}{2E}\mathrm{tr}\left[\left(\boldsymbol{H}^{\mathrm{p}}\boldsymbol{\sigma}_+^{\mathrm{D}}\boldsymbol{H}^{\mathrm{p}}\right)^2 + h_{\mathrm{a}}\left(\boldsymbol{H}^{\mathrm{n}}\boldsymbol{\sigma}_-^{\mathrm{D}}\boldsymbol{H}^{\mathrm{n}}\right)^2\right] \\ &+ \frac{3(1-2\nu)}{2E}\left[\frac{\langle\sigma_{\mathrm{H}}\rangle^2}{(1-\eta D_{\mathrm{H}})^2} + h_{\mathrm{a}}\frac{\langle-\sigma_{\mathrm{H}}\rangle^2}{(1-\eta h_{\mathrm{a}}D_{\mathrm{H}})^2}\right].\end{aligned} \tag{2.103}$$

Still for the sake of relative simplicity, the dilatancy terms are dropped out next.

2.2.5.1 Discretization by the θ-method

The previous set of nonlinear equations is discretized by use of the θ-method (with $\theta = 1$ for the yield criterion). This defines the local residual for plasticity coupled with damage as

2.2 Fully-Coupled Analysis

$$\{R_{\text{loc}}\} = \begin{cases} R_{\epsilon^e} = \Delta\epsilon^e - \Delta\epsilon + \Delta r\, n_{n+\theta}\,, \\ R_\alpha = \Delta\alpha - \Delta r\, m_{n+\theta}\,, \\ R_r = f_{n+1} = (s_{n+1} - X_{n+1})_{\text{eq}} - R(r_{n+1}) - \sigma_y\,, \\ R_D = \begin{cases} \Delta D - \left(\dfrac{Y_{n+\theta}}{S}\right)^s \dfrac{\Delta r}{1 - D_{n+\theta}}\, \mathbf{1} & \text{for isotropic damage,} \\ \Delta D - \left(\dfrac{\overline{Y}_{n+\theta}}{S}\right)^s |n_{n+\theta}|\, \Delta r & \text{for anisotropic damage,} \end{cases} \\ R_\sigma = \epsilon^e_{n+\theta} - \underline{\underline{E}}^{-1} : \tilde{\sigma}_{n+\theta} \end{cases}$$

(2.104)

and the incremental form of the plasticity loading/unloading conditions still reads as $\Delta r \geq 0$, $f_{n+1} \leq 0$, and $\Delta r f_{n+1} = 0$.

The main difference with the case without the microdefects closure effect is that the relationship $\sigma(\epsilon^e, D)$ is now implicit and given by the numerical inverting of the elasticty law.

For visco-plasticity coupled with damage, we consider the Norton law for the residual R_r^v defined by (2.72) instead of R_r.

2.2.5.2 Jacobian Matrix

The Newton iterative scheme is used again here and the derivatives of the residuals with respect to the increments $\Delta\epsilon^e$, $\Delta\alpha$, Δr, ΔD, and $\Delta\sigma$ are needed.

First, we have

$$\begin{aligned} \frac{\partial n^X}{\partial s} &= -\frac{\partial n^X}{\partial X} = \frac{1}{(s-X)_{\text{eq}}}\left[\frac{3}{2}I - \frac{1}{2}\mathbf{1}\otimes\mathbf{1} - n^X\otimes n^X\right], \\ \frac{\partial n^X}{\partial \epsilon^e} &= 0, \\ \frac{\partial n^X}{\partial r} &= 0, \\ \frac{\partial n^X}{\partial \alpha} &= \frac{2}{3}C\frac{\partial n^X}{\partial X}, \\ \frac{\partial n^X}{\partial D} &= \frac{\partial n^X}{\partial s}:\frac{\partial s}{\partial D}, \\ \frac{\partial n^X}{\partial \sigma} &= \frac{\partial n^X}{\partial s}:\frac{\partial s}{\partial \sigma}, \end{aligned}$$

(2.105)

with

$$\begin{aligned} \frac{\partial s}{\partial D} &= \frac{\partial H^p \sigma^D H^p}{\partial H^p}:\frac{\partial H^p}{\partial D} \\ \frac{\partial s}{\partial \sigma} &= \underline{\underline{H^p \otimes H^p}} - \frac{1}{3}H^{p2}\otimes\mathbf{1}, \end{aligned}$$

(2.106)

where $\frac{\partial \boldsymbol{H}^{\mathrm{p}} \boldsymbol{\sigma}^{\mathrm{D}} \boldsymbol{H}^{\mathrm{p}}}{\partial \boldsymbol{H}^{\mathrm{p}}}$ and $\frac{\partial \boldsymbol{H}^{\mathrm{p}}}{\partial \boldsymbol{D}}$ are defined by (2.79) with $\boldsymbol{H}$ changed into $\boldsymbol{H}^{\mathrm{p}}$ and $\boldsymbol{n}^{\mathrm{X}}$ into $\boldsymbol{\sigma}^{\mathrm{D}}$.

Again, we set up $\boldsymbol{n} = (\boldsymbol{H}^{\mathrm{p}} \boldsymbol{n}^{\mathrm{X}} \boldsymbol{H}^{\mathrm{p}})^{\mathrm{D}} = \underline{\boldsymbol{Q}} : \boldsymbol{n}^{\mathrm{X}}$ with $\underline{\boldsymbol{Q}}$ and $\frac{\partial (\boldsymbol{H}^{\mathrm{p}} \, \boldsymbol{n}^{\mathrm{X}} \boldsymbol{H}^{\mathrm{p}})}{\partial \boldsymbol{H}^{\mathrm{p}}}$ obtained by replacing $\boldsymbol{H}$ by $\boldsymbol{H}^{\mathrm{p}}$ in (2.77) and (2.79). Then, we obtain

$$\begin{aligned}
\frac{\partial \boldsymbol{n}}{\partial s} &= -\frac{\partial \boldsymbol{n}}{\partial \boldsymbol{X}} = \underline{\boldsymbol{Q}} : \frac{\partial \boldsymbol{n}^{\mathrm{X}}}{\partial s}, \\
\frac{\partial \boldsymbol{n}}{\partial \boldsymbol{\epsilon}^{\mathrm{e}}} &= 0, \\
\frac{\partial \boldsymbol{n}}{\partial r} &= 0, \\
\frac{\partial \boldsymbol{n}}{\partial \boldsymbol{\alpha}} &= \frac{2}{3} C \frac{\partial \boldsymbol{n}}{\partial \boldsymbol{X}} = \underline{\boldsymbol{Q}} : \frac{\partial \boldsymbol{n}^{\mathrm{X}}}{\partial \boldsymbol{\alpha}}, \\
\frac{\partial \boldsymbol{n}}{\partial \boldsymbol{D}} &= \underline{\boldsymbol{Q}} : \frac{\partial \boldsymbol{n}^{\mathrm{X}}}{\partial \boldsymbol{D}} + \frac{\partial (\boldsymbol{H}^{\mathrm{p}} \, \boldsymbol{n}^{\mathrm{X}} \boldsymbol{H}^{\mathrm{p}})}{\partial \boldsymbol{H}^{\mathrm{p}}} : \frac{\partial \boldsymbol{H}^{\mathrm{p}}}{\partial \boldsymbol{D}} - \frac{1}{3} \left(\mathbf{1} \otimes \boldsymbol{n}^{\mathrm{X}}\right) : \left(\boldsymbol{H}^{\mathrm{p}2} \underline{\otimes} \boldsymbol{H}^{\mathrm{p}2}\right), \\
\frac{\partial \boldsymbol{n}}{\partial \boldsymbol{\sigma}} &= \underline{\boldsymbol{Q}} : \frac{\partial \boldsymbol{n}^{\mathrm{X}}}{\partial \boldsymbol{\sigma}}.
\end{aligned} \quad (2.107)$$

Finally, the derivatives of Y and $\overline{Y}$ are needed. For isotropic damage, closed-form expressions are obtained:

$$\begin{aligned}
\frac{\partial Y}{\partial D} =& \frac{1+\nu}{E} \left[\frac{\langle \boldsymbol{\sigma}\rangle^+ : \langle \boldsymbol{\sigma}\rangle^+}{(1-D)^3} + h^2 \frac{\langle \boldsymbol{\sigma}\rangle^- : \langle \boldsymbol{\sigma}\rangle^-}{(1-hD)^3}\right] \mathbf{1} \\
& - \frac{\nu}{E} \left[\frac{\langle \mathrm{tr}\,\boldsymbol{\sigma}\rangle^2}{(1-D)^3} + h^2 \frac{\langle -\mathrm{tr}\,\boldsymbol{\sigma}\rangle^2}{(1-hD)^3}\right] \mathbf{1} \\
\frac{\partial Y}{\partial \boldsymbol{\sigma}} =& \frac{1+\nu}{E} \left[\frac{\langle \boldsymbol{\sigma}\rangle^+}{(1-D)^2} + \frac{h\langle \boldsymbol{\sigma}\rangle^-}{(1-hD)^2}\right] \\
& - \frac{\nu}{E} \left[\frac{\langle \mathrm{tr}\,\boldsymbol{\sigma}\rangle}{(1-D)^2} - \frac{h\langle -\mathrm{tr}\,\boldsymbol{\sigma}\rangle}{(1-hD)^2}\right] \mathbf{1}.
\end{aligned} \quad (2.108)$$

For anisotropic damage, only the hydrostatic terms (those responsible for strain localization) can be obtained in a closed-form. For an estimation of the last Jacobian terms, let us consider the approximation

$$\overline{Y} \approx \frac{1+\nu}{3E} \frac{\sigma_{\mathrm{eq}}^2}{(1-D_{\mathrm{H}})^2} + \frac{3(1-2\nu)}{2E} \left[\frac{\langle \sigma_{\mathrm{H}}\rangle^2}{(1-\eta D_{\mathrm{H}})^2} + \frac{h_{\mathrm{a}}\langle -\sigma_{\mathrm{H}}\rangle^2}{(1-\eta h_{\mathrm{a}} D_{\mathrm{H}})^2}\right], \quad (2.109)$$

which leads to

$$\begin{aligned}
\frac{\partial \overline{Y}}{\partial D} &\approx \left\{\frac{2(1+\nu)}{9E} \frac{\sigma_{\mathrm{eq}}^2}{(1-D_{\mathrm{H}})^3} + \frac{\eta(1-2\nu)}{E} \left[\frac{\langle \sigma_{\mathrm{H}}\rangle^2}{(1-\eta D_{\mathrm{H}})^3} + \frac{h_{\mathrm{a}}^2 \langle -\sigma_{\mathrm{H}}\rangle^2}{(1-\eta h_{\mathrm{a}} D_{\mathrm{H}})^3}\right]\right\} \mathbf{1} \\
\frac{\partial \overline{Y}}{\partial \boldsymbol{\sigma}} &\approx \frac{1+\nu}{E} \frac{\boldsymbol{\sigma}^{\mathrm{D}}}{(1-D_{\mathrm{H}})^2} + \frac{1-2\nu}{E} \left[\frac{\langle \sigma_{\mathrm{H}}\rangle}{(1-\eta D_{\mathrm{H}})^2} - \frac{h_{\mathrm{a}}\langle -\sigma_{\mathrm{H}}\rangle}{(1-\eta h_{\mathrm{a}} D_{\mathrm{H}})^2}\right] \mathbf{1}
\end{aligned} \quad (2.110)$$

Table 2.3 gives a summary of the Jacobian terms. For visco-plasticity, the derivatives for $\frac{\partial R_r^v}{\partial \Delta \mathcal{W}}$ are still given by (2.82) and (2.83) as in the case without microdefects closure effects. Note that in last equation of Table 2.3, $\mathcal{H}$ is the Heaviside function, $\mathcal{H}(x) = 1$ for $x \geq 0$, $\mathcal{H}(x) = 0$ for $x < 0$.

Table 2.3. Jacobian terms for damage models with microdefects closure effect

Derivatives of $\boldsymbol{R}_{\epsilon^e}$	
$\dfrac{\partial \boldsymbol{R}_{\epsilon^e}}{\partial \Delta \boldsymbol{\epsilon}^e} = \boldsymbol{I}$	$\dfrac{\partial \boldsymbol{R}_{\epsilon^e}}{\partial \Delta r} = \boldsymbol{n}_{n+\theta}$
$\dfrac{\partial \boldsymbol{R}_{\epsilon^e}}{\partial \Delta \boldsymbol{\alpha}} = \dfrac{2}{3} C\theta \Delta r \, \underline{\boldsymbol{Q}}_{n+\theta} : \left.\dfrac{\partial \boldsymbol{n}^{\mathrm{X}}}{\partial \boldsymbol{X}}\right|_{n+\theta}$	$\dfrac{\partial \boldsymbol{R}_{\epsilon^e}}{\partial \Delta \boldsymbol{D}} = \theta \Delta r \left.\dfrac{\partial \boldsymbol{n}}{\partial \boldsymbol{D}}\right|_{n+\theta}$
$\dfrac{\partial \boldsymbol{R}_{\epsilon^e}}{\partial \Delta \boldsymbol{\sigma}} = \theta \Delta r \, \underline{\boldsymbol{Q}}_{n+\theta} : \left.\dfrac{\partial \boldsymbol{n}^{\mathrm{X}}}{\partial \boldsymbol{\sigma}}\right|_{n+\theta}$	

Derivatives of $\boldsymbol{R}_\alpha$	
$\dfrac{\partial \boldsymbol{R}_\alpha}{\partial \Delta \boldsymbol{\epsilon}^e} = 0$	$\dfrac{\partial \boldsymbol{R}_\alpha}{\partial \Delta r} = -\boldsymbol{m}_{n+\theta}$
$\dfrac{\partial \boldsymbol{R}_\alpha}{\partial \Delta \boldsymbol{\alpha}} = (1+\gamma \theta \Delta r)\,\boldsymbol{I} - \dfrac{2}{3} C\theta \Delta r \left.\dfrac{\partial \boldsymbol{n}^{\mathrm{X}}}{\partial \boldsymbol{X}}\right|_{n+\theta}$	$\dfrac{\partial \boldsymbol{R}_\alpha}{\partial \Delta \boldsymbol{D}} = 0$
$\dfrac{\partial \boldsymbol{R}_\alpha}{\partial \Delta \boldsymbol{\sigma}} = -\theta \Delta r \left.\dfrac{\partial \boldsymbol{n}^{\mathrm{X}}}{\partial \boldsymbol{\sigma}}\right|_{n+\theta}$	

Derivatives of R_r (take $\theta = 1$ for the plastic case)	
$\dfrac{\partial R_r}{\partial \Delta \boldsymbol{\epsilon}^e} = 0$	$\dfrac{\partial R_r}{\partial \Delta r} = -\theta\, R'(r_{n+\theta})$
$\dfrac{\partial R_r}{\partial \Delta \boldsymbol{\alpha}} = -\dfrac{2}{3} C\theta\, \boldsymbol{n}^{\mathrm{X}}_{n+\theta}$	$\dfrac{\partial R_r}{\partial \Delta \boldsymbol{D}} = \theta\, \boldsymbol{n}^{\mathrm{X}}_{n+\theta} : \left.\dfrac{\partial s}{\partial \boldsymbol{D}}\right|_{n+\theta}$
$\dfrac{\partial R_r}{\partial \Delta \boldsymbol{\sigma}} = \theta\, \boldsymbol{n}_{n+\theta}$	

(continue next page)

2.2.5.3 Updating the Thermodynamics Variables

Once $\boldsymbol{\epsilon}^e_{n+1}$, $\boldsymbol{\alpha}_{n+1}$, r_{n+1}, $\boldsymbol{D}_{n+1}$, and $\boldsymbol{\sigma}_{n+1}$ are known, the remaining thermodynamics variables are updated explicitly:

- Plastic strain tensor: $\boldsymbol{\epsilon}^p_{n+1} = \boldsymbol{\epsilon}_{n+1} - \boldsymbol{\epsilon}^e_{n+1}$

Table 2.3. (continued)

Derivatives of $\boldsymbol{R}_\mathrm{D}$

Isotropic damage	Anisotropic damage
$\dfrac{\partial \boldsymbol{R}_\mathrm{D}}{\partial \Delta \boldsymbol{\epsilon}^\mathrm{e}} = 0$	$\dfrac{\partial \boldsymbol{R}_\mathrm{D}}{\partial \Delta \boldsymbol{\epsilon}^\mathrm{e}} \approx -\dfrac{s\theta \overline{Y}_{n+\theta}^{s-1} \Delta r}{S^s} \|\boldsymbol{n}_{n+\theta}\| \otimes \left.\dfrac{\partial \overline{Y}}{\partial \boldsymbol{\epsilon}^\mathrm{e}}\right|_{n+\theta}$
$\dfrac{\partial \boldsymbol{R}_\mathrm{D}}{\partial \Delta \alpha} = 0$	$\dfrac{\partial \boldsymbol{R}_\mathrm{D}}{\partial \Delta \boldsymbol{\alpha}} \approx 0$
$\dfrac{\partial \boldsymbol{R}_\mathrm{D}}{\partial \Delta r} = -\left(\dfrac{Y_{n+\theta}}{S}\right)^s \dfrac{1}{1-D_{n+\theta}} \mathbf{1}$	$\dfrac{\partial \boldsymbol{R}_\mathrm{D}}{\partial \Delta r} = -\left(\dfrac{\overline{Y}_{n+\theta}}{S}\right)^s \|\boldsymbol{n}_{n+\theta}\|$
$\dfrac{\partial \boldsymbol{R}_\mathrm{D}}{\partial \Delta D} = \boldsymbol{I} - \left(\dfrac{Y_{n+\theta}}{S}\right)^s$	$\dfrac{\partial \boldsymbol{R}_\mathrm{D}}{\partial \Delta \boldsymbol{D}} \approx \boldsymbol{I}$
$\qquad \times \dfrac{\theta \Delta r}{(1-D_{n+\theta})^2} \dfrac{1}{3} \mathbf{1} \otimes \mathbf{1}$	$\qquad -\dfrac{s\theta \overline{Y}_{n+\theta}^{s-1} \Delta r}{S^s} \|\boldsymbol{n}_{n+\theta}\| \otimes \left.\dfrac{\partial \overline{Y}}{\partial \boldsymbol{D}}\right|_{n+\theta}$
$\qquad -\dfrac{s\theta Y_{n+\theta}^{s-1}\Delta r}{S^s(1-D_{n+\theta})} \mathbf{1} \otimes \left.\dfrac{\partial Y}{\partial D}\right|_{n+\theta}$	
$\dfrac{\partial \boldsymbol{R}_\mathrm{D}}{\partial \Delta \boldsymbol{\sigma}} \approx -\dfrac{s\theta Y_{n+\theta}^{s-1}\Delta r}{S^s} \|\boldsymbol{n}_{n+\theta}\| \otimes \left.\dfrac{\partial Y}{\partial \boldsymbol{\sigma}}\right|_{n+\theta}$	$\dfrac{\partial \boldsymbol{R}_\mathrm{D}}{\partial \Delta \boldsymbol{\sigma}} \approx -\dfrac{s\theta \overline{Y}_{n+\theta}^{s-1}\Delta r}{S^s} \|\boldsymbol{n}_{n+\theta}\| \otimes \left.\dfrac{\partial \overline{Y}}{\partial \boldsymbol{\sigma}}\right|_{n+\theta}$

Derivatives of $\boldsymbol{R}_\sigma$

$$\dfrac{\partial \boldsymbol{R}_\sigma}{\partial \Delta \boldsymbol{\epsilon}^\mathrm{e}} = \theta \, \boldsymbol{I} \qquad \dfrac{\partial \boldsymbol{R}_\sigma}{\partial \Delta r} = 0 \qquad \dfrac{\partial \boldsymbol{R}_\sigma}{\partial \Delta \alpha} = 0$$

$$\dfrac{\partial \boldsymbol{R}_\sigma}{\partial \Delta \boldsymbol{D}} \approx -\dfrac{\theta(1+\nu)}{E}\left[\boldsymbol{I} - \dfrac{1}{3}\mathbf{1}\otimes\mathbf{1}\right] : \left.\dfrac{\partial \boldsymbol{s}}{\partial \boldsymbol{D}}\right|_{n+\theta}$$
$$-\dfrac{\eta\theta(1-2\nu)}{3E}\left[\dfrac{\langle\sigma_{\mathrm{H}\ n+\theta}\rangle}{(1-\eta D_{\mathrm{H}\ n+\theta})^2} - \dfrac{h_\mathrm{a}\langle-\sigma_{\mathrm{H}\ n+\theta}\rangle}{(1-\eta h_\mathrm{a} D_{\mathrm{H}\ n+\theta})^2}\right]\mathbf{1}\otimes\mathbf{1}$$

$$\dfrac{\partial \boldsymbol{R}_\sigma}{\partial \Delta \boldsymbol{\sigma}} \approx -\dfrac{\theta(1+\nu)}{E}\left[\boldsymbol{I} - \dfrac{1}{3}\mathbf{1}\otimes\mathbf{1}\right] : \left.\dfrac{\partial \boldsymbol{s}}{\partial \boldsymbol{\sigma}}\right|_{n+\theta}$$
$$-\dfrac{\theta(1-2\nu)}{3E}\left[\dfrac{\mathcal{H}(\sigma_{\mathrm{H}\ n+\theta})}{1-\eta D_{\mathrm{H}\ n+\theta}} + \dfrac{\mathcal{H}(-\sigma_{\mathrm{H}\ n+\theta})}{1-\eta h_\mathrm{a} D_{\mathrm{H}\ n+\theta}}\right]\mathbf{1}\otimes\mathbf{1}$$

- Strain hardening variables: $R_{n+1} = R(r_{n+1})$, $\boldsymbol{X}_{n+1} = \dfrac{2}{3}C\boldsymbol{\alpha}_{n+1}$
- Accumulated plastic strain: $p_{n+1} = p_n + \sqrt{\dfrac{2}{3}\boldsymbol{n}_{n+\theta} : \boldsymbol{n}_{n+\theta}}\,\Delta r$
- Stored energy:

$$w_{s\,n+1} = AR(r_{n+1})r_{n+1}^{1/m} + \frac{1}{3}C\,\boldsymbol{\alpha}_{n+1}:\boldsymbol{\alpha}_{n+1} - \hat{\omega}_{n+1},$$

$$\hat{\omega}_{n+1} = \hat{\omega}_n + \frac{A}{2}\left[R'(r_n)r_n^{1/m} + R'(r_{n+1})r_{n+1}^{1/m}\right]\Delta r,$$

with $R'(r) = bR_\infty \exp(-br)$ for exponential isotropic hardening
- Damage:

$$\boldsymbol{H}^{\mathrm{p}}_{n+1} = (\mathbf{1} - \boldsymbol{D}_{n+1})^{-\frac{1}{2}}, \qquad \boldsymbol{H}^{\mathrm{n}}_{n+1} = (\mathbf{1} - h_{\mathrm{a}}\boldsymbol{D}_{n+1})^{-\frac{1}{2}} \quad \text{and}$$
$$Y_{n+1} = Y(\boldsymbol{D}_{n+1}, \boldsymbol{\sigma}_{n+1}) \quad \text{or} \quad \overline{Y}_{n+1} = \overline{Y}(\boldsymbol{D}_{n+1}, \boldsymbol{\sigma}_{n+1}).$$

2.2.5.4 Consistent Tangent Operator

The variables $\{\boldsymbol{\epsilon}^{\mathrm{e}}, \boldsymbol{\alpha}, r, \boldsymbol{D}, \boldsymbol{\sigma}\}$ have been considered for the local integration of the constitutive equations. The consistent tangent operator is then a block of the Jacobian matrix at the convergence of the local Newton scheme as

$$\Delta\boldsymbol{\sigma} = [\mathrm{Jac}]^{-1}_{\boldsymbol{\sigma},\boldsymbol{\epsilon}^{\mathrm{e}}}\Delta\boldsymbol{\epsilon} \implies \underline{\boldsymbol{L}}^{\mathrm{c}} = [\mathrm{Jac}]^{-1}_{\boldsymbol{\sigma},\boldsymbol{\epsilon}^{\mathrm{e}}}. \qquad (2.111)$$

2.2.6 Performing FE Damage Computations♂

Once the damage models are available as part of the FE code material library, their use is identical to the use of elasto-(visco-)plasticity models. The material parameters need to be supplied in an input data file with eventually tables of the temperature dependency information.

Users prefer the anisotropic damage model because it demands the same computation effort than the isotropic model but takes into account the induced anisotropy and gives good results for the strain-damage localization with a realistic mesh size. Recall that due to the strong nonlinearity encountered, the computations are costly in time. When a whole structure is under consideration, we advise starting with a purely elasto-(visco-)plastic analysis before running the fully coupled computation (set for example, the damage threshold to a very large value) and performing an uncoupled analysis of damage as detailed in Sect. 2.1.3. Systematic comparisons between both uncoupled and coupled analysis are encouraged to build up the engineering experience.

The outputs of a fully coupled damage analysis are the history of the nodal displacements and, at each Gauss point, the stresses, strains, and state and internal variables, particularly the damage field history. The damage maps identifies the structure's weak points. There is crack initiation when the damage reaches the critical value D_{c}.

Different options are then possible

- To stop the computation when strain-damage localization takes place. This is the most commonly used option as the FE solution is mesh-independent up to this point.

- To continue the continuum mechanics computation with a given mesh size in the zone of intense damage. The strain-damage localization pattern gives an idea of the further development of cracks. The FE solution is mesh-dependent and should only be used to compare different design solutions.
- To continue the computation with a given mesh size and releasing nodes to model the crack initiation and propagation (see Table 1.8 of Sect. 1.6.3 for the orientation and the size of the initial mesocrack). This option needs the difficult operation of release of the adequate nodes to create a crack with the right size and orientation. The FE solution is still mesh-dependent; it has no absolute value but has again a relative value for the comparison of different design solutions.
- To initiate and propagate mesocracks with sizes and orientations independent of the mesh nodes by use of enriched finite elements with embedded displacements discontinuities. New classes of efficient methods emerged recently: manifold method (G. Shi 1992) or more generally the partition-of-unity method (J.M. Melenk and I. Babuska 1996, N. Moës, J. Dolbow, and T. Belytschko 1999). But the extension to damage behavior with softening, strain localization and mesh dependency is still under study.

2.2.7 Localization Limiters☺☺

When performing the FE analysis of a structure made of damageable materials, the instabilities encountered are due to the numerical representation of the physical occurrence of shear bands where the initial fracture mechanics mesocrack will initiate. The good thing is that damage mechanics reproduces such a phenomenon. As we just pointed out, the drawback is the mesh dependency of the solution and the shear band thickness. This emphasizes that there is missing information in the formulation. For an arbitrary small mesh size, the computed shear band may be much too thin, corresponding to a non-physical, unlimited, strain localization. Localization limiters are then necessary to introduce by a way or another a characteristic length.

Initially proposed for the crack band model (S.T. Pietruszczak and Z. Mroz 1981, Z.P. Bažant and B.H. Oh 1983), the simplest localization limiter is to set the mesh size to a constant value. Some authors propose considering the mesh size as a material parameter (and identifying it) in order to compute the crack propagation within damage mechanics framework (G. Rousselier 1987). The characteristic length introduced may be seen as the size of the RVE. But the "correct" mesh size for a good quality computation seems to be related to the shape, the size of the whole structure and to the loading!

As 21st century localization limiters, a material internal length is introduced directly into the constitutive laws, the main idea being that plasticity and/or damage evolution at a structure point $\boldsymbol{x}$ depend on the value of the thermodynamics variables at the considered point as well as a small domain around it. One speaks then of non-local constitutive models.

2.2.7.1 Strongly-Non-local Theories
(Z.P. Bažant and G. Pijaudier-Cabot 1987)

Each local thermodynamics variable $\mathcal{V}$ (or some of them) may be replaced by an "average" $\mathcal{V}^{\text{nl}}$ of the variable around the considered point. The delocalization of $\mathcal{V}$ is described by

$$\mathcal{V}^{\text{nl}} = \frac{1}{V_r} \int_V W(\boldsymbol{x} - \boldsymbol{s}) \mathcal{V}(\boldsymbol{s}) \mathrm{d}V \quad \text{and} \quad V_r = \int_V W(\boldsymbol{x} - \boldsymbol{s}) \mathrm{d}V, \quad (2.112)$$

where $W(\boldsymbol{x} - \boldsymbol{s})$ is a generalized weight function and V is the structure volume. The non-local weight function is a new parameter of the model, chosen as the Gauss distribution function for instance:

$$W(\boldsymbol{x}) = W_0 \exp\left(-\frac{\|\boldsymbol{x}\|^2}{2\delta_c^2}\right) \quad (2.113)$$

or the polynomial bell-shaped function with finite support (Z.P. Bažant and J. Ožbolt 1990):

$$W(\boldsymbol{x}) = W_0' \left\langle 1 - \frac{\|\boldsymbol{x}\|^2}{\delta_c'^2} \right\rangle^2, \quad (2.114)$$

with W_0 and W_0' as the normalization factors and δ_c, δ_c' as "the" internal lengths of the medium.

To perform the variables delocalization makes the numerical implementation of the constitutive equations difficult in classical FE computer codes. The local integration subroutines now need the value of the thermodynamics variables at many Gauss points, eventually located far outside the considered element.

2.2.7.2 Gradient Theories
(E. Aifantis 1987)

An alternative way that is easier to implement is to consider gradient theories for which plasticity and/or damage evolution at a structure point depend on the value of the thermodynamics variables as well as their first, second, ... n-th gradients. Replace then each local variable $\mathcal{V}$ by the expansion

$$\mathcal{V}^{\text{nl}} = \mathcal{V} + c_1 \cdot \nabla \mathcal{V} + c_2 \nabla^2 \mathcal{V} + \ldots, \quad (2.115)$$

often with c_2 as the only non-vanishing term (second gradient theories). It is also possible to define $\mathcal{V}^{\text{nl}}$ as the implicit solution of

$$\mathcal{V}^{\text{nl}} - c_1 \cdot \nabla \mathcal{V}^{\text{nl}} - c_2 \nabla^2 \mathcal{V}^{\text{nl}} - \ldots = \mathcal{V}, \qquad (2.116)$$

where the "internal lengths" are introduced through the parameters c_k.

Note that the gradients theories are recovered by introducing the Taylor expansion

$$\mathcal{V}(\boldsymbol{x} + \boldsymbol{s}) = \mathcal{V}(\boldsymbol{x}) + \nabla \mathcal{V} \cdot \boldsymbol{s} + \frac{1}{2!}\nabla^{(2)}\mathcal{V} \cdot (\boldsymbol{s} \otimes \boldsymbol{s}) + \ldots \qquad (2.117)$$

in the average integrals of the non-local theories.

Many non-local or gradient models may be constructed, depending then on the choice of the variables to be delocalized and the delocalization procedure (T. Belytschko 1988, Z.P. Bažant and F.B. Lin 1988, R. de Borst and H.B. Mulhaus 1991, S. Forest and E. Lorentz 2004). They all introduce an internal length. We will present in the next section a simple way to make the damage models of Chap. 1 non-local.

2.2.7.3 Simple Non-local Damage Model

One possible way to study damage models is to delocalize the damage variable and use

$$\boldsymbol{D}^{\text{nl}} = \frac{1}{V_r} \int_V W(\boldsymbol{x} - \boldsymbol{s})\boldsymbol{D}(\boldsymbol{s})\mathrm{d}V \qquad (2.118)$$

instead of $\boldsymbol{D}$ in the constitutive equations.

But from the remark of Sect. 1.6 that the strain localization is strongly related to the value of the parameter η, which defines an hydrostatic damage d_{H} (fully acting when strain localization occurs) as

$$d_{\text{H}} = \eta D_{\text{H}} \quad \text{where} \quad \tilde{\sigma}_{\text{H}} = \frac{\sigma_{\text{H}}}{1 - d_{\text{H}}}, \qquad (2.119)$$

the simplest is to make the previous relationship non-local and to calculate d_{H} as

$$d_{\text{H}} = \frac{1}{V_r} \int_V W(\boldsymbol{x} - \boldsymbol{s})\eta D_{\text{H}}(\boldsymbol{s})\mathrm{d}V, \quad \text{still with} \quad \tilde{\sigma}_{\text{H}} = \frac{\sigma_{\text{H}}}{1 - d_{\text{H}}}, \qquad (2.120)$$

or to consider the gradient form $d_{\text{H}} - c_{\text{D}}\nabla^2 d_{\text{H}} = \eta D_{\text{H}}$, with c_{D} as a material parameter. The last equation written either in an integral or a gradient form is the only change when compared to the local damage model.

2.3 Locally-Coupled Analysis

In **quasi-brittle failure** and **high-cycle fatigue** applications, the damage is very localized in such a way that the damaged material occupies a small volume in comparison with the macroscale of the structural component and even with the mesoscale of the RVE. This is due to the high sensitivity of damage to stress concentrations at the mesoscale and defects at the microscale. According to the hypothesis of a weak, damageable inclusion embedded in a meso-RVE, such a state of localized damage is represented by the **two-scale damage model** of Sect. 1.5.5 (see also Fig. 2.5).

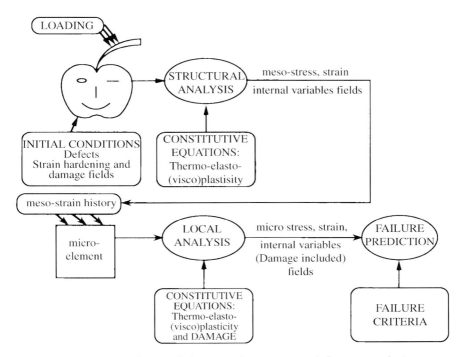

Fig. 2.5. Locally-coupled approach to structural damage analysis

By considering damage on the microscale only, the two-scale model predicts crack initiation when the microdamage reaches the critical value D_c. The crack initiation condition $D = D_c$ on the microscale coincides with the initiation of a mesocrack. The model is 3D and applies to complex loadings under isothermal conditions. It needs robust algorithms of resolution and an implicit scheme for its numerical implementation is described here.

2.3.1 Post-Processing a Reference Structure Calculation♂

The damage is determined by post-processing a reference structure calculation, performed in elasticity for quasi-brittle and high-cycle fatigue applications and in elasto-plasticity for more ductile conditions. The reference calculation provides information on the history of the stresses, strains, and plastic strains with no damage at the mesolevel of classical continuum mechanics. Considered altogether with the Eshelby–Kröner localization law and isotropic elasticity,

$$\epsilon^\mu = \epsilon + \beta(\epsilon^{\mu p} - \epsilon^p) \quad \text{and} \quad \beta = \frac{2}{15}\frac{4-5\nu}{1-\nu}, \qquad (2.121)$$

are inputs for the time integration of elasto-(visco-)plasticity fully coupled with damage constitutive equations on the microscale.

The numerical scheme used to solve those equations is the one described in Sect. 2.2.4 for the local integration of the fully-coupled analysis with only slight differences:

- For the fully-coupled damage analysis, the full calculation is performed by iterations made of global FE resolutions followed by local integrations at the structure Gauss points. Here, the time integration of the constitutive equations is performed only one time as a post-processor of the reference calculation. It may then be performed at the most loaded point only.
- In nonlinear FE computations, the local time integrations are made at constant strain. Here, the localization law has to be considered and the calculation is made at constant $(\epsilon^\mu - \beta\epsilon^{\mu p})$.
- The microscale constitutive equations are those of elasto-(visco-)plasticity coupled with damage, but with linear kinematic hardening and isotropic damage only. At points where $f^\mu > 0$, a set of 4 nonlinear differential equations is solved in the same manner as in Sect. 2.2.4 (μ-superscripts standing for "microscale" are added):

$$\dot{\epsilon}^{\mu e} + (1-\beta)\dot{r}^\mu n^\mu - \dot{\epsilon} + \beta\dot{\epsilon}^p = 0,$$
$$\dot{\alpha}^\mu - \dot{r}^\mu m^\mu = 0,$$
$$f^\mu - \sigma_v^\mu = (\tilde{\sigma}^\mu - X^\mu)_{eq} - \sigma_f^\infty - \sigma_v^\mu = 0, \qquad (2.122)$$
$$\dot{D} - \left(\frac{Y^\mu}{S}\right)^s \dot{p}^\mu = 0 \quad \text{if} \quad p^\mu > p_D,$$

where $\tilde{\sigma}^\mu = \sigma^\mu/(1-D)$ is the microscale effective stress, the yield stress σ_y^μ is equal to the fatigue limit σ_f^∞, σ_v^μ, if needed, is given by a viscosity law and where (isothermal case)

$$m^\mu = \frac{3}{2}\frac{\tilde{\sigma}^{\mu D} - X^\mu}{(\tilde{\sigma}^{\mu D} - X^\mu)_{eq}},$$
$$n^\mu = \frac{m^\mu}{1-D},$$
$$\tilde{\sigma}^\mu = \frac{\sigma^\mu}{1-D} = \underline{E} : \epsilon^{\mu e},$$

$$\boldsymbol{X}^{\mu} = \frac{2}{3}C_{y}\boldsymbol{\alpha}^{\mu},$$

$$Y^{\mu} = \frac{1+\nu}{2E}\left[\frac{\langle\boldsymbol{\sigma}^{\mu}\rangle^{+}:\langle\boldsymbol{\sigma}^{\mu}\rangle^{+}}{(1-D)^{2}} + h\frac{\langle\boldsymbol{\sigma}^{\mu}\rangle^{-}:\langle\boldsymbol{\sigma}^{\mu}\rangle^{-}}{(1-hD)^{2}}\right] \quad (2.123)$$

$$-\frac{\nu}{2E}\left[\frac{\langle\operatorname{tr}\boldsymbol{\sigma}^{\mu}\rangle^{2}}{(1-D)^{2}} + h\frac{\langle-\operatorname{tr}\boldsymbol{\sigma}^{\mu}\rangle^{2}}{(1-hD)^{2}}\right].$$

2.3.2 Implicit Scheme for the Two-Scale Model♂♂
(I. Doghri and R. Billardon 1988)

The reference computation gives the history of stresses, strains, and plastic strains on the mesoscale as discrete values at every instant $t = 0, \ldots, t_k, \ldots t_n$, $t_{n+1}, \ldots$ With the microscale variables at time t_n, the mesoscale variables at times t_n, and t_{n+1} being known, the microscale variables at time t_{n+1} are determined following a two step procedure:

1. A **local elastic prediction**, which assumes an elastic behavior with constant plastic strains $\boldsymbol{\epsilon}^{\mu p} = \boldsymbol{\epsilon}_n^{\mu p}$, constant kinematic hardening $\boldsymbol{X}^{\mu} = \boldsymbol{X}_n^{\mu}$, and constant damage $D = D_n$, gives a first estimate of the strains, elastic strains, and effective stresses on a microscale at time t_{n+1}:

$$\begin{aligned}
\boldsymbol{\epsilon}^{\mu} &= \boldsymbol{\epsilon} + \beta\left(\boldsymbol{\epsilon}_n^{\mu p} - \boldsymbol{\epsilon}_{n+1}^{p}\right), \\
\boldsymbol{\epsilon}^{\mu e} &= \boldsymbol{\epsilon}^{\mu} - \boldsymbol{\epsilon}_n^{\mu p}, \\
\tilde{\boldsymbol{\sigma}}^{\mu} &= \boldsymbol{\underline{E}} : \boldsymbol{\epsilon}^{\mu e} = \boldsymbol{\sigma} - 2G(1-\beta)\left(\boldsymbol{\epsilon}_n^{\mu p} - \boldsymbol{\epsilon}_{n+1}^{p}\right).
\end{aligned} \quad (2.124)$$

2. A **local plastic correction** of the state and internal variables $\boldsymbol{\epsilon}^{\mu}$, $\boldsymbol{\epsilon}^{\mu e}$, and $\tilde{\boldsymbol{\sigma}}$ made at constant $\boldsymbol{\epsilon}^{\mu} - \beta\boldsymbol{\epsilon}^{\mu p} = \boldsymbol{\epsilon} - \beta\boldsymbol{\epsilon}^{p}$ gives the strain $\boldsymbol{\epsilon}_{n+1}^{\mu}$, elastic strain $\boldsymbol{\epsilon}_{n+1}^{\mu e}$, plastic strain $\boldsymbol{\epsilon}_{n+1}^{\mu p}$, stress $\boldsymbol{\sigma}_{n+1}$, kinematic hardening $\boldsymbol{X}_{n+1}$, and damage D_{n+1} at time t_{n+1}.

To detail the second step, consider that the elastic predictor has given $\tilde{\boldsymbol{\sigma}}^{\mu}$, $\boldsymbol{\epsilon}^{\mu}$, and $\boldsymbol{\epsilon}^{\mu e}$ as initial estimates of the stresses, strains and elastic strains. If the yield condition $f^{\mu} \leq 0$ is satisfied, then $\tilde{\boldsymbol{\sigma}}_{n+1}^{\mu} = \tilde{\boldsymbol{\sigma}}^{\mu}$, $\boldsymbol{\epsilon}_{n+1}^{\mu} = \boldsymbol{\epsilon}^{\mu}$, and $\boldsymbol{\epsilon}_{n+1}^{\mu e} = \boldsymbol{\epsilon}^{\mu e}$ is set. If not, the Newton iterative process starts. For simplicity, the damage is assumed to remain constant over a time increment (in fatigue it may even be assumed constant over a whole cycle) and it is then the same as to integrate a law of elasto-plasticity with linear kinematic hardening only.

The nonlinear equations to be solved in a coupled manner are (implicit Euler scheme is used)

$$\begin{cases}
\Delta\boldsymbol{\epsilon}^{\mu e} + (1-\beta)\Delta r^{\mu}\,\boldsymbol{n}_{n+1}^{\mu} - \Delta\boldsymbol{\epsilon} + \beta\Delta\boldsymbol{\epsilon}^{p} = 0, \\
\Delta\boldsymbol{\alpha}^{\mu} - \Delta r^{\mu}\,\boldsymbol{m}_{n+1}^{\mu} = 0, \\
f_{n+1}^{\mu} = \left(\tilde{\boldsymbol{\sigma}}_{n+1}^{\mu} - \boldsymbol{X}_{n+1}^{\mu}\right)_{\text{eq}} - \sigma_{\text{f}}^{\infty}.
\end{cases} \quad (2.125)$$

They may be solved straightforwardly by use of the Newton iterative scheme as in Sects. 2.2.4 and 2.2.5 for the fully-coupled analysis, but it is advantageous here to write them first in terms of the plastic strain p^μ and the variable $s^\mu = \tilde{\sigma}^\mu - X^\mu$ as a set of two equations, as follows:

$$\begin{cases} R_s = s_{n+1}^\mu + \frac{2}{3}\mathcal{G}m_{n+1}^\mu \Delta p^\mu - \underline{E} : \epsilon + 2G\beta\epsilon^p + 2G(1-\beta)\epsilon_n^{\mu p} + X_n^\mu = 0, \\ R_p = (s_{n+1}^\mu)_{\text{eq}} - \sigma_f^\infty = 0, \end{cases} \quad (2.126)$$

where $\mathcal{G} = 3G(1-\beta) + C_y(1-D_n)$, $m^\mu = \frac{3}{2}\frac{s^{\mu D}}{s_{\text{eq}}^\mu}$, and the mesoscale strains $\epsilon = \epsilon_{n+1}$ and $\epsilon^p = \epsilon_{n+1}^p$ are known from the initial FE analysis.

For each iteration q of the Newton scheme, the solution $s^{\mu(q+1)}$ and $p^{\mu(q+1)}$, or in an equivalent manner, the "corrections" $C_s = s^{\mu(q+1)} - s^{\mu(q)}$, $C_p = p^{\mu(q+1)} - p^{\mu(q)}$ to be applied at each step to the previous iterated $s^{\mu(q)}$ and $p^{\mu(q)}$ are given by

$$\begin{cases} R_s + \dfrac{\partial R_s}{\partial s^\mu} : C_s + \dfrac{\partial R_s}{\partial p} C_p = 0, \\ R_p + \dfrac{\partial R_p}{\partial s^\mu} : C_s = 0, \end{cases} \quad (2.127)$$

where R_s, R_p, and their partial derivatives are taken at time t_{n+1} and the iteration q. The starting solution $s^{\mu(0)} = \tilde{\sigma}^\mu - X_n^\mu$ and $p^{\mu(0)} = p_n^\mu$ corresponds to the elastic predictor (2.124).

The set of equations (2.127) are then written as

$$\begin{cases} R_s + \left[\underline{I} + \frac{2}{3}\mathcal{G}\Delta p^\mu \dfrac{\partial m^\mu}{\partial s^\mu}\right] : C_s + \frac{2}{3}\mathcal{G}m^\mu C_p = 0, \\ R_p + m^\mu : C_s = 0, \end{cases} \quad (2.128)$$

where $\underline{I}$ is the fourth order identity tensor and

$$\dfrac{\partial m^\mu}{\partial s^\mu} = \dfrac{1}{s_{\text{eq}}^\mu}\left[\dfrac{3}{2}\underline{I} - \dfrac{1}{2}\mathbf{1} \otimes \mathbf{1} - m^\mu \otimes m^\mu\right]. \quad (2.129)$$

Since $m^\mu : m^\mu = \dfrac{3}{2}$ and $m^\mu : \dfrac{\partial m^\mu}{\partial s^\mu} = 0$, the system explicitly gives the corrections for p^μ and s^μ as

$$\begin{aligned} C_p &= \dfrac{R_p - m^\mu : R_s}{\mathcal{G}} \\ C_s &= \dfrac{2}{3}\left(m^\mu : R_s - R_p\right)m^\mu - \dfrac{R_s s_{\text{eq}}^\mu + \frac{2}{3}\mathcal{G}\Delta p^\mu (m^\mu : R_s)m^\mu}{s_{\text{eq}}^\mu + \mathcal{G}\Delta p^\mu}. \end{aligned} \quad (2.130)$$

For the solution to any system of equations, the procedure is then an implicit scheme with the advantage of an explicit one: the unknowns are updated explicitly by use of the closed-form formulae (2.130) for C_p and C_s.

Once the convergence is reached, $p_{n+1}^\mu = p^{\mu(q+1)}$ and $s_{n+1}^\mu = s^{\mu(q+1)}$ is set and the remaining variables are updated as

- Normal $\boldsymbol{m}^\mu$: $\boldsymbol{m}^\mu = \dfrac{3}{2} \dfrac{\boldsymbol{s}_{n+1}^{\mu\mathrm{D}}}{\sigma_{\mathrm{f}}^\infty}$
- Plastic strain: $\boldsymbol{\epsilon}_{n+1}^{\mu\mathrm{p}} = \boldsymbol{\epsilon}_n^{\mu\mathrm{p}} + \boldsymbol{m}^\mu \Delta p^\mu$
- Kinematic hardening: $\boldsymbol{X}_{n+1}^\mu = \dfrac{2}{3} C_{\mathrm{y}} (1 - D_n) \boldsymbol{\epsilon}_{n+1}^{\mu\mathrm{p}}$
- Effective stress: $\tilde{\boldsymbol{\sigma}}_{n+1}^\mu = \boldsymbol{s}_{n+1}^\mu + \boldsymbol{X}_{n+1}^\mu$
- Elastic strain: $\boldsymbol{\epsilon}_{n+1}^{\mu\mathrm{e}} = \underline{\boldsymbol{E}}^{-1} : \tilde{\boldsymbol{\sigma}}_{n+1}^\mu$
- Damage: $D_{n+1} = D_n + \left(\dfrac{Y_{n+1}^\mu}{S}\right)^s \Delta p^\mu$ if $p > p_{\mathrm{D}}$, with

$$Y_{n+1}^\mu = \dfrac{1+\nu}{2E} \left[\langle \tilde{\boldsymbol{\sigma}}_{n+1}^\mu \rangle^+ : \langle \tilde{\boldsymbol{\sigma}}_{n+1}^\mu \rangle^+ + h \left(\dfrac{1-D_n}{1-hD_n}\right)^2 \langle \tilde{\boldsymbol{\sigma}}_{n+1}^\mu \rangle^- : \langle \tilde{\boldsymbol{\sigma}}_{n+1}^\mu \rangle^- \right]$$
$$- \dfrac{\nu}{2E} \left[\langle \mathrm{tr}\, \tilde{\boldsymbol{\sigma}}_{n+1}^\mu \rangle^2 + h \left(\dfrac{1-D_n}{1-hD_n}\right)^2 \langle -\mathrm{tr}\, \tilde{\boldsymbol{\sigma}}_{n+1}^\mu \rangle^2 \right]$$
(2.131)

- Stress tensor: $\boldsymbol{\sigma}_{n+1}^\mu = (1 - D_{n+1}) \tilde{\boldsymbol{\sigma}}_{n+1}^\mu$

2.3.3 DAMAGE 2000 Post-processor⌀

The locally-coupled analysis of damage is made as a post-processing computation of an initial elastic or elasto-(visco-)plastic analysis. For simple geometries and/or loading types (such as uniaxial or proportional), the most loaded point is simply determined as the point where the damage equivalent stress $\sigma^\star = \sigma_{\mathrm{eq}} R_\nu^{1/2}$ is maximum (see also Sect. 2.1.3). It is sufficient to perform the post-processing at this single point to calculate the failure condition by the mesocrack initiation condition $D = D_{\mathrm{c}}$.

For complex geometries and/or loading types (such as non-proportional), the most loaded point may vary from one time increment to another or from one cycle (or block of cycles) to another. The calculation of the damage accumulation at the possibly weak points of the structure must be handled properly. This includes post-processing at all structure Gauss points and drawing damage maps. The result of the locally-coupled analysis will be the numerical prediction of a mesocrack initiation at the point where $D = D_{\mathrm{c}}$ is reached for the corresponding time t_{R} or number of cycles N_{R} to supture.

The computer code DAMAGE 2000 allows us to solve the two-scale model constitutive equations on the basis of the numerical scheme of Sect. 2.3.2. The input data are the material parameters and the history of the strain components ϵ_{ij} and $\epsilon_{ij}^{\mathrm{p}}$. The outputs are the evolutions of the damage D, the accumulated plastic strain p^μ, and the microscale stress components σ_{ij}^μ up to crack initiation. DAMAGE 2000 can handle any kind of proportional (or non-proportional), monotonic or fatigue loading, fatigue periodic by block, or random fatigue. For the fatigue loadings, the jump-in-cycles procedure of Sect. 2.1.5 may be activated in order to save much computer time. Examples are given in Sects. 6.4.1 and 6.4.2.

To finish, note that:

- The case $\beta = 0$ and $\sigma_f^\infty = \sigma_y$ corresponds to the single-scale damage post-processing with the uncoupled analysis of Sect. 2.1.
- The case $\beta = 0$ corresponds to the use of the Lin–Taylor localization law (instead of the Elshelby–Kröner law). Coupled with the consideration of $C_y = 0$, $m = 1$, and $s = 1$ it corresponds to the initial two-scale damage model described in the book *A course on Damage Mechanics* by J. Lemaitre, in which the FORTRAN 77 listing of the DAMAGE 90 post-processor is given.
- The effect of free edges and free surfaces on mesocrack initiation in fatigue may be represented by considering an adequate localization law (see Sect. 6.4.6) programmed as a possible option of DAMAGE 2000.

2.4 Precise Identification of Material Parameters

Contrary to linear elasticity for which the Young's modulus and Poisson ratio have almost the same value for each class of materials (steel, aluminum alloys, concrete...), nonlinear behaviors introduce parameters whose values strongly depend on the material. Due to the diversity of the engineering materials and the strong influence of the microstructure on the failure properties, the material parameters need to be identified directly from specific experiments despite the existence of handbooks of parameters. The table in Sect. 2.6 gives only the order of magnitude of parameters for some materials.

This section deals with precise identification procedures. They require computer tools which at least allow for the calculation of the material response in uniaxial tension-compression under isothermal conditions. They are

- **Mathematical softwares** that perform the time integration of the constitutive laws.
- **FE computer codes** that may include numerical optimization procedures.
- Specific identification by **optimization softwares** such as SIDOLO (P. Pilvin 1983).

Three levels of tests are generally used:

- **Qualitative tests** to choose the proper state variables to measure, to determine how they act, and to determine the couplings between the phenomena represented.

2.4 Precise Identification of Material Parameters

- **Quantitative tests** to identify the numerical values of the material parameters for each material at each temperature considered.
- **Tests of validation** to check the ability of the constitutive equations to represent more complex situations than those used for the parameters identification.

This section completes the **fast procedures** of Sect. 1.4.4 used to obtain an approximate initial set of parameters to be improved. If anisothermal constitutive equations are considered, the identification procedure consists of identifying the material parameters at each temperature and considering by interpolation all the parameters as functions of the temperature.

2.4.1 Formulation of an Identification Problem♂♂

Identifying the parameters of a model consists of finding the set of material parameters which gives the best representation of a maximum volume of experimental data and information on a given material. The data comes from

- Simple laboratory tests as monotonic (with repeated unloadings) or cyclic tension and compression experiments, monotonic or cyclic shear experiments in torsion on tubes, as well as creep and relaxation experiments performed after either a monotonic loading or a cyclic loading.
- More elaborate laboratory tests in the bending of beams, in bi-axial loading on cross-shaped specimens or in tri-axial loading on pseudo-cubic specimens. As the damage is not uniform over the whole geometry, such experiments need structure computations to determine the history of stress and strain on the RVE where the mesocrack is initiated.
- Any failure of a component, provided the structure calculation is available. An inverse method is not advised at this level because the damage is always localized and has an effect that is too small on control global variables. It may be possible only if local and very accurate measurements are performed.
- Any qualitative or quantitative information from the "state of the art" of experienced engineers and technicians!

Note that as damage governs the mesocrack initiation, it is essential to run all the laboratory tests up to the point of failure. Never interrupt a test which runs too long. Increase the applied load or the applied strain range instead in order to break the specimen more quickly. The corresponding data may then be used to validate the identification.

The database is then a finite set of load-displacement or stress-strain charts, or it could be a set of load, displacement, stress or strain versus time charts. The first thing to do is to extract as much information as possible from each test and to determine the history of all the state variables (plastic strain, accumulated plastic strain, hardenings, damage ...) or at least some components of the internal variables. This operation defines for each experiment an array, $\mathcal{Z}^{\text{exp}}(t)$, of the collected history at discrete times, t_i, of the variables.

The identification problem consists of determining the material parameters denoted formally as $\mathcal{A} = \{E, \nu, \sigma_y \ldots\}$ which minimize the difference between the observations $\mathcal{Z}^{\text{exp}}(t)$ and the numerical simulation of the model that is denoted as $\mathcal{Z}(t, \mathcal{A})$. From a classical mathematical point of view, a functional $\mathcal{L}(\mathcal{A})$ is introduced to measure the accumulated error between observations and simulations. The term $\mathcal{L}$ is the sum of functionals $\mathcal{L}_k$ for each test k such that

$$\mathcal{L} = \sum_k \mathcal{L}_k \quad \text{and} \quad \mathcal{L}_k = \frac{1}{N_k^{\text{exp}}} \sum_i \|\mathcal{Z}_k^{\text{exp}}(t_i) - \mathcal{Z}_k(t_i, \mathcal{A})\|^2, \tag{2.132}$$

where N_k^{exp} is the number of experimental points considered for the k-th experiment.

At this stage, one would expect the precise identification to end ($\mathcal{L} \approx 0$) because such a formulation may be handled automatically by use of numerical minimization procedures such as the Simplex method (not described in this book), the Newton or Quasi-Newton methods, the Levenberg–Marquardt method, or the Sequential Quadratic Programming (SQP or Projected Lagragian Method), using evolution or genetic algorithms or combinations of different minimization procedures (e.g., the identification software SIDOLO or FE computer code optimizers). In fact, the question concerning the choice of the norm on the space of thermodynamics variables is somewhat subjective. It depends on the relative accuracy the experimental results and the domains of application in which a good accuracy is necessary. Note that a ponderation (weight) matrix $\boldsymbol{P}$ needs to be defined in order to make $\|\mathcal{Z}^{\text{exp}}(t) - \mathcal{Z}(t, \mathcal{A})\|$ dimensionless (P. Pilvin 1988):

$$\|\mathcal{Z}^{\text{exp}}(t) - \mathcal{Z}(t, \mathcal{A})\|^2 = \{\mathcal{Z}^{\text{exp}}(t) - \mathcal{Z}(t, \mathcal{A})\}^{\text{T}} \boldsymbol{P} \{\mathcal{Z}^{\text{exp}}(t) - \mathcal{Z}(t, \mathcal{A})\}. \tag{2.133}$$

One may refer to Sect 2.4.6 on "Sensitivity Analysis" of the following chapters as an aid to avoid an arbitrary choice of the ponderation coefficients.

In order to avoid giving too much weight to the comparison times t_i (for non-regular time discretization, for instance), the expression

$$\mathcal{L}_k = \frac{1}{T_k^{\text{exp}}} \int_{T_k^{\text{exp}}} \|\mathcal{Z}_k^{\text{exp}}(t) - \mathcal{Z}_k(t, \mathcal{A})\|^2 dt \tag{2.134}$$

may be used instead of (2.132) for some experiments or some part of experiments (with T_k^{exp} as the observation time of the k-th experiment). The integral is numerically calculated, for example by use of trapezes method.

An example of precise identification is given in Sect. 2.4.3.

Some numerical minimization procedures for least squares nonlinear problems are described in next section. Due to the strong nonlinearity of the identification problem, local minima may be obtained and as in general, the solution of the optimization problem is not unique. It is of first importance to have a starting solution $\mathcal{A}^0$ that is "not too far" from "the" solution. When the convergence fails or becomes slow, it is important to be able to switch automatically from one method to another.

2.4.2 Minimization Algorithm for Least Squares Problems♂♂♂

The optimum set of material parameters $\mathcal{A}^{\text{opt}}$ minimizes the error $\mathcal{L}(\mathcal{A})$ that is mathematically called the objective function. The term $\mathcal{L}$ may generally be written as

$$\mathcal{L}(\mathcal{A}) = \frac{1}{2}\sum_j r_j^2(\mathcal{A}) = \frac{1}{2}\|\vec{r}(\mathcal{A})\|^2, \qquad (2.135)$$

where the residuals r_j or their vector form $\vec{r}$ are defined according to (2.132) and (2.134). This is a least squares problem that is strongly nonlinear for the nonlinear constitutive equations considered and is non-convex that is with possibly several local minima.

The mathematical tools to handle such a problem exist and are well described in the applied mathematics literature. For further details, readers can refer, for instance, to the book *Numerical optimization* by J. Nocedal and S.J. Wright (1999).

There are two classes of procedures:

- The unconstrained optimization procedures with no conditions (constraints) on the material parameters. From this class the Newton method, Gauss–Newton method, BFGS method, and Levenberg–Marquardt method are described below.
- The constrained optimization procedures with conditions (constraints) on the material parameters to force them to remain bound by some "reasonable" values: see for instance the SQP method described below.

For the identification problems, it is often efficient to disregard the constraints and assume that they have no effect on the optimal solution, or to replace them by additional penalization terms in the objective function (use logarithmic penalization for inequality constraints). These terms have the effect of discouraging constraints violation. One can then use the unconstrained optimization procedures that are simpler to implement.

For all the methods (except for the genetic algorithms), the gradient of the objective function is needed. For the least squares problem (2.135), we need

$$\nabla\mathcal{L}(\mathcal{A}) = J^{\text{T}}(\mathcal{A})\vec{r}(\mathcal{A}); \qquad (2.136)$$

for some of them the Hessian is needed:

$$\nabla^2\mathcal{L}(\mathcal{A}) = J(\mathcal{A})J^{\text{T}}(\mathcal{A}) + \sum_j r_j(\mathcal{A})\nabla^2 r_j, \qquad (2.137)$$

where the Jacobian of $\vec{r}$, $J(\mathcal{A}) = \left[\dfrac{\partial r_j}{\partial \mathcal{A}_i}\right]$, is introduced. Note that knowing the Jacobian allows us to compute the first part of the Hessian for free. Near convergence, the second term in (2.137) may be neglected and the approximation

$$\nabla^2 \mathcal{L}(\mathcal{A}) \approx J(\mathcal{A}) J^{\mathrm{T}}(\mathcal{A}) \tag{2.138}$$

may be used.

2.4.2.1 Generalities on Iterative Minimization Procedures

The numerical minimization procedures for nonlinear problems are iterative. The solution at the iteration k is calculated as

$$\mathcal{A}_{k+1} = \mathcal{A}_k + \rho_k w_k, \tag{2.139}$$

where w_k is the search direction and ρ_k the step length. The expressions for the search direction depend on the method and w_k is usually calculated first. The step length is initially taken to be equal to unity but one may optimize its value in order to get a substantial reduction, $\mathcal{L}(\mathcal{A}_k) - \mathcal{L}(\mathcal{A}_{k+1})$, of the objective function. The idea is to minimize at low cost $\mathcal{L}(\mathcal{A}_k + \rho w_k)$ with respect to ρ and an iterative process may again take place.

An efficient possibility for convergence reasons is to choose ρ_k which satisfy the Armijo–Wolfe condition

$$\begin{cases} \mathcal{L}(\mathcal{A}_k + \rho_k w_k) \leq \mathcal{L}(\mathcal{A}_k) + c_1 \rho_k \nabla \mathcal{L}_k^{\mathrm{T}} w_k, \\ \nabla \mathcal{L}^{\mathrm{T}}(\mathcal{A}_k + \rho_k w_k) w_k \geq c_2 \nabla \mathcal{L}_k^{\mathrm{T}} w_k, \end{cases} \tag{2.140}$$

with $0 < c_1 < c_2 < 1$ as constants of the method.

For the sake of simplicity the expression, $\nabla \mathcal{L}(\mathcal{A}_k) = \nabla \mathcal{L}_k$, has been set up and we shall also denote $\vec{r}_k = \vec{r}(\mathcal{A}_k)$ and $\nabla^2 \mathcal{L}(\mathcal{A}_k) = \nabla^2 \mathcal{L}_k$. With such notations, equations (2.136) and (2.138) stand as

$$\nabla \mathcal{L}_k = J_k^{\mathrm{T}} \vec{r}_k \quad \text{and} \quad \nabla^2 \mathcal{L}_k \approx J_k J_k^{\mathrm{T}}. \tag{2.141}$$

2.4.2.2 Newton Method

The search direction is the solution of

$$\nabla^2 \mathcal{L}_k w_k^{\mathrm{N}} = -J_k^{\mathrm{T}} \vec{r}_k \tag{2.142}$$

and $\rho_k^{\mathrm{N}} = 1$.

2.4.2.3 Gauss–Newton Method

The search direction is the solution of

$$J_k J_k^{\mathrm{T}} w_k^{\mathrm{GN}} = -J_k^{\mathrm{T}} \vec{r}_k \tag{2.143}$$

where the approximation (2.138) of the Hessian is used.

We then find a step length ρ_k^{GN} which satisfies the Armijo–Wolfe condition (try $\rho_k^{\mathrm{GN}} = 1$ first).

2.4.2.4 Broyden–Fletcher–Golfarb–Shanno or BFGS Method

The search direction is
$$w_k^{\text{BFGS}} = -\mathcal{H}_k J_k^{\text{T}} \vec{r}_k, \tag{2.144}$$

with $\mathcal{H}_k$ the approximation of the inverse of the Hessian calculated iteratively,

$$\mathcal{H}_{k+1} = \left(I - \frac{s_k y_k^{\text{T}}}{y_k^{\text{T}} s_k}\right) \mathcal{H}_k \left(I - \frac{y_k s_k^{\text{T}}}{y_k^{\text{T}} s_k}\right) + \frac{s_k s_k^{\text{T}}}{y_k^{\text{T}} s_k}, \tag{2.145}$$

where $s_k = \mathcal{A}_{k+1} - \mathcal{A}_k$, $y_k = \nabla \mathcal{L}_{k+1} - \nabla \mathcal{L}_k$, and I is the identity matrix. Set the initial $\mathcal{H}_0$ to be equal or proportional to the identity matrix or set it to the inverse of an approximate Hessian $\nabla^2 \mathcal{L}^{-1}$ at $\mathcal{A} = \mathcal{A}^0$ (calculated, for example, by finite differences).

The step length ρ_k^{BFGS} is chosen to satisfy the Armijo–Wolfe condition.

2.4.2.5 Levenberg–Marquardt Method

The Levenberg–Marquardt method is a robust, first order method especially meant for least squares optimization.

The search direction is the solution of

$$\left(J_k J_k^{\text{T}} + \lambda_k I\right) w_k^{\text{LM}} = -J_k^{\text{T}} \vec{r}_k, \tag{2.146}$$

which corrects the possible rank deficiency of the Jacobian. The scalar λ_k may be interpreted as a Lagrange multiplier.

Take $\rho_k^{\text{LM}} = 1$. As long as $\mathcal{L}(\mathcal{A}_k + w_k) \geq \mathcal{L}(\mathcal{A}_k)$, multiply the previous try for λ_k by a factor of 10 ($\lambda_k \leftarrow 10\lambda_k$). When $\mathcal{L}(\mathcal{A}_k + w_k) < \mathcal{L}(\mathcal{A}_k)$, take $\lambda_{k+1} = \lambda_k/10$.

2.4.2.6 Sequential Quadratic Programming (SQP)

The SQP method explicitly handles equality and inequality constraints.

2.4.2.6.1 Equality Constraints Minimization

To set up the SQP framework, consider first the problem:

$$\min \mathcal{L}(\mathcal{A}) \quad \text{subject to} \quad g_i(\mathcal{A}) = 0 \quad \text{where} \quad i = 1, \ldots m, \tag{2.147}$$

which allows us to define the Lagrangian as $\overline{\mathcal{L}}(\mathcal{A}, \lambda) = \mathcal{L} - \lambda^{\text{T}} g$.

The minimization problem at the k-th iteration is written as the minimization of a quadratic problem under linear constraints:

$$\min_w \frac{1}{2} w^{\text{T}} W_k w + \nabla \mathcal{L}_k^{\text{T}} w \quad \text{subject to} \quad A_k w + g_k = 0, \tag{2.148}$$

where $W_k = \nabla^2_{\mathcal{A}\mathcal{A}}\overline{\mathcal{L}}(\mathcal{A}_k,\lambda_k)$, $g_k = g(\mathcal{A}_k)$, and $\nabla g_i = \nabla g_i(\mathcal{A}_k)$, $A_k = [\nabla g_1,\ldots,\nabla g_m]^{\mathrm{T}}$. The term λ_k is the Lagrange multiplier vector of the k-th iteration.

The solution set $\mathcal{A}_{k+1} = \mathcal{A}_k + w_k$ and $\lambda_{k+1} = \lambda_k + w_\lambda$ solves the Karush–Kuhn–Tucker (KKT) system,

$$\begin{bmatrix} W_k & -A_k^{\mathrm{T}} \\ A_k & 0 \end{bmatrix} \begin{bmatrix} w_k \\ w_\lambda \end{bmatrix} = \begin{bmatrix} -\nabla \mathcal{L}_k + A_k^{\mathrm{T}} \lambda_k \\ -g_k \end{bmatrix}, \qquad (2.149)$$

which may also be written as

$$\begin{bmatrix} W_k & -A_k^{\mathrm{T}} \\ A_k & 0 \end{bmatrix} \begin{bmatrix} w_k \\ \lambda_{k+1} \end{bmatrix} = \begin{bmatrix} -\nabla \mathcal{L}_k \\ -g_k \end{bmatrix}. \qquad (2.150)$$

2.4.2.6.2 Inequality Constraints Minimization

The nonlinear minimization problem,

$$\min \mathcal{L}(\mathcal{A}) \quad \text{subject to} \quad g_i(\mathcal{A}) \geq 0 \quad \text{where} \quad i = 1,\ldots m, \qquad (2.151)$$

defines the linearized problem at the k-th iteration,

$$\min_w \frac{1}{2} w^{\mathrm{T}} W_k w + \nabla \mathcal{L}_k^{\mathrm{T}} w \quad \text{subject to} \quad A_k w + g_k \geq 0, \qquad (2.152)$$

which is solved iteratively. The method defines then the auxiliary equality constraints sub-problem:

$$\min_{x_w} \frac{1}{2} x_w^{\mathrm{T}} W_k x_w + [W_k w^{(r)} + \nabla \mathcal{L}_k]^{\mathrm{T}} x_w \quad \text{subject to} \quad A_k^{(r)} x_w = 0, \quad (2.153)$$

where $x_w = w - w^{(r)}$ is set.

The SQP method proceeds as follows:
1. Set $\mathcal{C}_0$ as the subset of the active constraints at $\mathcal{A} = \mathcal{A}^{(0)} = \mathcal{A}_k$ (if g_i is active $g_i(\mathcal{A}) = 0$).
2. In order to find x_w, solve the auxiliary KKT problem associated with (2.153) with the equality constraints $\nabla g_i^{\mathrm{T}} x_w = 0$ ($i \in \mathcal{C}_r$) rewritten as $A_k^{(r)} x_w = 0$:

$$\begin{bmatrix} W_k & -A_k^{(r)\mathrm{T}} \\ A_k^{(r)} & 0 \end{bmatrix} \begin{bmatrix} x_w \\ \lambda \end{bmatrix} = \begin{bmatrix} -W_k w^{(r)} - \nabla \mathcal{L}_k \\ -g^{(r)} \end{bmatrix}, \qquad (2.154)$$

with $g^{(r)}$ as a vector with components $g_i(\mathcal{A}_k)$, $i \in \mathcal{C}_r$.
3. a) If $x_w \neq 0$, compute the auxiliary step length as

$$\rho_r = \min\left[1, \min_{(i \notin \mathcal{C}_r, \nabla g_i^{\mathrm{T}} x_w < 0)} \left(-\frac{\nabla g_i^{\mathrm{T}} w^{(r)} + g_i}{\nabla g_i^{\mathrm{T}} x_w}\right)\right] \qquad (2.155)$$

2.4 Precise Identification of Material Parameters

and try for the new search direction

$$w^{(r+1)} = w^{(r)} + \rho_r x_w \,. \tag{2.156}$$

Take then first $\mathcal{C}_{r+1} = \mathcal{C}_r$. If there are blocking constraints (i.e., constraints i for which the minimum in (2.155) is achieved), update $\mathcal{C}_{r+1}$ by adding one of these constraints to the subset of the active constraints $\mathcal{C}_{r+1}$.

b) If $x_w = 0$, the current value $\hat{w}$ of $w^{(r)}$ minimizes the objective function over the current working set $\hat{\mathcal{C}} = \mathcal{C}_r$. Compute the Lagrange multipliers that satisfy

$$\sum_{i \in \hat{\mathcal{C}}} \nabla g_i \hat{\lambda}_i = W_k \hat{w} + \nabla \mathcal{L}_k \,. \tag{2.157}$$

If all the $\hat{\lambda}_i$ are positive, stop with the solution $w_k = \hat{w}$. If some $\hat{\lambda}_j$ are strictly negative, subtract the constraint j corresponding to the larger $|\hat{\lambda}_j|$ from the subset of the active constraints ($\hat{\mathcal{C}} = \mathcal{C}_{r+1} \leftarrow \mathcal{C}_r \setminus \{j\}$).

4. Take $\rho^{\text{SQP}} = 1$ and $\mathcal{A}_{k+1} = \mathcal{A}_k + w_k$.

Note that the calculation of W_k needs the calculation of the second derivatives of $\mathcal{L}$. The method is then similar to a Newton scheme. Again, approximate expressions for $\nabla^2 \mathcal{L}$ may be used, the most popular being the BFGS approximation of the Hessian.

2.4.3 Procedure for Numerical Identification ♂♂

Consider that the constitutive laws are available in a computer code which offers optimization facilities. The identification is then performed automatically by the minimization of the functional $\mathcal{L}(\mathcal{A})$ but some practical (convergence) difficulties may be encountered.

As a guide to optimize the full identification process, a 5-step procedure may then be followed:

1. **Choose** the plasticity, visco-plasticity and damage models, bearing in mind the experimental data available and the considered application. For example considering both isotropic and kinematic hardenings is useless for a monotonic application, considering classical plasticity models (those described in this book) gives a poor accuracy in modelling the ratcheting effect.

   And it is no secret that for low cycle fatigue applications, the best choice is to perform the model identification on low-cycle fatigue tests. In general, the identification is better as the identification tests are close to the cases of the application!

2. **Perform** the fast identification procedure of Sect. 1.4.4 to determine an initial set of material parameters.

3. **Estimate** qualitatively and quantitatively the sensitivity of the simulation to each parameter (see Sect. 2.4.6). This allows for a qualitative understanding of the model and a way to find out in which experiment or in which stage of the experiment each material parameter has a minor or major effect.
4. **Build** a chart of the experimental values of all the possible variables for discrete values of the time. The chart will compare the experimental values with the numerical simulation in order to calculate the error $\|\mathcal{Z}_k^{\text{exp}}(t_i) - \mathcal{Z}_k(t_i, \mathcal{A})\|^2$ or the integral $\int_{T_k^{\text{exp}}} \|\mathcal{Z}_k^{\text{exp}}(t) - \mathcal{Z}_k(t, \mathcal{A})\|^2 \mathrm{d}t$.
5. **Use** the set of initial parameters to perform the numerical minimization of the functional $\mathcal{L}$. Some FE codes propose this option. For better convergence, proceed step by step and try to identify first the plasticity parameters, second the viscosity parameters, and last the damage parameters. In a last step, perform a numerical identification of all the parameters at the same time.

Let us detail each point of this identification procedure on an example:

1. The whole set of operations is described here for the elasto-(visco-)plasticity coupled with the damage model of Table 1.3 for which the one-dimensional isothermal constitutive equations are recalled below for the Norton viscosity law:

$$\epsilon = \epsilon_\mathrm{e} + \epsilon_\mathrm{p}$$

$$\epsilon_\mathrm{e} = \frac{\sigma}{E(1-D)}$$

$$f = \left|\frac{\sigma}{1-D} - X\right| - R - \sigma_\mathrm{y} = \sigma_\mathrm{v}$$

$$\dot{\epsilon}_\mathrm{p} = \mathrm{sgn}(\sigma - X)\,\dot{p} = \mathrm{sgn}(\sigma - X)\,\frac{\dot{r}}{1-D}$$

$$R = R_\infty\left(1 - \exp(-br)\right)$$

$$\dot{X} = C(1-D)\dot{\epsilon}_\mathrm{p} - \gamma X \dot{r} \quad \text{with } C = \gamma X_\infty \qquad (2.158)$$

$$\dot{D} = \left(\frac{Y}{S}\right)^s \dot{p} \quad \text{if} \quad p > p_\mathrm{D},$$

mesocrack initiation if $D = D_\mathrm{c}$

$$Y = \frac{\sigma^2}{2E(1-D)^2}$$

$$\dot{p} = \left\langle\frac{\sigma_\mathrm{v}}{K_N}\right\rangle^N = \left\langle\frac{f}{K_N}\right\rangle^N$$

2. The initial set of material parameters is determined according to the fast identification procedures of Sects. 1.4.4 and 1.5.1. The precise identifi-

cation is performed for the steel alloy 2-1/4 CrMo at a temperature of 580°C and from the following initial data:
- Fatigue limit $\sigma_f^\infty = 60$ MPa
- Yield stress $\sigma_{y02} = 95$ MPa
- Ultimate stress (at $\dot{\epsilon} = 10^{-3}$ s^{-1}) $\sigma_u = 187$ MPa for an ultimate plastic strain $\epsilon_{pu} = 0.2$
- Rupture stress (at $\dot{\epsilon} = 10^{-3}$ s^{-1}) $\sigma_R = 150$ MPa for a rupture plastic strain $\epsilon_{pR} = 0.3$
- A few tension-compression tests at very low strain rate $\dot{\epsilon} = 10^{-6}$ s^{-1} give for the yield thresholds σ_s^+ in tension and σ_s^- in compression:

$$\begin{aligned}
\epsilon_p &= 0, & \sigma_s^+ &= \sigma_y = 80 \text{ MPa}, & \sigma_s^- &= -80 \text{ MPa}, \\
\epsilon_p &= 0.1\ 10^{-2}, & \sigma_s^+ &= 87 \text{ MPa}, & \sigma_s^- &= -77 \text{ MPa}, \\
\epsilon_p &= 0.4\ 10^{-2}, & \sigma_s^+ &= 99 \text{ MPa}, & \sigma_s^- &= -69 \text{ MPa}, \\
\epsilon_p &= 1.4\ 10^{-2}, & \sigma_s^+ &= 105 \text{ MPa}, & \sigma_s^- &= -65 \text{ MPa}.
\end{aligned} \quad (2.159)$$

They are used for the determination of both kinematic and isotropic hardenings as the laws $R = R_\infty[1 - \exp(-b\epsilon_p)]$ and $X = X_\infty[1 - \exp(-\gamma\epsilon_p)]$ for monotonic tension fit the discrete values

$$X = X(\epsilon_p) = \frac{\sigma_s^+ + \sigma_s^-}{2} \quad \text{and} \quad R = R(\epsilon_p) = \sigma_s^+ - \sigma_y - X. \quad (2.160)$$

- A few creep tests at different constant values of stress give the secondary plastic strain rate:

$$\begin{aligned}
\sigma &= 110 \text{ MPa}, & \dot{\epsilon}_p &= 10^{-6} \text{ s}^{-1}, \\
\sigma &= 120 \text{ MPa}, & \dot{\epsilon}_p &= 2\ 10^{-5} \text{ s}^{-1}, \\
\sigma &= 150 \text{ MPa}, & \dot{\epsilon}_p &= 2\ 10^{-4} \text{ s}^{-1}.
\end{aligned} \quad (2.161)$$

They are used to identify the Norton parameters K_N and N (with the hypothesis in of saturated hardenings) in

$$\dot{\epsilon}_p = \left(\frac{\sigma - \sigma_y - R_\infty - X_\infty}{K_N}\right)^N. \quad (2.162)$$

As the linear regression

$$\ln \dot{\epsilon}_p = N \ln(\sigma - \sigma_y - R_\infty - X_\infty) - \ln K_N^N \quad (2.163)$$

fits the experimental points in (2.161) in the $\ln \dot{\epsilon}_p$ vs $\ln(\sigma - \sigma_y - R_\infty - X_\infty)$ diagram.
- Furthermore, the thermo-elasticity parameters are known as
  - Elasticity modulus $E = 134000$ MPa
  - Poisson ratio $\nu = 0.3$

130   2 Numerical Analysis of Damage

– Dilatation coefficient $\alpha = 1.3 \cdot 10^{-5}/°C$

Finally, the fast identification leads to the following set ($\mathcal{A}^0$) of material parameters:
- Thermo-elasticity: $E = 134000 MPa$, $\nu = 0.3$, $\alpha = 1.3 \cdot 10^{-5}/°C$
- Yield stress: $\sigma_y = 80$ MPa
- Isotropic hardening: $R_\infty = 5$ MPa, $b = 300$
- Kinematic hardening: $X_\infty = 20$ MPa, $\gamma = 270$
- Damage: $S = 0.113$ MPa, $s = 5$, $D_c = 0.2$
- Damage threshold: $\epsilon_{pD} = 0.2$, $\sigma_u = 187$ MPa at $\dot\epsilon = 10^{-3}$ s^{-1}, $\sigma_f^\infty = 60$ MPa, $m = 4$
- Viscosity: $K_N = 1450$ MPa s$^{1/N}$, $N = 2.4$

3. For this didactic example we assume that all the parameters have the same sensitivity. Each particular case is examined in the following chapters.
4. The experiments used to improve the set of initial parameters are those of Fig. 2.6: two low cycle fatigue tests up to rupture, a cyclic hardening curve, and a relaxation test.

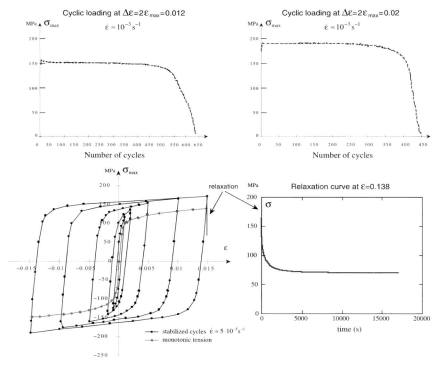

Fig. 2.6. Additional experiments for parameters identification of the 2-1/4 CrMo steel at 580°C (J.-P. Sermage 1998)

2.4 Precise Identification of Material Parameters

The charts $\mathcal{Z}_k^{\text{exp}}$ of the state variables needed to apply a numerical optimization procedure are:

- One file for each fatigue test $\sigma_{\max}(N)$ curve with the following columns:
  Col. 1: The time t_N corresponding to the maximum of the N-th cycle ($T = 2\Delta\epsilon/\dot\epsilon$ is the period of the loading). For the practical application, choose the points $N = 100, 200, 300, 400, 500, 550$, and 600 cycles for the curve at strain range $\Delta\epsilon = 1.2\ 10^{-2}$, and $N = 100, 200, 300, 350, 400$, and 425 cycles for $\Delta\epsilon = 2\ 10^{-2}$.
  Col. 2: The accumulated plastic strain at $t = t_N$ and $p \approx 2N\Delta\epsilon_{\text{p}}$. The plastic strain range $\Delta\epsilon_{\text{p}}$ is known either from the experiment or calculated as $\Delta\epsilon_{\text{p}}(t_N) = \Delta\epsilon - 2\sigma_{\max}(N)/E(1 - D(N))$.
  Col. 3: The damage $D(t_N) = D(N)$ calculated from the decrease of the maximum stress, $D(N) = 1 - \sigma_{\max}(N)/\sigma_{\max}(N = 100)$.
  The corresponding charts (or arrays of data) are used for a point to point comparison (application of eq. (2.132)).

- One file per strain range for the cyclic stress-strain curves of Fig. 2.6 at constant strain rate $\dot\epsilon = 10^{-3}$ s^{-1}. Use enough points as these files are used for the integral comparison (2.132) over each stabilized cycle, the observation time for the cycle with strain range $\Delta\epsilon_k$ where $T_k^{\text{exp}} = 2\Delta\epsilon_k/\dot\epsilon$. Place in columns:
  Col. 1: Time t_i corresponding to the point i of the stress-strain stabilized cycle,
  Col. 2: Stress $\sigma(t_i)$,
  Col. 3: Strain $\epsilon(t_i)$,
  Col. 4: Plastic strain (measured by fictitious unloadings) $\epsilon_{\text{p}}(t_i)$,
  Col. 5: Accumulated plastic strain $p(t_i) = p^A + |\epsilon_{\text{p}}(t_i) - \epsilon_{\text{p}}^A|$ during plastic unloading, $p(t_i) = p^B + \epsilon_{\text{p}}(t_i) - \epsilon_{\text{p}}^B$ during plastic loading with $p^A \approx 2N\Delta\epsilon_{\text{p}}$ and $p^B \approx (2N+1)\Delta\epsilon_{\text{p}}$ after a number of cycles N large enough to reach stabilization,
  Col. 6: Isotropic hardening $R(t_i)$,
  Col. 7: Kinematic hardening $X(t_i)$,
  Col. 8: Damage $D(t_i)$.

- One two-column file $\sigma(t_i)$ for the relaxation test of Fig. 2.6.

The last but not the least practical point concerns the choice of the ponderation matrix $\boldsymbol{P}$ which allows us to make some experiments preponderant over some others but also to make some parameters more sensible than some other. Unfortunately, many choices are possible. As proposed by P. Pilvin (1988) take the ponderation matrices for each variable as the inverse of the square of the accuracy times identity, i.e.,

- For the stresses, $\boldsymbol{P}_\sigma = \dfrac{1}{(1\ \text{MPa})^2}\boldsymbol{1}$

- For the strains or the plastic strains, $\boldsymbol{P}_\epsilon = \dfrac{1}{(10^{-5})^2}\boldsymbol{1}$

- For the accumulated plastic strain (fatigue applications),
$$P_p = \frac{1}{(10^{-4})^2}\mathbf{1}$$
- For the damage, $P_D = \dfrac{1}{(10^{-2})^2}\mathbf{1}$
- For the hardening, $P_R = P_X = \dfrac{1}{(2\text{ MPa})^2}\mathbf{1}$

5. The numerical optimization is finally performed with SIDOLO optimizer by keeping unchanged the thermo-elasticity parameters E, ν, α, and the damage threshold in pure tension ϵ_{pD}. The final set of parameters for the 2-1/4 CrMo steel at 580°C is then:
   - Thermo-elasticity: $E = 134000$ MPa, $\nu = 0.3$, $\alpha = 1.3 \cdot 10^{-5}/°C$
   - Fatigue limit: $\sigma_f^\infty = 60$ MPa
   - Yield stress: $\sigma_y = 85$ MPa instead of 80
   - Isotropic hardening: $R_\infty = 30$ MPa instead of 5, $b = 2$ instead of 300
   - Kinematic hardening: $X_\infty = 22$ MPa instead of 20, $\gamma = 250$ instead of 270
   - Damage: $S = 0.6$ MPa instead of 0.113, $s = 2$ instead of 5, $D_c = 0.2$
   - Damage threshold: $\epsilon_{pD} = 0.2$, $\sigma_u = 200$ MPa (at $\dot\epsilon = 10^{-3}$ s^{-1}) instead of 187, $\sigma_u = 137$ MPa at $\dot\epsilon \approx 0$, $m = 4$,
   - Viscosity: $K_N = 1220$ MPa s$^{1/N}$ instead of 1450, $N = 2.5$ instead of 2.4,
   - If necessary, the parameters A and w_D for the damage threshold are determined from the knowledge of the plastic strain range $\Delta\epsilon_p$ and the corresponding threshold p_D by use of (1.114)–(1.116).

The example has been derived for isotropic damage. If the anisotropic damage model is considered, apply the same procedure but change the D-columns to D_1-columns. The damage measurement on the maximum stress for cyclic loading is:

$$\sigma_{\max}(N) = \frac{3\sigma_{\max}(N = 100)}{\dfrac{2}{1-D_1} + \dfrac{1}{1-\dfrac{D_1}{2}}} \longrightarrow D_1(N). \tag{2.164}$$

2.4.4 Cross Identification of Damage Evolution Laws

For each application, different laws may model the damage evolution with either a phenomenological, thermodynamical, or micromechanical basis. One can consider:

- Damage as a function of the energy density release rate, i.e., $D = D(Y)$
- Damage as a function of the accumulated plastic strain, i.e., $D = D(p)$
- Damage as a function of the damage equivalent stress $D = D(\sigma^\star)$, in which the von Mises stress and triaxiality apply

2.4 Precise Identification of Material Parameters 133

- Damage as a function of the accumulated plastic strain and the triaxiality

Note that the previous laws $D(Y)$ and $D(\sigma^*)$ do not apply to fatigue. The law $D(p)$ may apply, but with different values of the damage parameters than for ductile failures!!

The initial difficulty for the numerical analysis of a structure is then the choice of a damage law and its material parameters. Each set of parameters is related to the damage law considered and even if the different models are available in a FE code, it may not be a small task to switch from one to an other. A solution may then be:

1. To build a **new data base** by numerical simulation with the law for which the numerical values of the parameters are known
2. To apply the identification procedures of Sects. 1.4.4 and 2.4 on this new database in order to find the values of the parameters related to the new law used.

An example of such a cross identification is given in Sect. 3.4.5 where the damage parameters ϵ_{pD}, S, s, and D_c are determined from the Gurson model parameters by a fastened procedure.

2.4.5 Validation Procedure♂

Any (good!) model must be able to represent situations other than those used for the identification of the material parameters. Furthermore, it is essential to know its domains of validity for safety reasons. This is why it is important to check the model against special tests designed for the validation. It can be:

- 2D or 3D state of stress measurements to check the criteria used
- Non-proportional tests to check the validity of the anisotropy representation
- Complex history of stress to check the representation of the damage accumulation
- Non-isothermal tests to check the temperature dependence
- Tests with gradient of stress to check the stress redistribution due to the coupling between damage and strain behaviors

As an example, the elasto-(visco-)plasticity coupled with the damage model identified on the 2-1/4 CrMo steel at 580°C in Sect. 2.4.3 has also been identified at room temperature, 300, 400, and 500°C. From the results, two series of validation have been performed:

1. One-dimensional non-isothermal cyclic tests are described in Figs. 2.7 and 2.8. The history of strain (or stress) and temperature are given for both tests. They are introduced as inputs for the calculation of the stress (or strain) and damage for comparison with the experiments. The graphs

correspond to the first cycle. The relative difference of about 25% on the strain, stress or number of cycles to rupture is on the order of magnitude expected in this kind of prediction.

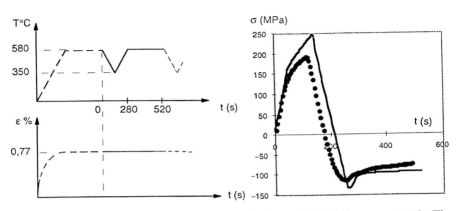

Fig. 2.7. Thermal stress cyclic test on 2-1/4 Cr steel (J.-P. Sermage 1998). The solid line represents the calculation and the dots represent experimental results

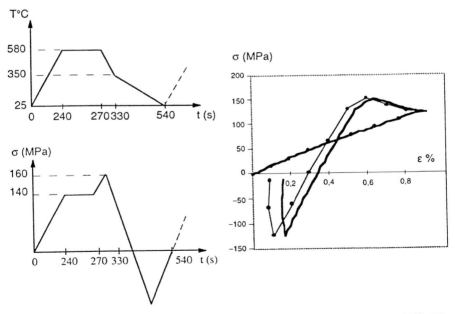

Fig. 2.8. Thermomechanical cyclic test on 2-1/4 Cr steel (J.-P. Sermage 1998). The solid line represents the calculation and the dots represent experimental results

2. Bidimensional non-isothermal cyclic tests up to crack initiation on Maltese cross-shaped specimens for which the numerical simulation are described in detail in Sect. 5.4.3 which is devoted to creep-fatigue damage. The difference between the number of cycles predicted and obtained by experiments is on the order of 25% to 100% depending on the method of calculation used.

The final set of temperature-dependent material parameters is available in the Table 5.1 in Sect. 5.4.3.

2.4.6 Sensitivity Analysis$^{\circ\circ}$

The accuracy of a structure failure prediction depends very much on the accuracy of the material parameters, but not all of the parameters have the same effect on the model predictions concerning the strain-stress response and the rupture of a Representative Volume Element.

The sensitivity of the calculated response to slight or main changes in each variable and material parameter value is of first importance at different stages of the mechanical analysis:

- **At the stage of parameters identification**, one has to minimize the error ($\mathcal{L}(\mathcal{A})$) between the numerical response of the model and the experimental database with respect to the set of material parameters ($\mathcal{A}$). The sensitivity of the functional $\mathcal{L}$ with respect to the material parameters is in fact the gradient ($\frac{\partial \mathcal{L}}{\partial \mathcal{A}}$) or its numerical approximation ($\frac{\Delta \mathcal{L}}{\Delta \mathcal{A}}$). In general, it is necessary to efficiently apply the minimization procedures leading to the best set of parameters. The expression for $\frac{\partial \mathcal{L}}{\partial \mathcal{A}}$ is quite complex and is numerically and automatically derived within the identification procedure. It is however a mathematical definition which has no clear mechanical meaning.

It is difficult to properly handle the contribution to the functional $\mathcal{L}$ due to damage. Following the work of G. Amar and J. Dufailly (1993), it is thus preferable to define the sensitivity to the material parameters as the relative sensitivity of the engineering variable considered for each specific application to the relative changes of each material parameter $\mathcal{A} = [\mathcal{A}_1, \mathcal{A}_2, \ldots \mathcal{A}_n]$ such as the plastic strain at failure p_R, the stress to failure σ_R, the number of cycles to failure N_R, or more generally the time-to-failure t_R. A dimensionless sensitivity matrix may then be introduced, such as

$$\begin{bmatrix} \frac{\delta p_R}{p_R} \\ \frac{\delta \sigma_R}{\sigma_R} \\ \frac{\delta N_R}{N_R} \\ \frac{\delta t_R}{t_R} \end{bmatrix} = \begin{bmatrix} S_{\mathcal{A}1}^{p_R} & S_{\mathcal{A}2}^{p_R} & \ldots & S_{\mathcal{A}n}^{p_R} \\ S_{\mathcal{A}1}^{\sigma_R} & S_{\mathcal{A}2}^{\sigma_R} & \ldots & S_{\mathcal{A}n}^{\sigma_R} \\ S_{\mathcal{A}1}^{N_R} & S_{\mathcal{A}2}^{N_R} & \ldots & S_{\mathcal{A}n}^{N_R} \\ S_{\mathcal{A}1}^{t_R} & S_{\mathcal{A}2}^{t_R} & \ldots & S_{\mathcal{A}n}^{t_R} \end{bmatrix} \cdot \begin{bmatrix} \frac{\delta \mathcal{A}_1}{\mathcal{A}_1} \\ \frac{\delta \mathcal{A}_2}{\mathcal{A}_2} \\ \ldots \\ \frac{\delta \mathcal{A}_n}{\mathcal{A}_n} \end{bmatrix} \quad (2.165)$$

Each line of the sensitivity matrix corresponds to a given case: monotonic loading for $S_{\mathcal{A}k}^{p_R}$ or $S_{\mathcal{A}k}^{\sigma_R}$ terms, fatigue loading for $S_{\mathcal{A}k}^{N_R}$ terms, more complex loading for $S_{\mathcal{A}k}^{t_R}$ terms. A few other lines may be added for experiments performed at different strain rates or at different temperatures, for example.

- **At the early stage of design of mechanical components**, the choice of the materials may not be definitive. A sensitivity analysis made on the failure properties may help for such a choice. Performed on the FE analysis of the components with real loading conditions, it will tell if it is better to
    - Choose a material with better plasticity properties through the hardening parameters σ_y, R_∞, b, X_∞, and γ
    - Choose a material with better creep properties through the viscosity parameters K_N, N.
- **At the stage of component manufacturing**, the previous sensitivity matrix may tell if a change of material supplier will have minor or major consequences for the component design and if further complementary studies are necessary to ensure the safety conditions.

Quantitative results of this method of sensitivity analysis are given for specific applications in the following chapters (see Sects. 3.2.3, 4.2.3, 5.2.3, 6.2.4, 7.2.5).

Note finally that if the expressions of the material parameters are known as (smooth) functions of the temperature ($\mathcal{A}_k = \mathcal{A}_k(T)$) or the microstructure through a few parameters $\mu_1, \mu_2 \ldots$, the knowledge of the sensitivity matrix allows us to determine the sensitivity of the failure properties with respect to a change in temperature as

$$\begin{cases} \dfrac{\partial p_R}{\partial T} = \sum_k S_{\mathcal{A}k}^{p_R} \dfrac{p_R}{\mathcal{A}_k} \dfrac{\mathrm{d}\mathcal{A}_k}{\mathrm{d}T} , \\ \dfrac{\partial \sigma_R}{\partial T} = \sum_k S_{\mathcal{A}k}^{\sigma_R} \dfrac{\sigma_R}{\mathcal{A}_k} \dfrac{\mathrm{d}\mathcal{A}_k}{\mathrm{d}T} , \\ \dfrac{\partial N_R}{\partial T} = \sum_k S_{\mathcal{A}k}^{N_R} \dfrac{N_R}{\mathcal{A}_k} \dfrac{\mathrm{d}\mathcal{A}_k}{\mathrm{d}T} , \\ \dfrac{\partial t_R}{\partial T} = \sum_k S_{\mathcal{A}k}^{t_R} \dfrac{t_R}{\mathcal{A}_k} \dfrac{\mathrm{d}\mathcal{A}_k}{\mathrm{d}T} , \end{cases} \qquad (2.166)$$

or with respect to a change in microstructure as

$$\begin{cases} \dfrac{\partial p_R}{\partial \mu_i} = \sum_k S_{\mathcal{A}k}^{p_R} \dfrac{p_R}{\mathcal{A}_k} \dfrac{\mathrm{d}\mathcal{A}_k}{\mathrm{d}\mu_i} , \\ \dfrac{\partial \sigma_R}{\partial \mu_i} = \sum_k S_{\mathcal{A}k}^{\sigma_R} \dfrac{\sigma_R}{\mathcal{A}_k} \dfrac{\mathrm{d}\mathcal{A}_k}{\mathrm{d}\mu_i} , \\ \dfrac{\partial N_R}{\partial \mu_i} = \sum_k S_{\mathcal{A}k}^{N_R} \dfrac{N_R}{\mathcal{A}_k} \dfrac{\mathrm{d}\mathcal{A}_k}{\mathrm{d}\mu_i} , \\ \dfrac{\partial t_R}{\partial \mu_i} = \sum_k S_{\mathcal{A}k}^{t_R} \dfrac{t_R}{\mathcal{A}_k} \dfrac{\mathrm{d}\mathcal{A}_k}{\mathrm{d}\mu_i} . \end{cases} \qquad (2.167)$$

In anticipation of the sensitivity analyses made in the following chapters for the accumulated plastic strain, stress, time, or number of cycles to rupture, we can make the following general conclusions:

- The most important parameter is the loading for which a relative error on the stress $\frac{\delta\sigma}{\sigma}$ influences the relative error on the results by a factor $S^{...}_\sigma$ of 5 to 10. If the loading is expressed in terms of **strain**, the factor $S^{...}_\epsilon$ may be reduced to 1.
- The material parameters which do not need a high degree of accuracy are the damage threshold in tension ϵ_{pD} and the critical damage D_c because their sensitivity factor $S^{...}_{\epsilon_{pD}}$ and $S^{...}_{D_c}$ are only of the order of 0.5.
- Due to the nonlinearities, all the sensitivity factors for the other parameters are larger than 1 and of the order of 1 to 5.
- Finally, if all the parameters are known with a relative accuracy of a few percent one may expect or fear (who knows?) an accuracy on the order of:
    - $\pm 20\%$ on the accumulated plastic strain to rupture in ductile failures
    - 0.5 to 2× on the number of cycles to rupture in low cycle fatigue failures
    - 0.5 to 2× on the time-to-rupture in creep failures
    - 0.2 to 5× on the number of cycles to rupture in creep-fatigue failures
    - 0.1 to 10× on the number of cycles to rupture in high cycle fatigue failures
    - $\pm 5\%$ on the stress to rupture in brittle failures

Large numbers indeed! This explains why the job of designers or engineers is still, at least partially, an art!

2.5 Hierarchic Approach and Model Updating$^{\sigma,\sigma\sigma\sigma}$

"It is always easy to make things complicated," tells a chinese proverb! This is particularly true when modelling materials and structures with computers which may give a multitude of useless output data.

To understand well what can happen on a structure in service, we strongly advise beginning predictions with simple models and increasing the accuracy with more sophisticated models progressively. This is the hierarchic approach (D. Marquis 1990). It is developed at the end of each applications chapter but some general conclusions are:

- $\sigma^* = \sigma_u$ is a good rupture criterion to begin with
- Isotropic hardening is sufficient for monotonic proportional loading
- Kinematic hardening is necessary to represent cyclic loading
- Linear kinematic hardening is often sufficient for damage purposes
- The energy density release rate $Y = constant$ in the damage law is a good approximation when the strain hardenings are large
- The damage threshold $p_D = constant$ is a good approximation in ductile failures

- But it is necessary to take into account the variation of p_D with the stress in low or high cycle fatigue
- An elastic calculation followed by the Neuber correction is a good approximation in small scale yielding

Updating a model consists of changing the model or the value of the material parameters each time new information is available. It can be

- More information concerning the material as the choices in early designs progress
- More accurate values of the damage parameters as a test program progress
- New phenomena which may take place during validation tests on samples or structures
- New in situ tests results on the structure itself to check if the real structure is made of exactly the same material as the one used for the laboratory identification tests
- Aging in service
- Injuries due to overloading or accident in service which may affect the state of strain hardening and damage governing the residual strength

In order to be prepared for model updating of a structure in service, two important points to remember are:

- Keep the initial calculations and the possibilities to change the inputs in proper archives. Unfortunately, it is not so simple over several decades with the change of computers and persons every few years!
- The possibility to practice in situ tests: the microhardness test and the digital image correlation are good candidates!

2.6 Table of Material Damage Parameters

To end this chapter, let us give sets of the numerical values of damage parameters for some materials. You should take the numbers only as an order of magnitude because the name of a material does not fully identify all the properties. Be careful to always check the values using appropriate tests or information in each particular case. Fortunately, you are clever and you know that most often you only need qualitative results or relative values to compare different design solutions for which approximate results are sufficient.

2.6 Table of Material Damage Parameters

Material	T °C	E GPa	ν	σ_y MPa	σ_u MPa	σ_f^∞ MPa	C_y MPa	S MPa	s	D_c	ϵ_{pD}	m	h	η
SOLDUR 355 steel	20	230	0.3	375	474		3000	0.57 / 0.43	4	0.5	0.025	2.5	h_a 1 / h_a 0	2.8
Weldox 460 E steel	20	200	0.33	490					1	0.3	0			-
Steel 2-1/4 CrMo	20 / 580	200 / 134	0.3 / 0.3	180 / 85	450 / 137	140 / 60	6000	2.8 / 0.6	2 / 2	0.2 / 0.2	0.12 / 0.2	2 / 4	0.2 / 0.2	- / -
AISI 316	20 / 600	200 / 140	0.3 / 0.32	260 / 6	700	180 / 3		7 / 0.2	1 / 1	0.15 / 0.5	0.1 / 0		0.2 / 0.5	- / -
AISI 1010	20	190	0.28	320	700	200		2.4	1	0.2	0.44		0.2	-
Refr. alloy IN 100	827	170	0.3	30				0.2	1	0.3	0.005			-
Al alloy 2024	20	72	0.32	300	500	250		1.7	1	0.23	0.03		0.2	-
Al alloy 6014 T7	20	70	0.33	240.6	275			1.22	2	0.36	0.05			-
Copper	20	100	0.33	190	300	100		0.4	1	0.85	0.35		0.6	-
Concrete	20	42	0.2	σ_y^+ 2.1 / σ_y^- 7.4	σ_u^+ 4 / σ_u^- 38			A $5\,10^3$	a $2.93\,10^{-4}$	0.99	κ_0 $5\,10^{-5}$			3
Al ceramic	20	400	0.2	306	500	300		6	1	0.99	0.1		0.2	-
Carbon-epoxy ply composite	20	E_1 311 / E_2 6.35 / E_3 20	ν_{12} 0.35 / ν_{13} 0.35 / ν_{23} 0.48	1760 / τ_y 17	1760			MPa$^{1/2}$ S_T 1.96 S_S 2.55	0.5	0.99	Y_{DT} 0 Y_{DS} 0			-
SB rubber	20	$2\,10^{-3}$	0.5	1.16				5.6	5	0.2	0			-

3

Ductile Failures

With this chapter, we begin our discussion of the applications of damage mechanics to specific structure failures classified by a phenomenon: here we are focusing on ductile failures which may occur in structures due to **overloading** and also during **metal forming**. Ductile damage without any rupture may also be produced by forming or mechanical processes and may modify the mechanical properties which must be taken into account in the constitutive equations used to evaluate the further strength of structures in service.

From the physical point of view, ductile damage is essentially atomic decohesions following dislocations piling in metals or growth and coalescence of **cavities** induced by large deformations in the vicinity of inclusions in both metals and polymers. From the micromechanical point of view, this is the growth of a spheric or elliptic hole in a plastic medium subjected to large strains and the problem can be solved analytically (F.A. Mac Clintoch 1968, J.R. Rice and D.M. Tracey 1969) or numerically (V. Tvergaard and A. Needleman 1982). From the Continuum Damage Mechanics point of view, this is a reduction of the resisting area in any plane of a Representative Volume Element that is governed by the **elastic energy** and the **accumulated plastic strain**, as explained in Chap. 1.

All of the following chapters on applications have the same format: some engineering considerations; description of "**fast methods**" which can be used in early design where the hypothesis of proportional loading and small scale yielding apply, with indications on the sensitivity analysis and possible safety margin; **closed-form solutions** of simple problems currently found in engineering such as structures with holes or notches, pressurized cylinders, and forming limits and post-buckling behavior; **specific cases treated by numerical analysis**, such as finite strains, **porous materials**, forming processes such as **deep drawing** and **extrusion** and **frames analysis** using Lumped Damage Mechanics. All applications are considered on materials at room temperature and low strain rates (the domain of application of plasticity). A ductile damage model for elastomers which also applies in fatigue is described in Chap. 4. Creep effects and also "dynamic plasticity" are relevant to visco-plasticity and are treated in Chap. 5. Problems of elastomers or polymers involving strain rate effects are approached at the end of Chap. 5.

3.1 Engineering Considerations

Ductile failures are most often related to large deformations, plastic strains in **metals**, and irreversible strains in **polymers**. These conditions can arise locally in the vicinity of notches or severe bendings. They can also occur on a large scale during forming processes but they are always associated with instabilities and the phenomenon of localization of strains and damage. Fortunately these phenomena occur at the latest stage of failures, allowing us to only look into their initiation.

The "in service" security of pressure vessels or mechanical members with holes or notches needs to be checked against ductile failures in case of overloads. The case of cyclic loadings is treated in Chap. 4 for low cycle fatigue and in Chap. 5 for elevated temperatures.

Another area of applications is analyzing the forming processes in order to choose the proper machine, to optimize the processes for economical energy consumption, avoid trouble such as necking and cracks, and determine the modifications of the constitutive equations after forming. **Be careful** when reading values of material parameters in a metallurgy catalog because they characterize materials in the state of the products sold (e.g., sheets) but not the products after metal forming (e.g., deep drawing). The sequences of machining like stamping may take advantage of non-proportional loadings to increase the forming limits.

In post-buckling analysis of crash problems, the energy involved is much dependent on the damage which reduces the strength and the energy absorbed. The dynamic case, which needs to take into account the strain rate effect, is described in Sects. 5.3.1 and 5.4.4.

3.2 Fast Calculation of Structural Failures

Fast calculation means obtaining the desired result using closed-form solutions or mathematical softwares on personal computers which do not need large finite element computations on big computers. Often, taking time for a clever qualitative analysis of the problem allows you to consider a simplified geometry of the structure together with a simplified loading, which then allows for a "fast calculation." For such cases, we advise using isotropic damage without the quasi-unilateral condition of microdefects closure.

3.2.1 Uniaxial Behavior and Validation of the Damage Law

The basic uniaxial test is the "simple" tension test which is not so simple in the range of large deformation where necking occurs. Figure 3.1 is an example of the same test represented in the engineering stress-strain reference of small strain theory and in the local "true" stress-"true" strain reference ($\sigma_{tr} = \sigma(1+\epsilon)$, $\epsilon_{tr} = \ln(1+\epsilon)$) of large strain theory. The difference in stress may reach 50% for strains around 0.5.

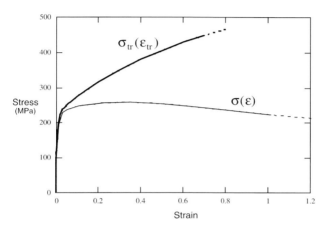

Fig. 3.1. Engineering and true stress-strain curves of Cu/Al at room temperature (J. Dufailly 1995)

Ductile properties of the material must take into consideration this big difference. The easiest way is to use the simple tension test only up to the strain corresponding to the ultimate stress ($\sigma = \sigma_u$) and to explore the damage during a low cycle fatigue test for which no necking appears. The ultimate stress (σ_u) is the true stress (F_u/S where S is the necking section) if a large deformation theory is used; it is the engineering stress otherwise (F_u/S_0 where S_0 is the initial section of the specimen). The method of identification of material parameters advised is the one discussed in Sect. 1.4.4 for a fast identification, improved with a numerical optimization if other tests are available as described in Sect. 2.4. Due to large deformations which correspond to saturated strain hardening at the ultimate stress σ_u, the plasticity criterion for monotonic tension is

$$f = \frac{\sigma}{1-D} - \sigma_u = 0, \tag{3.1}$$

the energy density release rate Y is almost constant:

$$R_\nu = 1 \quad \text{and} \quad Y \approx \frac{\sigma_u^2}{2E} \approx const, \tag{3.2}$$

and the unified damage law reduces to

$$\dot{D} = \left(\frac{Y}{S}\right)^s \dot{p} = \left(\frac{\sigma_u^2}{2ES}\right)^s |\dot{\epsilon}_p| \quad \text{if} \quad \epsilon_p > \epsilon_{pD}. \tag{3.3}$$

A simple integration gives a relation between the critical value of the damage D_c at mesocrack initiation and the plastic strain at rupture ϵ_{pR}:

$$D_c = \left(\frac{\sigma_u^2}{2ES}\right)^s (\epsilon_{pR} - \epsilon_{pD}). \tag{3.4}$$

This expression may be used to decrease the number of material parameters in the damage law, as then

$$\dot{D} = \frac{D_c}{\epsilon_{pR} - \epsilon_{pD}} |\dot{\epsilon}_p| \quad \text{if} \quad \epsilon_p > \epsilon_{pD}, \tag{3.5}$$

or by an obvious integration in monotonic loading:

$$\boxed{D = D_c \left\langle \frac{\epsilon_p - \epsilon_{pD}}{\epsilon_{pR} - \epsilon_{pD}} \right\rangle.} \tag{3.6}$$

The ductile damage is a linear function of the plastic strain. Checking this property by experiments is a validation of the damage law to be added to the procedures of identification of Sects. 1.4.4 and 2.4.

An approximate order of magnitude of the damage is obtained with $D_c \approx 0.5$, $\epsilon_{pD} \approx \epsilon_{pR}/2$, yielding

$$D \approx \left\langle \frac{\epsilon_p}{\epsilon_{pR}} - \frac{1}{2} \right\rangle \tag{3.7}$$

(recall that $\langle x \rangle$ denotes the positive part of x).

3.2.2 Case of Proportional Loading

As defined in Sect. 2.1.2, the proportional loading condition for a structure is when the loading induces principal directions of the stresses which are constant in time but which are eventually different in each point of the structure,

$$\sigma_{ij}(M,t) = \sigma_\Sigma(t)\Sigma_{ij}(M), \tag{3.8}$$

with the von Mises norm for easy applications

$$\frac{3}{2}\Sigma_{ij}^D \Sigma_{ij}^D = 1. \tag{3.9}$$

Let us recall that the main **sufficient condition** for a structure to be considered proportionally loaded is when there is only one applied load or when all the loads vary proportionally to one parameter.

The simplification which arises is the possible a priori integration of the damage law and the plastic constitutive equations as all the tensors are colinear:

$$\sigma^D_{ij} = \sigma_\Sigma \Sigma^D_{ij} \Rightarrow \sigma_{eq} = |\sigma_\Sigma|$$

$$\sigma_H = \sigma_\Sigma \Sigma_H \Rightarrow T_X = \frac{\sigma_H}{\sigma_{eq}} = \Sigma_H \cdot \operatorname{sgn}(\sigma_\Sigma) = \frac{1}{3}\Sigma_{kk} \cdot \operatorname{sgn}(\sigma_\Sigma) \approx T_{X\,elas}. \tag{3.10}$$

The value of the stress triaxiality in elasto-plasticity is usually close to the value obtained in a pure elastic calculation.

Using the isotropic damage law together with a monotonic loading, we have

$$\dot{D} = \left(\frac{\sigma_{eq}^2 R_\nu}{2ES(1-D)^2}\right)^s \dot{p} \quad \text{if} \quad p > p_D, \tag{3.11}$$

where

$$R_\nu = \frac{2}{3}(1+\nu) + 3(1-2\nu)\Sigma_H^2 = const \quad \text{and} \quad \sigma_{eq} = \sigma_\Sigma. \tag{3.12}$$

Considering natural initial conditions of no damage ($D = 0$ at the beginning of the loading), we have

$$\int_0^D (1-D)^{2s}\, dD = \left(\frac{R_\nu}{2ES}\right)^s \int_{t_D}^t \sigma_\Sigma^{2s}(t)\dot{p}(t)dt \tag{3.13}$$

$$\boxed{D = 1 - \left[1 - (2s+1)\left(\frac{R_\nu}{2ES}\right)^s \int_{t_D}^t \sigma_\Sigma^{2s}(t)\dot{p}(t)dt\right]^{\frac{1}{2s+1}},} \tag{3.14}$$

where t_D is the time for which the accumulated plastic strain is equal to the damage threshold $p = p_D$, i.e., for monotonic loading $p_D = \epsilon_{pD}$.

This integration may be performed once the time histories $\sigma_\Sigma(t)$ and $p(t)$ are determined, from an elasto-plastic computation or from the initial post-processing of an elastic computation using the Neuber method (see Sects. 2.1.4 and 3.2.4, respectively.)

If a saturated hardening is considered it is possible to simply relate the **accumulated plastic strain** at mesocrack initiation (p_R) to the **uniaxial plastic strain to rupture** (ϵ_{pR}):

$$\dot{D} = \left(\frac{\sigma_u^2 R_\nu}{2ES}\right)^s \dot{p} \quad \text{if} \quad p > p_D. \tag{3.15}$$

Integrating with the natural initial conditions and with $p_D = \epsilon_{pD}$ in monotonic loading leads to

$$D_c = \left(\frac{\sigma_u^2 R_\nu}{2ES}\right)^s (p_R - \epsilon_{pD}) \tag{3.16}$$

and dividing by the same equation in the uniaxial case gives

$$\boxed{\frac{p_R - \epsilon_{pD}}{\epsilon_{pR} - \epsilon_{pD}} = R_\nu^{-s} \quad \text{and} \quad R_\nu = \frac{2}{3}(1+\nu) + 3(1-2\nu)\Sigma_H^2,} \tag{3.17}$$

with $\Sigma_H = T_X$ (the triaxiality).

3.2.3 Sensitivity Analysis♂♀

Following the general method of Sect. 2.4.6, the sensitivity to loading conditions and material parameters is determined here on the predicted **plastic strain to rupture** using its closed-form expression in proportional loading (3.16),

$$D_c = \left(\frac{\sigma_u^2 R_\nu}{2ES}\right)^s (p_R - \epsilon_{pD}) \quad \text{or} \quad p_R = \epsilon_{pD} + \left(\frac{\sigma_u^2 R_\nu}{2ES}\right)^{-s} D_c, \tag{3.18}$$

with $R_\nu = \frac{2}{3}(1+\nu) + 3(1-2\nu)T_X^2$.

The logarithmic derivative allows us to determine the relative error on all the variables and parameters in order to build the scheme of the qualitative importance of each of them:

$$\ln(p_R - \epsilon_{pD}) = \ln D_c + s\left(\ln 2 + \ln E + \ln S - 2\ln \sigma_u - \ln R_\nu\right). \tag{3.19}$$

With the notation $\delta x = |dx|$ to ensure upper bounds on the errors (of unknown signs),

$$\frac{\delta(p_R - \epsilon_{pD})}{p_R - \epsilon_{pD}} = \frac{\delta p_R}{p_R}\frac{p_R}{p_R - \epsilon_{pD}} - \frac{\delta \epsilon_{pD}}{\epsilon_{pD}}\frac{\epsilon_{pD}}{p_R - \epsilon_{pD}} \tag{3.20}$$

and

$$\frac{\delta R_\nu}{R_\nu} = \frac{\delta \nu}{\nu}\frac{|6T_X^2 - \frac{2}{3}|\nu}{R_\nu} + \frac{\delta T_X}{T_X}\frac{6(1-2\nu)T_X^2}{R_\nu}, \tag{3.21}$$

writing $\frac{\delta p_R}{p_R} = \sum_k S_{Ak}^{p_R} \frac{\delta A_k}{A_k}$ where $S_{Ak}^{p_R}$ are the coefficients of the sensitivity matrix of Sect. 2.4.6, we obtain

$$S_{T_X}^{p_R} = \frac{p_R - \epsilon_{pD}}{p_R}\frac{6s(1-2\nu)T_X^2}{R_\nu},$$

$$S_{\sigma_u}^{p_R} = \frac{p_R - \epsilon_{pD}}{p_R} 2s,$$

3.2 Fast Calculation of Structural Failures

$$S_E^{p_R} = \frac{p_R - \epsilon_{pD}}{p_R} s,$$

$$S_\nu^{p_R} = \frac{p_R - \epsilon_{pD}}{p_R} \frac{|6T_X^2 - \frac{2}{3}|\nu s}{R_\nu},$$

$$S_{\epsilon_{pD}}^{p_R} = \frac{\epsilon_{pD}}{p_R},$$

$$S_S^{p_R} = \frac{p_R - \epsilon_{pD}}{p_R} s, \qquad (3.22)$$

$$S_s^{p_R} = \frac{p_R - \epsilon_{pD}}{p_R} s \left| \ln \frac{2ES}{\sigma_u^2 R_\nu} \right|,$$

$$S_{D_c}^{p_R} = \frac{p_R - \epsilon_{pD}}{p_R}.$$

Finally, the scheme of Fig. 3.2 gives an example of the values of the sensitivity parameters and shows the relative importance of each parameter given by the height of the frames around each sensitivity coefficient. It has more or less a general qualitative value for ductile failures as it has been obtained for mean values of the parameters: $T_X = 1$, $R_\nu = 2.07$, $\sigma_u = 500$ MPa, $E = 200000$ MPa, $\nu = 0.3$, $\epsilon_{pR}/\epsilon_{pD} = 2$, $S = 2$ MPa, and $s = 5$.

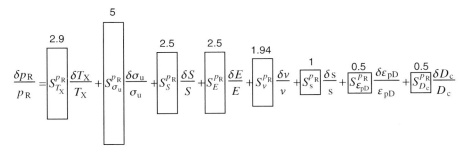

Fig. 3.2. Relative importance of each parameter in ductile failures

The loading is represented here by the triaxiality ratio T_X and by the stress through σ_u. Their influences are larger than those of the material parameters. The parameters which do not need a large accuracy are ϵ_{pD} and D_c. Nevertheless, if all the parameters were known with an error as small as 1%, the error on the accumulated plastic strain at failure could be of the order of 10 to 20%. Fortunately, in practice some errors are positive and some negative!

3.2.4 Stress Concentration and the Neuber Method

The case of small scale yielding allows us to avoid a whole elasto-plastic calculation of the structure by setting up a plastic correction of an elastic structure analysis. Small scale yielding refers to the cases where plastic strain is limited in energy and in geometrical space. To give a quantitative criterion, it is in the domain where $\sigma_{eq} > \sigma_y$ is below $\approx 1\%$ of the volume of the structure. This corresponds to the zones of stress concentrations as those close to holes or notches.

To evaluate the risk of a failure the proposed procedure is as follows:

- First, perform an elastic structure calculation
- Second, do a Neuber local plastic analysis as a correction of the elastic analysis. Other local energetic methods may be used. For instance, the strain energy density (SED) method described in Chap. 4 applies better at free edge points if the plasticity remains very localized. This is the case for cyclic loadings. For monotonic loadings, the Neuber method is generally preferred.
- Third, calculate the damage based on the results of the elastic and plastic analysis. The time integration of the damage evolution law is performed by using (3.6) or (3.14).

The Neuber method looks like an energy equivalence between the elastic and the elasto-plastic calculations of the same geometry submitted to the same loading but it is a heuristic.

For unidimensional states of stress, the product (stress × strain) calculated in elasticity is assumed to be locally identical to the same product calculated by means of an elasto-plastic analysis,

$$\sigma\epsilon = (\sigma\epsilon)_{elas} \quad \text{at the stress concentration point,} \qquad (3.23)$$

where $(.)_{elas}$ means "value determined from a strictly elastic computation," even if the stress is above the yield limit. The plastic state is then determined as the matching of the constitutive equation $\epsilon = \sigma/E + g(\sigma)$ with the equation of the hyperbole (stress × strain = the constant determined by the elastic calculation), where for monotonic loading isotropic hardening $R(\epsilon_p)$ alone has to be considered (Fig. 3.3). The function $g(\sigma) = p$ is $g(\sigma) = \left(\frac{\sigma - \sigma_y}{K_p^y}\right)^{M_y}$ for the hardening power law.

One may introduce the **stress concentration coefficient** in elasticity,

$$K_T = \frac{\sigma_{elas}}{\sigma_n} = \frac{\epsilon_{elas}}{\epsilon_n}, \qquad (3.24)$$

whose value may be found in handbooks (see bibliography) for common engineering geometries. The nominal stress σ_n is the nominal uniform stress field calculated from the external loading by the simple consideration of loads at equilibrium (see Fig. 3.4), $\epsilon_{elas} = \sigma_{elas}/E$, and $\epsilon_n = \sigma_n/E$. This allows us to write the Neuber heuristic as

$$\sigma\epsilon = K_T^2 \sigma_n \epsilon_n = \frac{K_T^2}{E}\sigma_n^2. \qquad (3.25)$$

3.2 Fast Calculation of Structural Failures   149

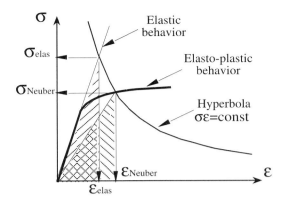

Fig. 3.3. The Neuber method

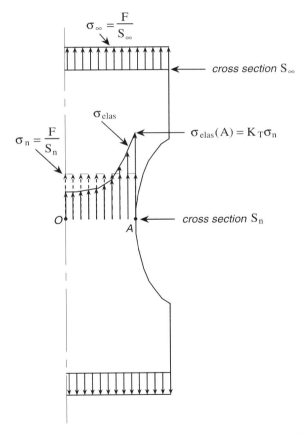

Fig. 3.4. Nominal stress (σ_n) and elastic stress concentration coefficient (K_T) for an applied load (F)

The result of the post-processing is the stress σ and the plastic strain ϵ_p determined graphically (Fig. 3.3) or analytically: use, for instance, (3.36) for linear hardening and solve (3.38) with $R_\nu = 1$ for nonlinear hardening. This defines the ratio

$$k_\mathrm{Neuber} = \frac{\sigma}{\sigma_\mathrm{elas}} \qquad (3.26)$$

and the stress concentration coefficient in plasticity as

$$\boxed{\frac{\sigma}{\sigma_\mathrm{n}} = k_\mathrm{Neuber} K_\mathrm{T}\,.} \qquad (3.27)$$

For three-dimensional states of stress, Neuber fundamental hypothesis may be written as

$$\boxed{\sigma_{ij}\epsilon_{ij} = (\sigma_{ij}\epsilon_{ij})_\mathrm{elas} \qquad \text{for monotonic loading,}} \qquad (3.28)$$

where the quantity $(\sigma_{ij}\epsilon_{ij})_\mathrm{elas}$ may also be written in terms of von Mises stress $\sigma_\mathrm{eq}^\mathrm{elas}$ and stress triaxiality $T_{\mathrm{X\,elas}}$ coming from the elastic reference computation,

$$(\sigma_{ij}\epsilon_{ij})_\mathrm{elas} = \frac{(\sigma_\mathrm{eq}^\mathrm{elas})^2 R_\nu^\mathrm{elas}}{E} \quad \text{and} \quad R_\nu^\mathrm{elas} = \frac{2}{3}(1+\nu) + 3(1-2\nu)T_{\mathrm{X\,elas}}^2\,. \qquad (3.29)$$

The plastic behavior is described by an integrated Hencky–Mises law for monotonic loading,

$$\sigma_{ij} = E_{ijkl}\epsilon_{kl} - 3G\frac{\sigma_{ij}^\mathrm{D}}{\sigma_\mathrm{eq}}p \quad \text{and} \quad p = R^{-1}(\sigma_\mathrm{eq} - \sigma_\mathrm{y})\,, \qquad (3.30)$$

where G is the shear modulus and $R(p)$ is the isotropic hardening law. For the power law, we have

$$p = \left(\frac{\sigma_\mathrm{eq} - \sigma_\mathrm{y}}{K_\mathrm{p}^\mathrm{y}}\right)^{M_\mathrm{y}}\,. \qquad (3.31)$$

Again, the values of the stress triaxiality T_X and the triaxiality function $R_\nu = \frac{2}{3}(1+\nu) + 3(1-2\nu)T_\mathrm{X}^2$ are needed in order to properly apply the Neuber method. Use the proportional loading assumption $T_\mathrm{X} \approx T_{\mathrm{X\,elas}}$ for general 3D cases. Use $T_\mathrm{X} = 1/3$ and $R_\nu = 1$ for stress concentration points located on free edges in plane stress. Use $T_\mathrm{X} \approx 0.58$ and $R_\nu \approx 1.27$ for stress concentration points located on free edges in plane strain (see Sect. 2.1.4).

3.2.4.1 Neuber Method with Linear Hardening, $R = C_\mathrm{y} p$

Assume that an elastic computation gives the product $(\sigma_{ij}\epsilon_{ij})_\mathrm{elas}$ as well as the von Mises stress $\sigma_\mathrm{eq}^\mathrm{elas}$ at a stress concentration point. The application of the Neuber method (3.28) leads to

3.2 Fast Calculation of Structural Failures 151

$$\sigma_{ij}\epsilon_{ij} = \frac{\sigma_{eq}^2 R_\nu}{E} + \frac{\sigma_{eq}\langle\sigma_{eq} - \sigma_y\rangle}{C_y} = (\sigma_{ij}\epsilon_{ij})_{elas} \qquad (3.32)$$

and then the evaluation of the von Mises stress in plasticity σ_{eq} if R_ν is known. The relationship $\sigma_{eq}(\sigma_{eq}^{elas})$ is explicit:

$$\boxed{\sigma_{eq} = \frac{\frac{E}{C_y}\sigma_y + \sqrt{\left(\frac{E}{C_y}\sigma_y\right)^2 + 4E\left(R_\nu + \frac{E}{C_y}\right)(\sigma_{ij}\epsilon_{ij})_{elas}}}{2\left(R_\nu + \frac{E}{C_y}\right)}} \qquad (3.33)$$

where we define

$$\boxed{k_{Neuber} = \frac{\sigma_{eq}}{\sigma_{eq}^{elas}} \quad \text{from the Neuber method}} \qquad (3.34)$$

and the accumulated plastic strain is

$$\boxed{p = \frac{\sigma_{eq} - \sigma_y}{C_y}.} \qquad (3.35)$$

This method has the main advantage of being fully explicit and giving closed-form expressions for σ_{eq} and p. The hardening parameters σ_y and C_y need to be identified from the tensile stress-strain curve within a range in accordance with the value of the plastic strain finally estimated.

For the 1D case,

$$\sigma = \sigma_n K_T k_{Neuber} = \sigma_n K_T \frac{1 + \sqrt{1 + 4\left(\frac{C_y}{E} + \frac{C_y^2}{E^2}\right)\frac{\sigma_{elas}^2}{\sigma_y^2}}}{2\left(\frac{C_y}{E} + 1\right)\frac{\sigma_{elas}}{\sigma_y}}, \qquad (3.36)$$

with $\sigma_{elas} = K_T \sigma_n$.

3.2.4.2 Neuber Method with Nonlinear Hardening, $R = R_\infty^y\left(1 - \exp\left(-b_y p\right)\right)$ ⚅⚅

The accumulated plastic strain is

$$p(\sigma_{eq}) = g(\sigma_{eq}) = -\frac{1}{b_y}\ln\left(\frac{R_\infty^y + \sigma_y - \sigma_{eq}}{R_\infty^y}\right) \qquad (3.37)$$

and the Neuber method now reads (R_ν known):

$$\sigma_{ij}\epsilon_{ij} = \frac{\sigma_{eq}^2 R_\nu}{E} + \sigma_{eq} p(\sigma_{eq}) = (\sigma_{ij}\epsilon_{ij})_{elas}, \qquad (3.38)$$

an equation which has to be solved numerically to estimate the von Mises stress in plasticity σ_{eq}, and then the accumulated plastic strain p and the auxiliary function $k_{Neuber} = \sigma_{eq}/\sigma_{eq}^{elas}$.

Considering a nonlinear hardening gives better results but a mathematical software is needed to solve the nonlinear equation (3.38).

3.2.4.3 Neuber Method Coupled with Damage

When damage occurs, the Neuber method can still be used but with the stress replaced by the effective stress, $\tilde{\sigma}_{\text{eq}} = \sigma_{\text{eq}}/(1-D)$.

Equation (3.33) for linear hardening becomes

$$\tilde{\sigma}_{\text{eq}} = \frac{\dfrac{E}{C_{\text{y}}}\sigma_{\text{y}} + \sqrt{\left(\dfrac{E}{C_{\text{y}}}\sigma_{\text{y}}\right)^2 + 4E\left(R_\nu + \dfrac{E}{C_{\text{y}}}\right)(\sigma_{ij}\epsilon_{ij})_{\text{elas}}}}{2\left(R_\nu + \dfrac{E}{C_{\text{y}}}\right)} \qquad (3.39)$$

and equation (3.38) for nonlinear hardening becomes

$$\frac{\tilde{\sigma}_{\text{eq}}^2 R_\nu}{E} + \tilde{\sigma}_{\text{eq}} p(\tilde{\sigma}_{\text{eq}}) = (\sigma_{ij}\epsilon_{ij})_{\text{elas}}, \qquad (3.40)$$

with the accumulated plastic strain

$$p = p(\tilde{\sigma}_{\text{eq}}) = R^{-1}(\tilde{\sigma}_{\text{eq}} - \sigma_{\text{y}}) = g\left(\frac{\sigma_{\text{eq}}}{1-D} - \sigma_{\text{y}}\right). \qquad (3.41)$$

The damage is equal to D_{c} at crack initiation.

3.2.5 Safety Margin and Crack Arrest ♂

The main parameter which allows us to evaluate how far or close a loaded structure is to a ductile mesocrack initiation is the maximum accumulated plastic strain p resulting from a structure analysis.

The plastic strain corresponding to a mesocrack initiation is given by the damage analysis:

$$p_{\text{R}} = p_{\text{D}} + D_{\text{c}}\left(\frac{\sigma_{\text{u}}^2 R_\nu}{2ES}\right)^{-s} = p_{\text{D}} + (\epsilon_{\text{pR}} - \epsilon_{\text{pD}})R_\nu^{-s}. \qquad (3.42)$$

If the number $Saf \geq 1$ is the safety factor to take into consideration in engineering, the safety margin is defined by

$$p < \frac{p_{\text{R}}}{Saf}, \qquad (3.43)$$

with $Saf = 5$, 2, or 1.1 depending on the compromise adopted between security and weight design.

Once the mesocrack is initiated, it is often important to check its rate of growth. As a finite crack exists, the simple Continuum Damage Mechanics is not appropriate any more and the concepts of Fracture Mechanics by cracking can be applied. The behavior of a ductile crack is approximately governed by a resistant curve (or R-curve) that depends on each material: an energy which depends on the area of the crack A,

$$R_c = R_c(A - A_0), \qquad (3.44)$$

where the function $R_c(A - A_0)$ is known from experiments (Fig. 3.5) and where A_0 is the area of the mesocrack initiated by damage (see Sect. 1.6.3),

$$A_0 = \frac{G_c^2}{\left(\dfrac{\sigma_u^2}{2E}D_c + \sigma_u \epsilon_{\mathrm{pR}}\right)^2}. \qquad (3.45)$$

The energy available for the crack growth is determined by the structure strain energy release rate (G) that depends on the geometry of the structure and the far field loading (σ_∞),

$$G = G(A, \sigma_\infty). \qquad (3.46)$$

It can be calculated from the Fracture Mechanics concepts and has the general form of

$$G = \kappa \frac{\sigma_\infty^2 \pi A^{1/2}}{E}, \qquad (3.47)$$

with κ a dimensionless shape parameter.

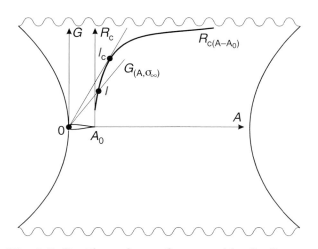

Fig. 3.5. Ductile crack growth governed by the R-curve

154   3 Ductile Failures

The strain energy release rate G is compared to the resistant curve (Fig. 3.5) as the fracture behavior is determined by

$$G(A, \sigma_\infty) = R_c(A - A_0). \tag{3.48}$$

- If this equality is satisfied, the intersection I of the two curves exists, there will be a crack arrest for the value of σ_∞ considered and a stable growth of the crack if σ_∞ increases.
- If this equality is not satisfied, which means $G > R_c$, there will be a fast crack propagation by instability as soon as the tangent point I_c is reached:

$$\boxed{\frac{\partial G}{\partial A} = \frac{\partial R_c}{\partial A}} \quad \text{at } I_c. \tag{3.49}$$

If the R-curve for the considered material is not available, there is always the possibility to compare the strain energy release rate for the value A_0 of the crack area with the toughness G_c: does $\kappa \sigma_\infty^2 \pi A_0^{1/2}/E$ remain much smaller than G_c? If yes, be happy!

3.3 Basic Engineering Examples

Some basic geometries and loadings are very common in engineering. For plates with holes or notches, for cylinders or for some plane stress problems there are approximate solutions which allow for fast calculations of the mesocrack initiation conditions. Furthermore, some results can be expressed in non dimensional form for use in large problem sets.

Here again, the damage is considered isotropic without the quasi-unilateral condition of microdefects closure.

3.3.1 Plates or Members with Holes or Notches♂

These are all cases where the critical RVE is submitted to a stress concentration, inducing a small scale local plasticity. Assume that the local elastic stress, ignoring the plasticity, is known for a monotonic loading either from a structure calculation or through the stress concentration coefficient (K_T) which may be found in handbooks (see bibliography):

$$\sigma_{eq}^{elas} = K_T \sigma_n. \tag{3.50}$$

with σ_n the nominal stress. When the load (σ_∞) increases, small scale local plasticity develops which decreases the stress concentration determined now from the 1D Neuber correction (as the stress concentration lies on a free surface, see Sect. 3.2.4). Then

$$\sigma_{eq} = k_{Neuber} \sigma_{eq}^{elas} = k_{Neuber} K_T \sigma_n. \tag{3.51}$$

This stress is used to calculate the accumulated plastic strain p which ensures the mesocrack initiation when it reaches ϵ_{pR} or $\epsilon_{\text{pR}}^\star$ given by the critical damage condition of Sect. 3.2.1,

$$D_{\text{c}} = \left(\frac{\sigma_{\text{u}}^2 R_\nu}{2ES}\right)^s (\epsilon_{\text{pR}} - \epsilon_{\text{pD}}) \quad \text{or} \quad \epsilon_{\text{pR}} = \epsilon_{\text{pD}} + D_{\text{c}} \left(\frac{2ES}{\sigma_{\text{u}}^2 R_\nu}\right)^s \quad (3.52)$$

with $R_\nu = 1$ for thin plates with the hypothesis of plane stresses and $R_\nu = 1.27$ for thick plates with the hypothesis of plane strains (stress triaxiality $T_X = 0.58$ from Sect. 2.1.4).

3.3.1.1 Linear Hardening

$$k_{\text{Neuber}} = \frac{\frac{E}{C_{\text{y}}}\sigma_{\text{y}} + \sqrt{\left(\frac{E}{C_{\text{y}}}\sigma_{\text{y}}\right)^2 + 4E\left(R_\nu + \frac{E}{C_{\text{y}}}\right)(\sigma_{ij}\epsilon_{ij})_{\text{elas}}}}{2\left(R_\nu + \frac{E}{C_{\text{y}}}\right)\sigma_{\text{eq}}^{\text{elas}}}, \quad (3.53)$$

with C_{y} the plastic modulus such that

$$p = \left\langle \frac{\frac{\sigma_{\text{eq}}}{1-D} - \sigma_{\text{y}}}{C_{\text{y}}} \right\rangle. \quad (3.54)$$

Writing the rupture condition as

$$\epsilon_{\text{pR}} = \left\langle \frac{\frac{\sigma_{\text{eq}}}{1-D_{\text{c}}} - \sigma_{\text{y}}}{C_{\text{y}}} \right\rangle = \epsilon_{\text{pD}} + D_{\text{c}}\left(\frac{2ES}{\sigma_{\text{u}}^2 R_\nu}\right)^s, \quad (3.55)$$

from which

$$\boxed{\sigma_{\text{eq}} = (1-D_{\text{c}})\left[\sigma_{\text{y}} + C_{\text{y}}\left(\epsilon_{\text{pD}} + D_{\text{c}}\left(\frac{2ES}{\sigma_{\text{u}}^2 R_\nu}\right)^s\right)\right]} \quad (3.56)$$

and the failure condition is

$$\boxed{\sigma_{\text{n}} = \frac{\tilde\sigma_{\text{eq}}}{k_{\text{Neuber}} K_{\text{T}}}, \quad \tilde\sigma_{\text{eq}} = \frac{\sigma_{\text{eq}}}{1-D_{\text{c}}}.} \quad (3.57)$$

3.3.1.2 Exponential Hardening, $R = R_\infty^{\text{y}}(1 - \exp(-b_{\text{y}}p))$

$k_{\text{Neuber}} = \tilde\sigma_{\text{eq}}/\sigma_{\text{eq}}^{\text{elas}}$ is determined by the Neuber method (3.38), with

$$p = -\frac{1}{b_{\text{y}}}\ln\left(\frac{R_\infty^{\text{y}} + \sigma_{\text{y}} - \tilde\sigma_{\text{eq}}}{R_\infty^{\text{y}}}\right). \quad (3.58)$$

156   3 Ductile Failures

Writing the rupture condition

$$p = -\frac{1}{b_y} \ln \left(\frac{R_\infty^y + \sigma_y - \dfrac{\sigma_{eq}}{1 - D_c}}{R_\infty^y} \right) = \epsilon_{pD} + D_c \left(\frac{2ES}{\sigma_u^2 R_\nu} \right)^s, \quad (3.59)$$

from which

$$\boxed{\sigma_{eq} = (1 - D_c) \left\{ \sigma_y + R_\infty^y \left[1 - \exp\left(-b_y \left(\epsilon_{pD} + D_c \left(\frac{2ES}{\sigma_u^2 R_\nu} \right)^s \right) \right) \right] \right\},} \quad (3.60)$$

the failure condition is

$$\boxed{\sigma_n = \frac{\tilde{\sigma}_{eq}}{k_{Neuber} K_T}, \quad \tilde{\sigma}_{eq} = \frac{\sigma_{eq}}{1 - D_c}.} \quad (3.61)$$

3.3.2 Pressurized Shallow Cylinders

A pressurized cylinder may explode if a crack is initiated through the phenomenon of ductility. It may happen in pipes or during the process of hydroforming. If the cylinder is considered long enough, with a thin thickness (t_{cyl}) compared to the mean radius (R_{cyl}, Fig. 3.6), the state of stress derives simply from overall equilibrium equations.

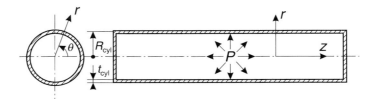

Fig. 3.6. Pressurized circular cylinder

In cylindrical coordinates (r, θ, z) with P as the internal pressure, we have

$$\sigma = \begin{bmatrix} \approx 0 & 0 & 0 \\ 0 & \dfrac{PR_{cyl}}{t_{cyl}} & 0 \\ 0 & 0 & \dfrac{PR_{cyl}}{2t_{cyl}} \end{bmatrix}, \quad (3.62)$$

from which it is easy to determine

3.3 Basic Engineering Examples

- The von Mises equivalent stress $\sigma_{eq} = \dfrac{\sqrt{3}}{2}\dfrac{PR_{cyl}}{t_{cyl}}$,
- The hydrostatic stress $\sigma_H = \dfrac{PR_{cyl}}{2t_{cyl}}$
- The triaxiality ratio $T_X = \dfrac{\sigma_H}{\sigma_{eq}} = \dfrac{1}{\sqrt{3}}$
- The triaxiality function

$$R_\nu = \frac{2}{3}(1+\nu) + 3(1-2\nu)T_X^2 = \frac{5-4\nu}{3} \qquad (3.63)$$

or $R_\nu = 1.27$ for $\nu = 0.3$.

A rough approximation consists of writing the condition of brittle fracture of Chap. 7 through the damage equivalent stress $\sigma^\star$ derived in Sect. 1.2.2 (1.34), using

$$\sigma^\star = \sigma_u \quad \text{or} \quad \sigma_{eq} R_\nu^{1/2} = \sigma_u, \qquad (3.64)$$

where σ_u is the ultimate stress in tension. The term P_c is the critical pressure at mesocrack initiation, where

$$\frac{\sqrt{3}}{2}\frac{P_c R_{cyl}}{t_{cyl}}\sqrt{\frac{5-4\nu}{3}} = \sigma_u \qquad (3.65)$$

gives

$$P_c = \frac{2}{\sqrt{5-4\nu}}\sigma_u \frac{t_{cyl}}{R_{cyl}}. \qquad (3.66)$$

The critical pressure is linear with the ultimate stress and the thickness-radius ratio.

A better analysis consists of the resolution of the unified damage law together with the plastic constitutive equations for which the hypothesis of linear hardening is sufficient:

$$\dot{p} = \left\langle \frac{\frac{\sigma_{eq}}{1-D} - \sigma_y}{C_y} \right\rangle. \qquad (3.67)$$

Since the loading is proportional and monotonic, considering the saturated strain hardening allows us to use the result of Sect. 3.2.5 (3.42):

$$p_R = p_D + D_c\left(\frac{\sigma_u^2 R_\nu}{2ES}\right)^{-s} = \epsilon_{pD} + (\epsilon_{pR} - \epsilon_{pD})R_\nu^{-s}. \qquad (3.68)$$

There will be a mesocrack initiation for the pressure corresponding to the accumulated plastic strain value equal to p_R, such that

$$\frac{\sqrt{3}P_c R_{cyl}}{2t_{cyl}(1-D_c)C_y} - \frac{\sigma_y}{C_y} = \epsilon_{pD} + \left(\frac{5-4\nu}{3}\right)^{-s}(\epsilon_{pR} - \epsilon_{pD}) \qquad (3.69)$$

or

$$P_c = \frac{2}{\sqrt{3}} \frac{t_{cyl}}{R_{cyl}} (1 - D_c) \left[\sigma_y + C_y \left(\epsilon_{pD} + \left(\frac{5 - 4\nu}{3} \right)^{-s} (\epsilon_{pR} - \epsilon_{pD}) \right) \right]. \tag{3.70}$$

The critical pressure is still linear with the ratio t_{cyl}/R_{cyl} but depends on the damage properties.

Example of a Pipe

For a pipe made of ferritic steel at room temperature:
$E = 200000$ MPa, $\nu = 0.3$, $\sigma_y = 375$ MPa, $\sigma_u = 474$ MPa, $\epsilon_{pu} = 0.15$,
$C_y \approx \dfrac{\sigma_u - \sigma_y}{\epsilon_{pu}} = 660$ MPa, $\epsilon_{pD} = 0.15$, $\epsilon_{pR} = 0.32$, $D_c = 0.3$, $s = 2.4$.
Equation (3.70) gives $P_c = 435 \dfrac{t_{cyl}}{R_{cyl}}$ (MPa). The condition of brittle fracture (3.66) leads to $P_c = 486 \dfrac{t_{cyl}}{R_{cyl}}$ (MPa) which overestimates the pressure by a factor of 12%.

3.3.3 Post-Buckling in Bending ♂

A situation which often occurs in accident evaluation is a post-buckling deformation following an instability due to buckling. The purpose here is to evaluate the energy absorbed by plastic deformation, taking into account the damage up to a mesocrack initiation for an elementary volume loaded in pure bending at slow strain rate.

The problem reduces to the analysis of a beam element of length l, height h and width b, loaded in circular bending of radius of curvature ρ up to a maximum plastic strain ϵ_{pR} or $\epsilon_{pR}^\star$ (Fig. 3.7).

The elementary beam theory is used together with the Bernoulli hypothesis applied on the plastic strain as the elastic strain can be neglected, yielding

$$\epsilon_p = \epsilon_{pM} \frac{2y}{h}, \tag{3.71}$$

where ϵ_{pM} is the maximum uniaxial plastic strain reached in the cross section and $\epsilon_{pM} = \epsilon_{pR}$ at mesocrack initiation. The density of energy to create a tension defined by a plastic strain ϵ_p is

$$w = \int_0^{\epsilon^p} \sigma d\epsilon^p. \tag{3.72}$$

For a material subjected to a power law hardening (with the yield stress taken as $\sigma_y = 0$) and the unified damage law (3.6) of Sect. 3.2.1 (with the damage threshold taken equal to $\epsilon_{pD} = 0$), we have

3.3 Basic Engineering Examples

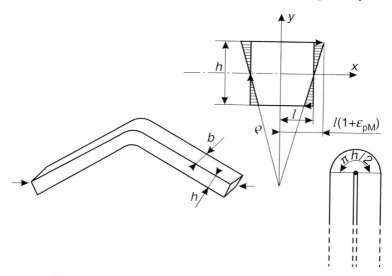

Fig. 3.7. Elementary bending of a rectangular beam

$$\epsilon_\mathrm{p} = \left[\frac{\sigma}{(1-D)K_\mathrm{p}^0}\right]^{M_0} \quad \text{and} \quad D = D_\mathrm{c}\frac{\epsilon_\mathrm{p}}{\epsilon_\mathrm{pR}}. \tag{3.73}$$

Then,

$$\begin{aligned} w &= \int_0^{\epsilon_\mathrm{p}} \left(1 - D_\mathrm{c}\frac{\epsilon_\mathrm{p}}{\epsilon_\mathrm{pR}}\right) K_\mathrm{p}^0 \epsilon_\mathrm{p}^{1/M_0} \mathrm{d}\epsilon_\mathrm{p} \\ &= \frac{M_0}{M_0+1} K_\mathrm{p}^0 \epsilon_\mathrm{p}^{\frac{M_0+1}{M_0}} - \frac{D_\mathrm{c} K_\mathrm{p}^0}{\epsilon_\mathrm{pR}} \frac{M_0}{2M_0+1} \epsilon_\mathrm{p}^{\frac{2M_0+1}{M_0}}. \end{aligned} \tag{3.74}$$

At this stage it is interesting to compare the energy involved without any damage at crack initiation ($\epsilon_\mathrm{p} = \epsilon_\mathrm{pR}$, $w = w_\mathrm{R}$):

$$w_0 = \int_0^{\epsilon_\mathrm{p}} K_\mathrm{p}^0 \epsilon_\mathrm{p}^{1/M_0} \mathrm{d}\epsilon_\mathrm{p} = \frac{M_0}{M_0+1} K_\mathrm{p}^0 \epsilon_\mathrm{p}^{\frac{M_0+1}{M_0}} \tag{3.75}$$

and

$$w_\mathrm{R} = w_0 \left(1 - D_\mathrm{c}\frac{M_0+1}{2M_0+1}\right). \tag{3.76}$$

This shows that a calculation made without consideration of damage leads to an overestimation of about $w_0 D_\mathrm{c}/2$, a relative error of 10 to 25%.

Back to the bending problem, w has to be integrated in the whole volume with the linear variation of ϵ_p in height $\epsilon_\mathrm{p} = \epsilon_\mathrm{pM}\frac{2y}{h}$:

$$W = 2bl \int_0^{h/2} \int_0^{\epsilon_{\mathrm{p}}(y)} \left(1 - D_{\mathrm{c}} \frac{\epsilon_{\mathrm{p}}}{\epsilon_{\mathrm{pR}}}\right) K_{\mathrm{p}}^0 \epsilon_{\mathrm{p}}^{1/M_0} \mathrm{d}\epsilon_{\mathrm{p}} \mathrm{d}y \qquad (3.77)$$

or

$$W = bhl K_{\mathrm{p}}^0 \frac{M_0}{2M_0+1} \epsilon_{\mathrm{p}}^{\frac{M_0+1}{M_0}} \left(\frac{M_0}{M_0+1} - D_{\mathrm{c}} \frac{M_0}{3M_0+1} \frac{\epsilon_{\mathrm{pM}}}{\epsilon_{\mathrm{pR}}}\right). \qquad (3.78)$$

If the damage is not taken into consideration, then

$$W_0 = bhl K_{\mathrm{p}}^0 \frac{M_0^2}{(2M_0+1)(M_0+1)} \epsilon_{\mathrm{p}}^{\frac{M_0+1}{M_0}} \qquad (3.79)$$

and for the mesocrack initiation condition ($\epsilon_{\mathrm{p}} = \epsilon_{\mathrm{pR}}$, $W = W_{\mathrm{R}}$),

$$\boxed{W_{\mathrm{R}} = W_0 \left(1 - D_{\mathrm{c}} \frac{M_0+1}{3M_0+1}\right).} \qquad (3.80)$$

When damage is not taken into account, the result is again an overestimation of the energy. The relative difference is slightly smaller and of about $D_{\mathrm{c}}/3$, that is 7 to 17%.

For the practical applications of bar structures collapsing with n_{h} identical **plastic hinges**, we estimate the bending length as a **full bending**: $l = \pi h/2$ (see Fig. 3.7) and take the following as a rough estimation of the energy absorbed:

$$W_{\mathrm{R}}(n_{\mathrm{h}}) = n_{\mathrm{h}} b \frac{\pi h^2}{2} K_{\mathrm{p}}^0 \frac{M_0^2}{(2M_0+1)(M_0+1)} \epsilon_{\mathrm{pR}}^{\frac{M_0+1}{M_0}} \left(1 - D_{\mathrm{c}} \frac{M_0+1}{3M_0+1}\right) \qquad (3.81)$$

or with $\epsilon_{\mathrm{pR}} = \left[\dfrac{\sigma_{\mathrm{R}}}{(1-D_{\mathrm{c}})K_{\mathrm{p}}^0}\right]^{M_0} \approx \left[\dfrac{\sigma_{\mathrm{u}}}{K_{\mathrm{p}}^0}\right]^{M_0}$,

$$\boxed{W(n_{\mathrm{h}}) = n_{\mathrm{h}} b \frac{\pi h^2}{2} \frac{M_0^2}{(2M_0+1)(M_0+1)} \sigma_{\mathrm{u}} \epsilon_{\mathrm{pR}} \left(1 - D_{\mathrm{c}} \frac{M_0+1}{3M_0+1}\right),} \qquad (3.82)$$

which only needs the knowledge of the geometry b, h, and the material parameters σ_{u}, ϵ_{pR}, M_0, and D_{c}.

3.3.4 Damage Criteria in Proportional Loading⌀

The case of proportional loading allows us to derive the mesocrack initiation condition as a function of the state of stresses or strains with the need of only a few material parameters. Equation (3.17) derived in Sect. 3.2.2 for saturated hardening is used here but its application is limited to positive stresses as the damage is considered isotropic, without any effect of microdefects closure in compression:

$$\boxed{\frac{p_{\mathrm{R}} - \epsilon_{\mathrm{pD}}}{\epsilon_{\mathrm{pR}} - \epsilon_{\mathrm{pD}}} = R_\nu^{-s}.} \qquad (3.83)$$

3.3.4.1 Three-Dimensional Loading

The large effect of the triaxiality ratio $T_X = \sigma_H/\sigma_{eq}$ on the accumulated plastic strain at mesocrack initiation p_R is shown in Fig. 3.8. The value of p_R may be deduced from the graphs if the values of the damage threshold ϵ_{pD}, the damage exponent s, the plastic strain to rupture in pure tension ϵ_{pR}, and the triaxiality ratio from a structure calculation are known.

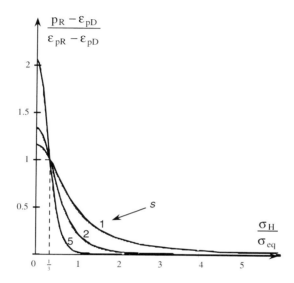

Fig. 3.8. Triaxiality makes materials more brittle!

3.3.4.2 Case of Plane Strains

The case of plain strains applies to thick sheets for which the strain within the thickness ϵ_3 is considered negligible in comparison to the in-plane principal strains ϵ_1 and ϵ_2:

$$\boldsymbol{\epsilon} = \begin{bmatrix} \epsilon_1 & 0 & 0 \\ 0 & \epsilon_2 & 0 \\ 0 & 0 & 0 \end{bmatrix} \quad \text{and} \quad \boldsymbol{\sigma} = \begin{bmatrix} \sigma_1 & 0 & 0 \\ 0 & \sigma_2 & 0 \\ 0 & 0 & \sigma_3 \end{bmatrix}. \quad (3.84)$$

Neglecting the elastic strains, the condition of plastic incompressibility gives

$$\epsilon_2 = -\epsilon_1. \quad (3.85)$$

The plastic constitutive equations integrated in case of proportional loading gives back the Hencky–Mises law with $p = g(\sigma_{eq}, D)$:

$$\epsilon_3 = \frac{3}{2}g(\sigma_{\text{eq}}, D)\frac{\sigma_{33}^{\text{D}}}{\sigma_{\text{eq}}}, \tag{3.86}$$

from which $\sigma_{33}^{\text{D}} = 0$,

$$\begin{aligned}\epsilon_1 &= \frac{3}{2}g(\sigma_{\text{eq}}, D)\frac{\sigma_{11}^{\text{D}}}{\sigma_{\text{eq}}}, \\ \epsilon_2 &= \frac{3}{2}g(\sigma_{\text{eq}}, D)\frac{\sigma_{22}^{\text{D}}}{\sigma_{\text{eq}}}.\end{aligned} \tag{3.87}$$

As $\epsilon_2 = -\epsilon_1$, $\sigma_{22}^{\text{D}} = -\sigma_{11}^{\text{D}}$, it follows that

$$\begin{aligned}\sigma_3 &= \frac{1}{2}(\sigma_1 + \sigma_2), \\ \sigma_{\text{H}} &= \frac{1}{2}(\sigma_1 + \sigma_2), \\ \sigma_{\text{eq}} &= \frac{\sqrt{3}}{2}|\sigma_1 - \sigma_2|.\end{aligned} \tag{3.88}$$

The accumulated plastic strain is

$$p = \sqrt{\frac{2}{3}(\epsilon_1^2 + \epsilon_2^2)} = \frac{2}{\sqrt{3}}|\epsilon_1| \tag{3.89}$$

and the crack initiation criterion becomes

$$\frac{\frac{2}{\sqrt{3}}|\epsilon_1| - \epsilon_{\text{pD}}}{\epsilon_{\text{pR}} - \epsilon_{\text{pD}}} = R_\nu^{-s} \tag{3.90}$$

or

$$\boxed{\begin{aligned}|\epsilon_1| &= |\epsilon_2| \\ &= \frac{\sqrt{3}}{2}\left[\epsilon_{\text{pD}} + (\epsilon_{\text{pR}} - \epsilon_{\text{pD}})\left[\frac{2}{3}(1+\nu) + (1-2\nu)\left(\frac{\sigma_1 + \sigma_2}{\sigma_1 - \sigma_2}\right)^2\right]^{-s}\right].\end{aligned}} \tag{3.91}$$

The case $\sigma_1 = \sigma_2$ corresponds to a pure hydrostatic state of stress $\sigma_1 = \sigma_2 = \sigma_3$ for which there is no plastic strain and no damage since it would need infinite strains.

3.3.4.3 Case of Plane Stresses

The case of plane stresses applies to thin sheets for which the stress within the thickness σ_3 is considered as negligible in comparison to the in-plane principal stresses σ_1 and σ_2.

3.3 Basic Engineering Examples   163

With the condition of incompressibility $\epsilon_{kk} = 0$, and neglecting the elastic strains,

$$\boldsymbol{\sigma} = \begin{bmatrix} \sigma_1 & 0 & 0 \\ 0 & \sigma_2 & 0 \\ 0 & 0 & 0 \end{bmatrix}, \quad \boldsymbol{\epsilon} = \begin{bmatrix} \epsilon_1 & 0 & 0 \\ 0 & \epsilon_2 & 0 \\ 0 & 0 & -(\epsilon_1 + \epsilon_2) \end{bmatrix} \quad (3.92)$$

To use the condition of mesocrack initiation (3.17)

$$\frac{p_{\mathrm{R}} - \epsilon_{\mathrm{pD}}}{\epsilon_{\mathrm{pR}} - \epsilon_{\mathrm{pD}}} = R_\nu^{-s} \quad (3.93)$$

one needs the accumulated plastic strain and the triaxiality function expressed either as functions of the strains or of the stresses.

3.3.4.3.1 Plane Stress Criterion as a Function of the Strains

$$p = \sqrt{\frac{2}{3}\epsilon_{ij}^{\mathrm{D}}\epsilon_{ij}^{\mathrm{D}}} = \frac{2}{\sqrt{3}}\sqrt{\epsilon_1^2 + \epsilon_2^2 + \epsilon_1\epsilon_2} \quad (3.94)$$

The triaxiality ratio $\sigma_{\mathrm{H}}/\sigma_{\mathrm{eq}}$ may be derived from the plastic constitutive equations (3.87). Both σ_{H} and σ_{eq} may be expressed as a function of σ_1 and ϵ_2/ϵ_1. First of all

$$\frac{\epsilon_2}{\epsilon_1} = \frac{\sigma_{22}^{\mathrm{D}}}{\sigma_{11}^{\mathrm{D}}}, \quad (3.95)$$

with

$$\sigma_{11}^{\mathrm{D}} = \sigma_1 - \frac{1}{3}(\sigma_1 + \sigma_2) \quad \text{and} \quad \sigma_{22}^{\mathrm{D}} = \sigma_2 - \frac{1}{3}(\sigma_1 + \sigma_2). \quad (3.96)$$

It follows that

$$\sigma_2 = \sigma_1 \frac{2\dfrac{\epsilon_2}{\epsilon_1} + 1}{2 + \dfrac{\epsilon_2}{\epsilon_1}}. \quad (3.97)$$

Then,

$$\sigma_{\mathrm{H}} = \frac{1}{3}(\sigma_1 + \sigma_2) = \frac{\sigma_1}{3}\left(1 + \frac{2\dfrac{\epsilon_2}{\epsilon_1} + 1}{2 + \dfrac{\epsilon_2}{\epsilon_1}}\right), \quad (3.98)$$

$$\sigma_{\mathrm{eq}} = \frac{1}{\sqrt{2}}\sqrt{(\sigma_1 - \sigma_2)^2 + \sigma_1^2 + \sigma_2^2} = \sigma_1\sqrt{\left(\frac{2\dfrac{\epsilon_2}{\epsilon_1} + 1}{2 + \dfrac{\epsilon_2}{\epsilon_1}}\right)^2 - \left(\frac{2\dfrac{\epsilon_2}{\epsilon_1} + 1}{2 + \dfrac{\epsilon_2}{\epsilon_1}}\right) + 1}, \quad (3.99)$$

and

$$T_{\mathrm{X}} = \frac{\sigma_{\mathrm{H}}}{\sigma_{\mathrm{eq}}} = \frac{\dfrac{\epsilon_2}{\epsilon_1} + 1}{\sqrt{3}\sqrt{\left(\dfrac{\epsilon_2}{\epsilon_1}\right)^2 + \dfrac{\epsilon_2}{\epsilon_1} + 1}}. \quad (3.100)$$

164   3 Ductile Failures

Finally at rupture, we have

$$\frac{\frac{2}{\sqrt{3}}\sqrt{\epsilon_1^2 + \epsilon_2^2 + \epsilon_1\epsilon_2} - \epsilon_{\text{pD}}}{\epsilon_{\text{pR}} - \epsilon_{\text{pD}}} = \left[\frac{2}{3}(1+\nu) + (1-2\nu)\frac{\left(\frac{\epsilon_2}{\epsilon_1}+1\right)^2}{\left(\frac{\epsilon_2}{\epsilon_1}\right)^2 + \frac{\epsilon_2}{\epsilon_1}+1}\right]^{-s}. \quad (3.101)$$

A resolution by any mathematics software with $\epsilon_{\text{pD}} = 0$ gives the graphs of Fig. 3.9 where the proportional loading is always a straight line in the plot of $\epsilon_2/\epsilon_{\text{pR}}$ vs $\epsilon_1/\epsilon_{\text{pR}}$. The admissible domain for the couple of strains ϵ_1 and ϵ_2 is smaller and smaller as the damage exponent s increases.

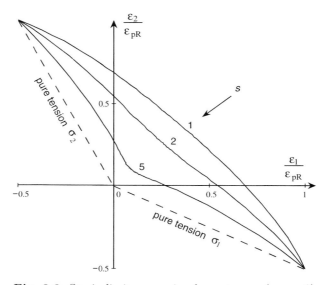

Fig. 3.9. Strain limit curves in plane stresses ($\epsilon_{\text{pD}} = 0$)

3.3.4.3.2 Plane Stress Criterion as a Function of Stresses

$$\sigma_{\text{H}} = \frac{1}{3}(\sigma_1 + \sigma_2)$$
$$\sigma_{\text{eq}} = \frac{1}{\sqrt{2}}\sqrt{(\sigma_1-\sigma_2)^2 + (\sigma_2-\sigma_3)^2 + (\sigma_3-\sigma_1)^2} = \sqrt{\sigma_1^2 + \sigma_2^2 - \sigma_1\sigma_2} \quad (3.102)$$

The plastic strain needs to be expressed as a function of the stresses. Using again the integrated plastic constitutive equations again, but with a zero yield stress, allows us to obtain a simple expression for the ratio $p_{\text{R}}/\epsilon_{\text{pR}}$.

3.3 Basic Engineering Examples

The hardening law reads:

$$p = g(\sigma_{\text{eq}}, D) = \left[\frac{\sigma_{\text{eq}}}{(1-D)K_{\text{p}}^0}\right]^{M_0}. \tag{3.103}$$

Close to crack initiation the hardening is close to saturation, leading to

$$p_{\text{R}} = \left[\frac{\sigma_{\text{eq}}}{(1-D_{\text{c}})K_{\text{p}}^0}\right]^{M_0} \quad \text{and} \quad \epsilon_{\text{pR}} = \left[\frac{\sigma_{\text{R}}}{(1-D_{\text{c}})K_{\text{p}}^0}\right]^{M_0} \approx \left[\frac{\sigma_{\text{u}}}{K_{\text{p}}^0}\right]^{M_0}. \tag{3.104}$$

Then

$$\frac{p_{\text{R}}}{\epsilon_{\text{pR}}} = \left[\frac{\sigma_{\text{eq}}}{\sigma_{\text{u}}(1-D_{\text{c}})}\right]^{M_0}, \tag{3.105}$$

where $\dfrac{p_{\text{R}}}{\epsilon_{\text{pR}}} = R_\nu^{-s}$ gives

$$\boxed{\frac{\sqrt{\sigma_1^2 + \sigma_2^2 - \sigma_1\sigma_2}}{(1-D_{\text{c}})\sigma_{\text{u}}} = \left[\frac{2}{3}(1+\nu) + \frac{1}{3}(1-2\nu)\frac{(\sigma_1+\sigma_2)^2}{\sigma_1^2+\sigma_2^2-\sigma_1\sigma_2}\right]^{-\frac{s}{M_0}}.} \tag{3.106}$$

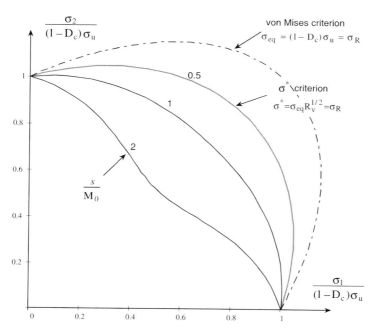

Fig. 3.10. Stress limit curves in plane stresses ($\epsilon_{\text{pD}} = 0$)

If the damage threshold is considered as $\epsilon_{\text{pD}} = 0$, the stress condition of mesocrack initiation depends only on the ultimate stress σ_{u}, the critical damage D_{c}, and the ratio of the damage and plastic exponents s/M_0, as shown in Fig. 3.10.

In the same figure, the von Mises criterion is plotted to show which mistake may be made in using it as a rupture criterion when the triaxiality ratio differs from $\approx 1/3$. The damage equivalent stress criterion $\sigma^{\star} = \sigma_{\text{eq}} R_{\nu}^{1/2} = \sigma_{\text{R}}$ corresponds to the curve $s/M_0 = 0.5$.

3.4 Numerical Failure Analysis

For complex geometries and loadings, closed-form solutions for the stress, strain, and damage fields usually do not exist and the loading is often non-proportional at most Gauss points of finite element analyses. No simplifying assumptions can be made, except the earlier hypothesis of localized plasticity (small scale yielding) which gives the possibility to efficiently use the Neuber method as described in Sect. 3.2.4. When plasticity takes place in a non-negligible domain of the structure (one speaks then of large scale yielding), stress redistribution occurs. It may have an accelerating effect on the failure conditions as it enhances the phenomenon of strain and damage localization.

We then use computer codes and perform FE analyses with the fully-coupled damage constitutive equations of Chap. 1. For ductile failures we use the damage initiation criterion written in terms of plastic strain where $p_{\text{D}} = \epsilon_{\text{pD}}$ (see Sect. 1.4.1). For monotonic loading, we use the single isotropic hardening R with either the exponential law $R = R_{\infty}^{\text{y}} (1 - \exp(-b_{\text{y}} r))$, the power law $R = K_{\text{p}}^{\text{y}} r^{1/M_{\text{y}}}$, or the linear law $R = C_{\text{y}} r$. When a numerical analysis is performed, there is no more reason to consider the unified damage law reduced to isotropy. We strongly advise using the anisotropic law to obtain a better accuracy. It requires only one additional material parameter which in practice may be taken as $\eta = 3$. The consideration of the quasi-unilateral conditions of microdefects closure is not necessary as far as monotonic loading is applied.

For more details see for instance Sect. 2.2.6 for the use of the fully-coupled scheme and more generally Sect. 2.2 for its numerical implementation, as well as Sects. 1.4.4 and 2.4 for the material parameters identification.

Other models concern ductile failures: an anisotropic plasticity model with the Hill yield surface that is compatible with the effective stress concept is described at the end of this chapter; the model extension to finite strains is briefly exposed in next section; and the Gurson model specific to ductile failure of porous materials is summarized in Sect. 3.4.5 and used for the cross identification of the unified damage law. Frames may be analyzed by limit analysis incorporing damage through Lumped Damage Mechanics of Sect. 3.4.6.

3.4.1 Finite Strains♂♂♂

When ductile failure occurs, the strains often become too large for the small strain perturbation hypothesis to remain valid: for metals the plastic strain at rupture is usually larger than 0.1 (and most of the time smaller than 1); for polymers it may reach much larger values corresponding sometimes to a few hundred percent. The large deformations framework has to be used and this can be done only with the help of computer codes.

Different solutions exist to extend nonlinear small strain constitutive equations to finite strains, mainly related to the material behavior description in either the reference initial configuration $\mathcal{C}_0$ or in the deformed or actual configuration $\mathcal{C}(t)$ at time t. The link between $\mathcal{C}_0$ and $\mathcal{C}(t)$ is the gradient of the transformation $\boldsymbol{F}$ and $\boldsymbol{F}^{\mathrm{T}}$ is its transpose.

One may consider different strain tensors, such as the **Green–Lagrange** tensor ($\boldsymbol{E}$) in the reference configuration, the Euler–Almansi tensor ($\boldsymbol{A}$) in $\mathcal{C}(t)$,

$$\boldsymbol{E} = \frac{1}{2}\left(\boldsymbol{F}^{\mathrm{T}}\boldsymbol{F} - \boldsymbol{1}\right) \quad \text{and} \quad \boldsymbol{A} = \frac{1}{2}\left(\boldsymbol{1} - \boldsymbol{F}^{-\mathrm{T}}\boldsymbol{F}^{-1}\right), \tag{3.107}$$

and even different strain rate tensors such as the velocity gradient $\boldsymbol{L}$ or its symmetric part: the strain rate tensor $\boldsymbol{\Delta}$,

$$\boldsymbol{L} = \dot{\boldsymbol{F}}\boldsymbol{F}^{-1} \quad \text{and} \quad \boldsymbol{\Delta} = \frac{1}{2}\left(\boldsymbol{L} + \boldsymbol{L}^{\mathrm{T}}\right). \tag{3.108}$$

But considering one definition instead of another does not provide an equivalent description and leads to different theories. Their use is also related to stress tensors definitions:

- The second Piola–Kirchhoff tensor $\boldsymbol{S}$ associated with $\boldsymbol{E}$ (defined in the reference configuration $\mathcal{C}_0$),
- The Cauchy stress tensor $\boldsymbol{\sigma}$ associated with $\boldsymbol{\Delta}$ (defined in $\mathcal{C}(t)$), and
- The first Piola–Kirchhoff stress tensor $\boldsymbol{\tau}$ associated with $\boldsymbol{F}$. The tensors $\boldsymbol{F}$ and $\boldsymbol{\tau}$ are not symmetric.

The following relations stand:

$$(\det \boldsymbol{F})\,\boldsymbol{\sigma} = \boldsymbol{\tau}\boldsymbol{F}^{\mathrm{T}} = \boldsymbol{F}\boldsymbol{S}\boldsymbol{F}^{\mathrm{T}}, \tag{3.109}$$

and one has the energetic equivalence

$$\frac{1}{\rho}\boldsymbol{\sigma} : \boldsymbol{\Delta} = \frac{1}{\rho_0}\boldsymbol{\tau} : \dot{\boldsymbol{F}} = \frac{1}{\rho_0}\boldsymbol{S} : \dot{\boldsymbol{E}}, \tag{3.110}$$

with ρ_0 as the initial density in $\mathcal{C}_0$ and $\rho = \rho_0/\det \boldsymbol{F}$ as the actual density in $\mathcal{C}(t)$.

There are also theoretical difficulties concerning the objectivity and framework independency of the time derivative of the stress tensor, the main question being, "In which framework or configuration do we take the derivatives?"

Different answers to this question lead to different definitions of the derivative (called objective), the most "popular" being the **Jaumann derivative**,

$$\frac{d_J}{dt}\boldsymbol{\sigma} = \dot{\boldsymbol{\sigma}} - \boldsymbol{W}\boldsymbol{\sigma} + \boldsymbol{\sigma}\boldsymbol{W} \quad \text{with} \quad \boldsymbol{W} = \frac{1}{2}\left(\boldsymbol{L} - \boldsymbol{L}^T\right). \tag{3.111}$$

The **Truesdell derivative** is also considered; it corresponds to the reactualized Lagrangian formulation,

$$\frac{d_{Tr}}{dt}\boldsymbol{\sigma} = \dot{\boldsymbol{\sigma}} - \boldsymbol{L}\boldsymbol{\sigma} - \boldsymbol{\sigma}\boldsymbol{L}^T + \text{tr}(\boldsymbol{L})\boldsymbol{\sigma}, \tag{3.112}$$

but other good (better?) definitions exist. We briefly present here three possible formulations and further details about objectivity or about plasticity and thermodynamics within the large deformation framework given in the books by G. Maugin (1992) and P. Ladevèze (1996). The thermodynamics framework of Chap. 1 then applies strictly if the constitutive equations are written in a "rotated" configuration.

In order to keep the presentation simple, only isotropic hardening (variables R, r) and isotropic damage (variables Y, D) are considered here.

3.4.1.1 Additive Split of the Strain Rate Tensor

The strain rates $\dot{\boldsymbol{\epsilon}}$, $\dot{\boldsymbol{\epsilon}}^e$, and $\dot{\boldsymbol{\epsilon}}^p$ of small deformation formulation are replaced by $\boldsymbol{\Delta}$, $\boldsymbol{\Delta}^e$, $\boldsymbol{\Delta}^p$, and the stress rate by Jaumann objective derivative (3.111). The strain rate partition reads

$$\boldsymbol{\Delta} = \boldsymbol{\Delta}^e + \boldsymbol{\Delta}^p. \tag{3.113}$$

The elasticity law is written in the rate form

$$\frac{d_J}{dt}\boldsymbol{\sigma} = \underline{\boldsymbol{a}} : \boldsymbol{\Delta}^e, \tag{3.114}$$

with $\underline{\boldsymbol{a}}$ an objective elasticity tensor.

The yield criterion is written in terms of the von Mises invariant of the Cauchy stress,

$$f = \frac{\sigma_{eq}}{1-D} - R - \sigma_y \quad \text{with} \quad R = R(r) \tag{3.115}$$

and the normality rule

$$\boldsymbol{\Delta}^p = \dot{\lambda}\frac{\partial f}{\partial \boldsymbol{\sigma}} \quad \text{with} \quad \begin{cases} \dot{\lambda} = \dot{r} = \dot{p}(1-D), \\ \dot{p} = \sqrt{\frac{2}{3}\boldsymbol{\Delta}^p : \boldsymbol{\Delta}^p} \end{cases} \tag{3.116}$$

which ensures the plastic incompressibility.

The damage evolution remains the same as for the small strain formulation of Chap. 1:

$$\dot{D} = \left(\frac{Y}{S}\right)^s \dot{p} \quad \text{if} \quad p > \epsilon_{pD}. \tag{3.117}$$

3.4.1.2 Multiplicative Decomposition of F

The multiplicative decomposition

$$\boldsymbol{F} = \boldsymbol{F}^\text{e} \boldsymbol{F}^\text{p} \tag{3.118}$$

defines an intermediate relaxed configuration which in small strain plasticity corresponds to the linearly-unloaded state. It defines an elastic strain tensor,

$$\boldsymbol{E}^\text{e} = \frac{1}{2} \left(\boldsymbol{F}^{\text{eT}} \boldsymbol{F}^\text{e} - \boldsymbol{1} \right), \tag{3.119}$$

associated with the stress tensor $\boldsymbol{S}^\text{e}$ in order to derive the elasticity law as

$$\boldsymbol{S}^\text{e} = \frac{\rho_i}{\rho} \boldsymbol{F}^{\text{e}-1} \boldsymbol{\sigma} \boldsymbol{F}^{\text{e}-1\text{T}} = \rho_i \frac{\partial \psi}{\partial \boldsymbol{E}^\text{e}}, \tag{3.120}$$

with ρ_i the density for the relaxed configuration.

The yield criterion is written as

$$f = \frac{\Sigma_\text{eq}}{1-D} - R(r) - \sigma_\text{y} \quad \text{with} \quad \boldsymbol{\Sigma} = \frac{1}{\rho_i} \boldsymbol{F}^{\text{eT}} \boldsymbol{F}^\text{e} \boldsymbol{S}^\text{e} \tag{3.121}$$

and the normality rule of standard generalized materials gives

$$\dot{\boldsymbol{F}}^\text{p} \boldsymbol{F}^{\text{p}-1} = \dot{\lambda} \frac{\partial f}{\partial \boldsymbol{\Sigma}} \quad \text{with} \quad \begin{cases} \dot{\lambda} = \dot{r} = \dot{p}(1-D) \\ \dot{p} = \sqrt{\frac{2}{3} \left(\dot{\boldsymbol{F}}^\text{p} \boldsymbol{F}^{\text{p}-1} \right) : \left(\dot{\boldsymbol{F}}^\text{p} \boldsymbol{F}^{\text{p}-1} \right)}, \end{cases} \tag{3.122}$$

$$\dot{D} = \left(\frac{Y}{S} \right)^s \dot{p} \quad \text{if} \quad p > \epsilon_\text{pD}.$$

3.4.1.3 Reactualized Lagrangian Formulation

This formulation extends the rate formulation $\dot{\boldsymbol{\sigma}} = \underline{\boldsymbol{L}} : \dot{\boldsymbol{\epsilon}}$ of small strain problems to large deformations

- By considering $\mathcal{C}(t)$ permanently as the reference configuration: take $\boldsymbol{F}(t) = \boldsymbol{1}$ at any time t (but $\dot{\boldsymbol{F}} \neq 0$) and
- By still considering the rate formulation, but on the deformed structure. It is then equivalent to $\dot{\boldsymbol{S}} = \underline{\boldsymbol{L}} : \dot{\boldsymbol{E}}$, with $\boldsymbol{S}$ and $\boldsymbol{E}$ defined with respect to $\mathcal{C}(t)$.

From (3.109) and (3.110), one can see that $\boldsymbol{S}$ and $\boldsymbol{E}$ are in fact replaced by $\boldsymbol{\sigma}$ and $\boldsymbol{\Delta}$ and after some calculations that $\dot{\boldsymbol{S}}$ is replaced by the Truesdell derivative of the Cauchy stress tensor $\frac{\text{d}_\text{Tr}}{\text{d}t} \boldsymbol{\sigma}$.

For numerical finite element analysis, the extension to finite strains is incremental and for the computation of the solution at time t_{n+1}, the deformed mesh $\mathcal{C}(t) = \mathcal{C}(t_n)$ is used.

3.4.2 Deep Drawing Limits♂♂

During deep drawing or forming processes to induce large plastic deformation of thin sheets, there are limitations in strains due to instabilities at small scale, instability of the phenomenon of strain localization, and crack initiation.

The first limitation governs the appearance of the surfaces. The second governs the local reduction of thickness; it appears sooner in the process of hydro-forming than in classical deep drawing. The third limitation is the local rupture of the sheet.

3.4.2.1 Forming Limits by Strain Localization

Knowing the loading path, the limit may be predicted by the strain damage localization criterion of Sect. 1.6.2:

$$\det(\vec{n}\underline{L}\vec{n}) = 0, \tag{3.123}$$

where $\underline{L}$ is the fourth order tangent operator derived from the elasto-plastic constitutive equations that may or may not be coupled with damage and $\vec{n}$ is the normal to the plane of localization.

Considering the problem of plane stress in proportional loading of Sect. 3.3.4, but without any simplified hypothesis here, it is possible to find at each increment the minimum of the criterion $\det(\vec{n}\underline{L}\vec{n})$ with respect to any normal $\vec{n}$ and to check when $\det(\vec{n}\underline{L}\vec{n}) = 0$ is satisfied.

Working with uniform fields of stresses and strains, we have

$$\boldsymbol{\sigma} = \begin{bmatrix} \sigma_1 & 0 & 0 \\ 0 & \sigma_2 & 0 \\ 0 & 0 & 0 \end{bmatrix}, \quad \boldsymbol{\epsilon} = \begin{bmatrix} \epsilon_1 & 0 & 0 \\ 0 & \epsilon_2 & 0 \\ 0 & 0 & -(\epsilon_1 + \epsilon_2) \end{bmatrix}. \tag{3.124}$$

and Fig. 3.11 gives an example for steel with the following parameters: $E = 200000$ MPa, $\nu = 0.3$, $\sigma_y = 200$ MPa, $K_p^y = 10000$ MPa, $M_y = 1.66$, $\epsilon_{pD} = 0$, $s = 1$, and $S = 0.5$ MPa. Numerical simulations show that the shape of the curve is very sensitive to the value of the plastic exponent M_y of the hardening law $R = K_p^y r^{1/M_y}$.

3.4.2.2 Forming Limits by Mesocrack Initiation

The problem of plane stress has already been solved in Sect. 3.3.4 but with a set of simplifications to yield a closed-form solution: proportional loading, elastic strain neglected, hardening saturated, and damage considered as isotropic.

With numerical simulations it is possible to be closer to reality where the forming processes are executed in several steps eventually on different

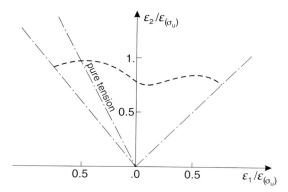

Fig. 3.11. Strain localization limit curves of deep drawing (I. Doghri 1989)

machines which correspond to non-proportional loadings, inducing effects of anisotropy. Furthermore, it can be interesting to beneficiate from particular loading paths to minimize the damage in order to increase the limit of forming.

Let us consider again the case of a plane stress problem and look for the strain forming limits in the plane (ϵ_1, ϵ_2). The limits are defined by the anisotropic critical damage criterion

$$\max D_I = D_c \qquad (3.125)$$

for different proportional and non-proportional loading paths represented in Fig. 3.12 as thin lines finishing by the point of mesocrack initiation.

The material is ARCELOR steel SOLDUR 355. The constitutive equations are those of elasto-plasticity coupled with anisotropic damage and microdefects closure effect of Table 1.6 (Sect. 1.5). A single isotropic hardening is considered with an exponential law $R = R_\infty^y (1 - \exp(-b_y r))$. Tensile tests performed on small specimens cut inside large plates damaged either in uniaxial tension or in plane tension, giving both the η-parameter and the anisotropic damage evolution for different stress triaxialities (see Sect. 1.3.3). The set of material parameters is: $E = 230000$ MPa, $\nu = 0.3$, $\sigma_y = 375$ MPa, $R_\infty^y = 120$ MPa, $b_y = 25$, $\epsilon_{pD} = 2.5 \cdot 10^{-2}$, $h_a = 0$, $S = 0.43$ MPa, $s = 4$, $\eta = 2.8$, and $D_c = 0.5$. Note that this corresponds to the same stress-strain response in tension as with $h_a = 1$ and $S = 0.57$ MPa for the model without microdefects closure effect. The computations are performed with the ZeBuLon FE code using the numerical implicit scheme of Sect. 2.2.5.

The numerical results for many monotonic proportional loading paths allow us to build the crack initiation limit curve of deep drawing (solid curve of Fig. 3.12) in which are also reported the results for some non-proportional paths (points •).

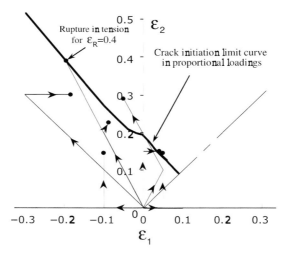

Fig. 3.12. Computed crack initiation limit curve and points of deep drawing of SOLDUR 355 (R. Desmorat and J. Besson 2004)

- It is seen that very large strains may be obtained for loadings close to pure shear $\epsilon_2 = -\epsilon_1$. This is the case of hydroforming of tube-like structures with axial compression added to pressure.
- The reality taking into account both localization and damage is a curve below the crack initiation curve of Fig. 3.12, with a flatter part for $\epsilon_1 > 0$ due to easier strain localization conditions.
- Some computed non-proportional loading paths exhibit a crack initiation below the proportional limit curve. Some specific non-proportional loadings increase the limit of forming as shown by F. Moussy and J.P. Cordebois (1990). This is mainly the case for the path $\epsilon_2 = 2\epsilon_1$ up to $\epsilon_1 = 0.1$, followed by $\dot\epsilon_2 = -\dot\epsilon_1/2$ (parallel to the tensile path) up to $\epsilon_2 = 0.29$ which is found to be very beneficial.

This kind of calculations offers the possibility to optimize loading paths to reach a state of strain with a minimum value of the damage.

3.4.3 Damage in Cold Extrusion Process♂♂
(K. Saanouni 2002)

Another important difficulty concerning the forming processes is the existence and then the determination of possible local instabilities. The global instabilities are well predicted from elasto-plastic structure computations. The local instabilities leading to damage initiation and to microcrack growth can only be studied by use of the damage mechanics constitutive equations. They correspond to the reach of the critical damage D_c for the local damage as in the following example or to the strain damage localization $\det(\vec{n}\underline{\boldsymbol{L}}\vec{n}) = 0$ as in the example of Sect. 3.4.4.

The forward cold extrusion process consists of obtaining a bar or a wire of diameter d from a bigger bar of diameter d_0 by pushing the material through a die. The quality of the result depends on several parameters: the ratio of the diameter d/d_0, the angle of the die α, the lubrication, and of course the ductility of the material. The optimization of the process consists of choosing these parameters in order to obtain bars with adequate properties and to avoid large damage leading to chevrons as in Fig. 3.13.

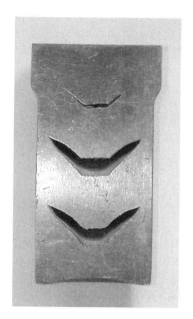

Fig. 3.13. Picture (M. Grange) and numerical simulations of damage and inside chevrons cracking during forward cold extrusion

It is possible to catch this periodic phenomenon using the finite strains framework and the constitutive equations of elasto-plasticity coupled with damage of Sect. 1.5.2, Table 1.3, or something similar. The corresponding finite element analysis does not ensure the convergence regarding the mesh size but in the following example, the smallest mesh size was adjusted to correctly fit a simple tensile test calculated by an axisymmetric FEA in the softening range up to failure. For the extrusion of the circular bar of diameter $d_0 = 35$ mm reduced to a diameter $d = 31$ mm, the meshing consists of 1312 triangular linear elements. The material is a ductile steel of yield stress $\sigma_y = 300$ MPa, ultimate stress $\sigma_u = 390$ MPa, and rupture strain $\epsilon_R = 0.7$. It is represented by the damage law $\dot{D} = (Y/S)^{s_1}\,\dot{p}/(1-D)^{s_2}$, with $S = 500$ MPa, $s_1 = 2.5$, and $s_2 = 7.5$. The axisymmetric calculation of the extrusion process corresponding to the picture of Fig. 3.13 was performed with FORGE 2 code using the reactualized Lagrangian formulation, the elements being removed when the damage reaches its critical value. The result is given in Fig. 3.13, with a close similitude with the experiment. The discrete appearance of the chevrons is due to the redistribution of stresses, a consequence of the coupling of damage with elasticity – if this coupling is neglected, the damaged zone is continuous all along the axis of the specimen.

As far as the specimen goes in the die, there is a large gradient of damage mainly governed by the stress triaxiality with a loss of rigidity. As the material moves on, a chevron crack forms at an angle determined by the angle of the die. At the same time the relaxation of stresses in the ligament behind the chevron does not produce any more damage until the material has moved on a certain distance to recover a damaging stress level. A new chevron crack then initiates ... and so on ...

From this kind of analysis performed on many cases, the following qualitative observations can help to decrease the risk of such damage and obtain a small damage state without any chevron:

- A small diameter reduction (e.g., 16%) needs a small die angle ($\approx 4°$)
- A large diameter reduction (e.g., 30%) may be obtained in good conditions with a large die angle ($\approx 15°$) which decreases the contact surface
- A large ductility of the material is better, of course
- A large friction (e.g., a friction coefficient of 0.2) can prevent the formation of chevrons but it increases the load to perform the process and damage may occur at the surface of the material due to the heavy contact with the die.

Nevertheless, these results may be helpful in the preparation of a precise numerical procedure of optimization of any extrusion process.

3.4.4 Crack Initiation Direction♂♂
(R. Billardon and I. Doghri 1989)

The question of the direction in which a mesocrack initiates is of first importance if a Fracture Mechanics analysis must be performed afterwards to determine the conditions of crack growth (see Sect. 3.2.5). An example is given below using the localization criterion of Sect. 1.6.2 for which the mathematical problem consists of determining the normal $\vec{n}$ (and the tangent operator $\underline{\boldsymbol{L}}$) which makes

$$\det(\vec{n}\underline{\boldsymbol{L}}\vec{n}) = 0. \qquad (3.126)$$

Consider the notched specimen of Fig. 3.14 with an out-of-axis hole of diameter $\phi = 1$mm, inducing a stress concentration area which may model an initial macroscopic defect of non-negligible size at the scale of the structure. The material is an aluminum alloy 2024 at room temperature.

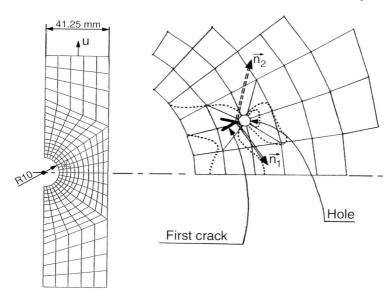

Fig. 3.14. Notched plate with a hole and iso-damage curves with the strain localization directions (R. Billardon and I. Doghri 1989)

A monotonic displacement $u(t)$ is applied. Before any crack initiates, the small hole ovalizes strongly and the specimen bends (without any buckling). The first crack initiates on the hole edge and propagates quickly toward the notched edge. A second crack initiates also on the hole and propagates through the whole specimen up to complete failure. The direction of the first crack is not easily guessed as it does not coincide with the shortest distance between the hole and the notch. Furthermore, the two possible directions for the crack initiated are observed on each side of the specimen, with the crack finally taking an intermediate direction.

The numerical simulation of this experiment has been performed in elasto-plasticity coupled with isotropic damage with ABAQUS code. The mesh is made of 8-node isoparametric elements with reduced integration. Plane stress conditions apply. The material parameters of the 2024 aluminum alloy considered are:

- $E = 72000$ MPa, $\nu = 0.33$ for elasticity
- $\sigma_y = 273.5$ MPa for the yield stress
- $R_\infty = 275$ MPa, $b = 1.86$ for the exponential isotropic hardening law
- $C = 4950$ MPa, $\gamma = 37.2$ for the nonlinear kinematic hardening law
- $p_D = \epsilon_{pD} = 0$, $S = 1.3$ MPa, $s = 1$ for damage evolution

Figure (3.14) also gives the iso-damage curves around the hole at strain-damage localization where $\det(\vec{n}\boldsymbol{L}\vec{n}) = 0$. In agreement with the experiments, two normals of localization ($\vec{n}_1$ and $\vec{n}_2$) are obtained showing the ability of the criterion to predict the direction of the cracks initiated drawn in solid lines on the figure.

3.4.5 Porous Materials – the Gurson Model♂♂
(A.L. Gurson 1977, V. Tvergaard and A. Needleman 1984)

Porous materials submitted to heavy loadings are subjected to volume or density change. Furthermore, the cavities can grow due to a ductile damage process. This is the case for geomaterials, powders, and also for certain steels where cavities may nucleate at the neighborhood of defects, grow by plastic deformation, and finally give a mesocrack by coalescence. The early Gurson model mathematically describes mainly the second mechanism. The Gurson–Tvergaard–Needleman (GTN) model describes the three mechanisms. This model may be used in numerical structure simulations to predict flow localization or final failure. It is restricted to proportional loading and applies badly when the loading induces mainly shear.

The basic micromechanics cell of Gurson analysis contains a hollow sphere and the damage variable is the volume fraction of cavities or the porosity f_v. At microlevel the material surrounding the cavities is considered elasto-plastic with isotropic hardening R and a yield stress $\sigma_s = \sigma_y + R$. At mesolevel the homogenized behavior is plasticity with volume change (i.e., $\epsilon^p_{kk} \neq 0$).

The plastic potential F_{Gurson} is obtained for proportional loading by considering an appropriate velocity field in the matrix,

$$F_{\text{Gurson}} = \frac{\sigma_{\text{eq}}^2}{\sigma_s^2} + 2q_1 f_v^\star \cosh\left(\frac{3}{2} q_2 \frac{\sigma_H}{\sigma_s}\right) - 1 - (q_1 f_v^\star)^2, \qquad (3.127)$$

with q_1 and q_2 as the two material parameters added by V. Tvergaard and A. Needleman to avoid an overestimation of the rupture strain which occurs for $f_v = f_R < 1$. Introduced by the same authors, the effective porosity $f_v^\star$ models the fast coalescence of voids once a threshold f_c is reached:

$$f_v^\star = \begin{cases} f_v, & \text{if } f_v < f_c, \\ f_c + \delta_{\text{GTN}} \cdot (f_v - f_c) & \text{if } f_v \geq f_c, \end{cases} \qquad (3.128)$$

where $\delta_{\text{GTN}} = \dfrac{q_1^{-1} - f_c}{f_R - f_c}$ is also a material parameter.

When $f_v^\star = f_v = 0$, the plastic potential (3.127) reduces to the von Mises criterion $\sigma_{\text{eq}} - (\sigma_y + R) = 0$.

Remark – *The Rousselier model (1981) uses the plastic potential coupled with porosity damage f_v:*

$$F_{\text{Rousselier}} = \frac{\sigma_{\text{eq}}}{\rho} - R(p) - \sigma_y + d_R f_v \sigma_1 \exp\left(\frac{\sigma_H}{\rho \sigma_1}\right) = 0, \qquad (3.129)$$

with $\rho = \frac{1-f}{1-f_0}$ as the relative density and f_0 as the initial porosity, and where d_R and σ_1 are material parameters.

The normality rule applied on F_{Gurson} gives the plastic strain rate as

$$\dot{\epsilon}^{\text{P}}_{ij} = \dot{\lambda}\left[\frac{3\sigma^{\text{D}}_{ij}}{\sigma^2_{\text{s}}} + \frac{q_1 q_2 f^\star_{\text{v}}}{\sigma_{\text{s}}}\sinh\left(\frac{3}{2}q_2\frac{\sigma_{\text{H}}}{\sigma_{\text{s}}}\right)\delta_{ij}\right], \quad (3.130)$$

where the plastic multiplier $\dot{\lambda}$ is determined from the consistency condition $F_{\text{Gurson}} = 0$ and $\dot{F}_{\text{Gurson}} = 0$. The hardening law in the yield criterion is a power or exponential law function $R(p)$ of the equivalent plastic strain p defined by the energetic equivalence

$$\sigma_{ij}\dot{\epsilon}^{\text{P}}_{ij} = (1-f_{\text{v}})\sigma_{\text{s}}\dot{p} \quad (3.131)$$

and p differs from the von Mises accumulated plastic strain $\int\sqrt{\frac{2}{3}\dot{\epsilon}^{\text{P}}_{ij}\dot{\epsilon}^{\text{P}}_{ij}}\,\mathrm{d}t$.

Next, the porosity evolution law

$$\dot{f}_{\text{v}} = (1-f_{\text{v}})\dot{\epsilon}^{\text{P}}_{kk} + \dot{f}_{\text{n}} \quad (3.132)$$

states that the porosity rate is the sum of the void growth rate $(1-f_{\text{v}})\dot{\epsilon}^{\text{P}}_{kk}$ determined by the plastic incompressibility of the material surrounding the cavities and the voids nucleation rate $\dot{f}_{\text{n}}$. Phenomenological evolution laws are used to model the nucleation phenomenon assumed either controlled

- by the stress,

$$\dot{f}_{\text{n}} = A_\sigma \dot{\sigma}_{\text{s}} + B_\sigma \frac{\dot{\sigma}_{kk}}{3}, \quad (3.133)$$

- or by the plastic strain,

$$\boxed{\dot{f}_{\text{n}} = A_{\text{n}}\dot{p},} \quad (3.134)$$

with A_σ, B_σ, and A_{n} as material constants to be identified or, better, functions of the strains and of the stress triaxiality.

The second law ($\dot{f}_{\text{n}} = A_{\text{n}}\dot{p}$) is often used with the statistically-based expression

$$\boxed{A_{\text{n}} = \frac{f_{\text{N}}}{S_{\text{N}}\sqrt{2\pi}}\exp\left(-\frac{1}{2}\left(\frac{p-\epsilon_{\text{N}}}{S_{\text{N}}}\right)^2\right)} \quad (3.135)$$

proposed by C.C. Chu and V. Tvergaard in 1980. Three additional material parameters are introduced:

- f_{N} is the volume fraction of voids which may nucleate,

$$f_{\text{N}} = \int_{-\infty}^{\infty} A_{\text{n}}(p)\mathrm{d}p \approx \int_{0}^{p_R} A_{\text{n}}(p)\mathrm{d}p \ll 1 \quad (3.136)$$

- ϵ_N is the mean strain at nucleation
- S_N is the standard deviation of the rupture strain

Finally, the conditions of ductile failure are obtained through a numerical analysis of elasto-plasticity coupled with porosity. It gives the condition of localization of plastic strain together with the final rupture by coalescence of cavities and initiation of a mesocrack for $f_v^\star = 1/q_1$ that is for a porosity:

$$f_R = f_c + \frac{1}{\delta_{GTN}} \left(\frac{1}{q_1} - f_c \right). \quad (3.137)$$

3.4.5.1 Material Parameter Identification of the GTN Model

Each application with a specific material needs:

- The elasticity parameters E and ν.
- The yield stress σ_y and either the power hardening parameters K_p^y and M_y or the exponential hardening parameters R_∞^y and b_y (see Sect. 1.5.1) determined on specimens without any damage.
- The parameters q_1 and q_2 governing the voids' growth are determined from numerical micromechanics analysis. In most cases good results for metals are obtained with $q_1 \approx 1.5$ and $q_2 \approx 1$.
- It is the same for the coalescence parameters f_R and f_c where estimations from numerical simulations and experiments on metals are $f_R = 0.25$ and $f_c = 0.09$ ($\delta_{GTN} = 3.6$).
- The initial void volume fraction and the nucleation parameters f_N, ϵ_N, and S_N can be estimated only from micrographs analysis at different states of deformation. $\epsilon_N = 0.3$ and $S_N = 0.1$ are reasonable values for metals.

3.4.5.2 Cross Identification of the Unified Damage Law $\dot{D}(Y, \dot{p})$

It is the purpose here to determine the material parameters of the damage law ($\dot{D} = (Y/S)^s \dot{p}$) from the knowledge of the GTN parameters following the method described in Sect. 2.4.4. It is also the purpose to estimate the damage threshold in monotonic tension (ϵ_{pD}). Recall that for Continuum Damage Mechanics, the isotropic damage (D) may either be evaluated on unloading through the elasticity change or on the stress softening part of the stress-strain curve. As for porosity damage, the GTN model neglects the coupling of elasticity with damage, allowing us to use the stress softening effect to make a link between both theories up to a fast identification procedure based on the expressions of the yield surfaces of the damaged material.

For the GTN model, the plastic potential is rewritten as

$$F_{Gurson} = \frac{\sigma_{eq}^2}{(\sigma_y + R)^2} + 2q_1 f_v^\star \cosh\left(\frac{3}{2} q_2 T_X\right) - 1 - (q_1 f_v^\star)^2 = 0, \quad (3.138)$$

with T_X as the stress triaxiality $\frac{1}{3}\frac{\sigma_{kk}}{\sigma_s}$.

For the damage model of Chap. 1, the yield criterion is (isotropic damage)

$$f = \frac{\sigma_{eq}}{1-D} - R - \sigma_y = 0 \qquad (3.139)$$

so that for saturated hardening $\sigma_y + R = \sigma_u$ (the ultimate stress), the damage D measured in tension is directly related to the void volume fraction f_v by simple substitution of (3.139) in (3.138),

$$D = 1 - \sqrt{1 + (q_1 f_v^\star)^2 - 2q_1 f_v^\star \cosh\left(\frac{3}{2} q_2 T_X\right)}. \qquad (3.140)$$

Next, we plot the damage D versus the accumulated plastic strain p calculated from the GTN model.

As $D = \left(\frac{\sigma_u^2 R_\nu}{2ES}\right)^s (p - \epsilon_{pD})$ for a monotonic tensile loading, a curve fitting gives, for a given s, both the damage threshold ϵ_{pD} and the ratio $\sigma_u^2/2ES$. If no other information is available, take $s = 1$. If results for different stress triaxiality are available, plot $\frac{dD}{dp}$ vs $\ln R_\nu$ (with $R_\nu = \frac{2}{3}(1+\nu) + 3(1-2\nu)T_X^2$).
As

$$\ln \frac{dD}{dp} = \ln \left(\frac{\sigma_u^2}{2ES}\right)^s + s \ln R_\nu, \qquad (3.141)$$

s is the slope of the curve obtained.

To illustrate the procedure, consider here that $f_v^\star = f_v$ (acceleration factor $\delta_{GTN} = 1$) and the porosity remains small. With the additional assumption of a constant stress triaxiality T_X (equal to 1/3 in tension), a closed-form expression may be derived from the GTN model for D vs p. The porosity evolution law becomes then

$$\dot{f}_v \approx \left[\frac{3}{2} q_1 q_2 f_v \sinh\left(\frac{3}{2} q_2 T_X\right) + A_n(p)\right] \dot{p}, \qquad (3.142)$$

where $\dot{p}$ is nearly the von Mises accumulated plastic strain rate. The solution for f_v is

$$f_v = \exp\left[\frac{3}{2} q_1 q_2 \sinh\left(\frac{3}{2} q_2 T_X\right) p\right] \cdot \int_0^p A_n(p) \exp\left[-\frac{3}{2} q_1 q_2 \sinh\left(\frac{3}{2} q_2 T_X\right) p\right] dp \qquad (3.143)$$

and the damage D is calculated by use of (3.140). The $D(p)$ curves obtained for the nucleation law $A_n(p)$ (3.135) are plotted in Fig. 3.15 for different values of the parameter f_N. Note that these curves could be plotted with a better accuracy (and also for larger values of the acceleration factor δ_{GTN}) from computations using Gurson or GTN model implemented in a finite element code.

180   3 Ductile Failures

Linear regressions of the $D(p)$ curves correspond to the integrated unified damage law

$$D = \left(\frac{\sigma_u^2 R_\nu}{2ES}\right)^s (p - \epsilon_{\text{pD}}). \tag{3.144}$$

With $R_\nu = 1$ in tension, they give for each f_N the damage threshold ϵ_{pD} and the slope $\left(\frac{\sigma_u^2 R_\nu}{2ES}\right)^s$. The damage exponent is determined from the stress triaxiality effect on damage growth ((3.141) with D given by (3.140)): the derivative $\frac{dD}{dp}$ is performed numerically and linear regression on $\ln\frac{dD}{dp}|_{p=p_{\text{given}}}$ vs $\ln R_\nu$ gives s as the slope of the curve (a given accumulated plastic strain $p = p_{\text{given}}$ is considered: $p_{\text{given}} = 0.25$ for $f_N = 0.01, 0.05, 0.1$, and 0.2).

The results are summarized in Table 3.1. An interesting point, a damage threshold of 0.15 is found for metals: physical existence of porosity at very small strains does not correspond to mechanical damage occurrence. Knowing the ultimate stress and the Young's modulus finally gives S.

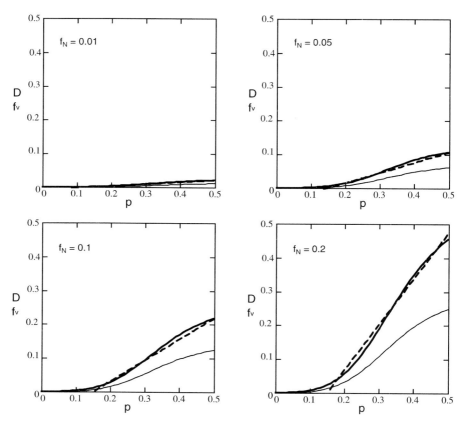

Fig. 3.15. Cross identification of the unified damage law ($\epsilon_N = 0.3$, $S_N = 0.1$). Damage D in thick line, porosity f_v in thin line, damage law in dotted line

Table 3.1. Results of cross identification for $q_1 = 1.5$, $q_2 = 1$, $\delta_{\text{GTN}} = 1$, $\epsilon_N = 0.3$, and $S_N = 0.1$.

f_N	0.01	0.05	0.1	0.2
ϵ_{pD}	0.15	0.15	0.15	0.15
s	2	2.5	2.5	2.5
$\dfrac{\sigma_u^2}{2ES}$	0.25	0.62	0.83	1.13

Last, consider that the reach of the critical damage D_c in tension corresponds to the rupture porosity $f_v^\star = f_R$. With $T_X \approx 1/3$, the critical damage is then

$$D_c = 1 - \sqrt{1 + (q_1 f_R)^2 - 2q_1 f_R \cosh \frac{q_2}{2}} \qquad (3.145)$$

and ranges between 0.1 and 0.5 (nothing new!) as $f_R = 0.1$ gives $D_c = 0.17$, $f_R = 0.2$ gives $D_c = 0.36$, and $f_R = 0.25$ gives $D_c = 0.46$.

Example: for the values of the GTN parameters of Table 3.1 for $f_N = 0.1$ and $f_R = 0.2$, the unified damage law reads:

$$\boxed{\begin{aligned} \dot{D} &= (0.83 \cdot R_\nu)^{2.5} \dot{p} \quad \text{if} \quad p_D > 0.15\,, \\ D_c &= 0.36 \longrightarrow \text{mesocrack initiation.} \end{aligned}} \qquad (3.146)$$

Now we have a law which can be adopted to obtain results for steel when material data are ignored!

3.4.6 Frame Analysis by Lumped Damage Mechanics
(J. Flórez-López 1998)

The analysis of inelastic frames is an important subject in civil engineering for metallic as well as concrete reinforced structures. The plastic hinge concept and limit analysis are efficient tools to design metallic frames and the conventional theory is based on perfect-plasticity. The extension to damage is possible by concentrating the damage in the hinges and is called Lumped Damage Mechanics. It applies to ductile monotonic failures as well as to cyclic and dynamic (seismic) loadings.

The model of a frame member between nodes i and j is made of two inelastic hinges i and j and an elastic column whose shape and size depend on the structure studied (but of elastic behavior Fig. 3.16). Generalized stresses $\{M\} = (M_i, M_j, N)^{\text{T}}$ and deformations $\{\Phi\} = (\phi_i, \phi_j, \delta)^{\text{T}}$ are used to write the frame behavior instead of stress and strain tensors in continuum mechanics. For instance, elasticity reads

$$\{\Phi\} = [S]\{M\} \quad \text{where} \quad [S] \text{ is the symmetric compliance matrix.} \qquad (3.147)$$

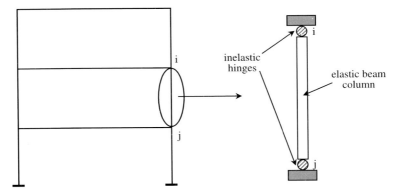

Fig. 3.16. Frame and lumped dissipation model of a frame member

We define the complementary energy for a damaged frame member as

$$W^\star = \frac{1}{2}\{M\}^{\mathrm T}[\tilde S]\{M\} \quad \text{and} \quad [\tilde S] = \begin{bmatrix} \dfrac{S_{11}}{1-d_i} & S_{12} & 0 \\ S_{21} & \dfrac{S_{22}}{1-d_j} & 0 \\ 0 & 0 & \dfrac{S_{33}}{1-d_n} \end{bmatrix} \quad (3.148)$$

where the components of $\{d\} = \{d_i, d_j, d_n\}^{\mathrm T}$ are the lumped damage variables. Note that the form for $W^\star$ is fully similar to the form for the complentary strain energy for anisotropic composite layers (see Sect. 7.4.1).

Inelastic frame deformation $\{\Phi^{\mathrm p}\}$ occurs due to plasticity and damage in the hinges so that the generalized state laws for elasticity coupled with damage are

$$\{\Phi^{\mathrm e}\} = \{\Phi - \Phi^{\mathrm p}\} = \frac{\partial W^\star}{\partial \{M\}},$$
$$\{y\} = \frac{\partial W^\star}{\partial \{d\}} \quad (3.149)$$

or

$$\phi_i^{\mathrm e} = \phi_i - \phi_i^{\mathrm p} = S_{11}\frac{M_i}{1-d_i} + S_{12}M_j,$$
$$\phi_j^{\mathrm e} = \phi_j - \phi_j^{\mathrm p} = S_{22}\frac{M_j}{1-d_j} + S_{21}M_i, \quad (3.150)$$
$$\delta^{\mathrm e} = \delta - \delta^{\mathrm p} = S_{33}\frac{N}{1-d_n}.$$

For the generalized strain energy release rates, the state laws are

$$y_i = \frac{S_{11}M_i^2}{2(1-d_i)^2},$$
$$y_j = \frac{S_{22}M_j^2}{2(1-d_j)^2}, \quad (3.151)$$
$$y_n = \frac{S_{33}N^2}{2(1-d_n)^2}.$$

To complete the model and write the generalized evolution laws, we will neglect the inelastic axial effect ($\delta \approx \delta^{\mathrm{e}}$, $d_n \approx 0$). The yield function for the damaged plastic hinge i is then simply written as

$$f_i = \left| \frac{M_i}{1-d_i} - x_i \right| - m_{\mathrm{y}} \leq 0, \quad (3.152)$$

where m_{y} is the yield momentum and x_i stands for linear kinematic hardening (parameter c):

$$\dot{x}_i = c(1-d_i)\dot{\phi}_i^{\mathrm{p}} \quad \text{(no summation)}. \quad (3.153)$$

Finally, the generalized damage evolution law is

$$\dot{d}_i = \left(\frac{y_i}{S}\right)^s \dot{\pi}_i \quad \text{if} \quad \pi_i > \pi_{\mathrm{D}}, \quad (3.154)$$

where $\pi = \int_0^t |\dot{\phi}^{\mathrm{p}}|\mathrm{d}t$ is the generalized accumulated plastic deformation and π_{D}, S, and s are the frame member damage parameters.

Remark – *For reinforced concrete frames, use the damage framework of Sect. 7.4.1, and consider the following as damage law:*

$$d_i = \kappa^{-1}(y_{i\,\mathrm{max}}) \quad \text{and} \quad y_{i\,\mathrm{max}} = \sup_{\tau \in [0,t]} y_i(\tau) \quad (3.155)$$

instead of (3.154). J. Flórez-López proposes modelling the crack-resistance-like phenomenon by using

$$\kappa(d_i) = y_{i\,\mathrm{max}} \quad \text{and} \quad \kappa(d) = A + B\frac{\ln(1-d)}{1-d} \quad (3.156)$$

with A and B as frame member damage parameters.

The frame analysis is then performed by assembling the frame members (use beam boundary conditions) and by computing the structure response to applied nodal loads or displacements in case of ductile failures or to inertia forces in case of dynamic failures. There are not many nodal quantities for a real structure and the method is usually of low cost (cheaper than the multifiber beam modelling of Sect. 7.4.3). Some difficulty may be encountered in the identification of the frame members parameters: S_{11}, $S_{12} = S_{21}$, S_{22}, and S_{33} for elasticity; m_{y} and c for plasticity; π_{D}, S, and s for damage (or A and B for quasi-brittle damage). These parameters depend on the cross section of the member, which can have any shape, size, or reinforcement. The good thing is that damage is localized at the ends of the member frames and no instabilities are encountered, such as strain-damage localization in Continuum Damage Mechanics.

As an illustration, Fig. 3.17 shows the kind of results expected on a frame subjected to seismic loadings. The location and size of the dots represent the damage field. A design criterion such as $D < D_c$ or $D < D_{\text{given}}$ tells us when and where to repair the structure. Following conventional engineering criteria, the damages encountered in frame (b) have very high values and the frame should not be repaired.

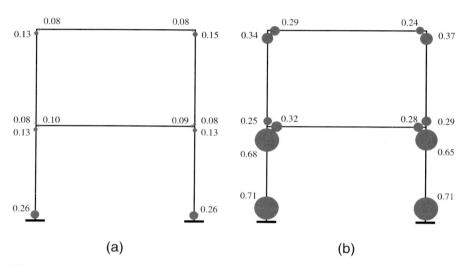

Fig. 3.17. Damage distribution in a frame (M.E. Marante and J. Flórez-López 1998) (**a**) after a low intensity shake and (**b**) after a severe shake

3.4.7 Predeformed and Predamaged Initial Conditions[♂♂]
(R. Billardon 1988)

The initial state of a material is a question with several answers! Most of the time, it is the state in which the material is received, but it can be before a forming process in order to calculate the process or after the process to determine the component's strength in service. It can be also the state of the material after an accident in service in order to determine the residual strength.

In all cases of structure calculation, the initial state is the state corresponding to the specimens used for the identification of the material parameters. A question arises immediately: how do we modify the constitutive equations to take into account a change without performing a completely new identification procedure?

This is the power of the thermodynamics approach to work with differential evolution laws of state variables which may have initial values. If an identification of a set of an elasto-(visco-)plastic coupled with damage constitutive equations has been performed on a material in its origin state (plane sheet for example), the initial conditions to use in a structure calculation after a forming process (a deep drawing for example), or any rheological modification, are the values of the internal variables at the end of the process:

- The accumulated plastic strain p_0
- The corresponding isotropic hardening R_0
- The corresponding kinematic hardening X_{ij}^0
- The eventual damage D_0 or, better, D_{ij}^0

Introducing these initial values in a numerical analysis consists of adding to the set of constitutive equations of Sect. 1.5 the fields of initial conditions:

$$\begin{aligned} R(t=0, M) &= R_0(M), \\ X_{ij}(t=0, M) &= X_{ij}^0(M), \\ p(t=0, M) &= p_0(M) \quad \text{in accordance with } R_0 \text{ and } X_{ij}^0, \\ D(t=0, M) &= D_0(M) \quad \text{or} \quad D_{ij}(t=0, M) = D_{ij}^0(M). \end{aligned} \quad (3.157)$$

In a finite element analysis they must be considered at each Gauss point, which is not a major problem. The main difficulty is to know that their values are often perturbed by an other initial condition coming from a structural effect: a field of residual self-equilibrated stresses, described by

$$\sigma_{ij}(t=0, M) = \sigma_{ij}^0(M). \quad (3.158)$$

The only way to properly determine these fields of initial conditions is to numerically simulate the process of their creation. Some experiments like the in situ microhardness or an X-ray analysis or a digital image correlation qmay give some indications but never enough information.

Concerning the ductile failure, the influence of non-zero initial conditions in proportional loading are:

- A modification of the yield stress (an increase if the original damage was zero or small),

$$\boxed{\sigma_y^0 = (\sigma_y + R_0 + X_{eq}^0)(1 - D_0)} \quad \text{with} \quad X_{eq}^0 = \sqrt{\frac{3}{2} X_{ij}^0 X_{ij}^0} \quad (3.159)$$

- A decrease of the plastic strain to rupture p_R:
  Back to the unified damage law of

$$\dot{D} = \left(\frac{\sigma_u^2 R_\nu}{2ES}\right)^s \dot{p} \quad \text{if} \quad \epsilon_p > p_D, \quad (3.160)$$

186   3 Ductile Failures

the integration with zero initial conditions gives

$$D_c = \left(\frac{\sigma_u^2 R_\nu}{2ES}\right)^s (p_R - p_D). \qquad (3.161)$$

The integration with initial conditions D_0 and $p_0 > p_D$ gives

$$D_c - D_0 = \left(\frac{\sigma_u^2 R_\nu}{2ES}\right)^s (p_R - p_0) \qquad (3.162)$$

from which the remaining plastic strain to rupture is

$$\boxed{p_R - p_0 = (p_R - p_D)\frac{D_c - D_0}{D_c}.} \qquad (3.163)$$

- Therefore some anisotropy can also be observed.

As an example, we will describe here the method used for the determination of the remaining strength of a pressurized gas pipe that has been damaged by a shock from a digging machine (Fig. 3.18):

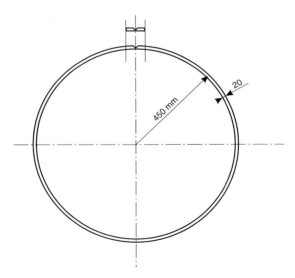

Fig. 3.18. Injured gas pipe and "equivalent" flat specimen

1. Careful measurement of the local geometry of the indented pipe in order to make a flat specimen ready for microhardness tests (Fig. 3.18) from the same material. Its notch is made as identical as possible to the accident notch by an indentation process.

2. Careful measurement of the microhardness all around the notch in order to determine the damage field and the extra hardening field induced by the indentation ((1.85) and (1.81) of Sect. 1.3.4):

$$D_0 = 1 - \frac{\tilde{H}}{H}\frac{\sigma_y}{\sigma_u} \quad \text{and} \quad R_0 = \frac{\sigma_s}{1-D_0} - \sigma_y. \quad (3.164)$$

The accumulated plastic strain field is deduced from the plastic constitutive equation:

$$p_0 = \left(\frac{R_0}{K_p^y}\right)^{M_y}. \quad (3.165)$$

3. These fields are given in Fig. 3.19 for the following notch geometry and material parameters corresponding to a low carbon laminated steel:
   - A notch with a depth of 2 mm made by an indenter with an angle of 60°
   - Parameters of the low carbon laminated steel determined from the initial virgin material: $\sigma_y = 575$ MPa, $K_p^y = 552$ MPa, and $M_y = 2.04$

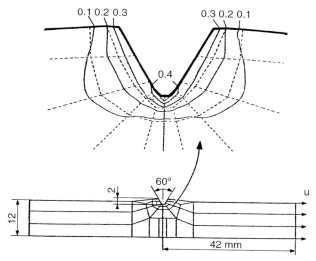

Fig. 3.19. Fields of accumulated plastic strain and damage in the neighboring of the notch (R. Billardon 1988)

4. Assuming these fields are representative of what happened in the gas pipe, they are introduced as initial conditions by interpolation at each Gauss point in a finite element analysis performed with the code ABAQUS in elasto-plasticity coupled with damage. The analysis corresponds to the in-service pressure increased up to crack initiation at the notch. In comparison to the strength of the undamaged gas pipe, the result is

a reduction of the pressure at crack initiation by an amount of about 50%. Regarding the fatigue strength due to a cyclic variation of the pressure the reduction is a factor of about 100 on the number of cycles to crack initiation.

3.4.8 Hierarchic Approach up to Full Anisotropy♂,♂♂♂

From the simplest analysis that require a low level of calculation and a small number of material parameters to advanced numerical methods with the highest degree of accuracy involving a large set of experimental data, the possibility to prevent ductile failures are as follows:

- Knowing only the sate of stress, use the damage equivalent stress criterion

$$\sigma^\star < \sigma_u. \tag{3.166}$$

- Knowing the state of the accumulated plastic strain, use the result obtained in Sect. 3.2.2 in proportional loading with the triaxiality deduced from an elastic calculation,

$$p < p_R \quad \text{with} \quad p_R = \epsilon_{\text{pD}} + (\epsilon_{\text{pR}} - \epsilon_{\text{pD}}) R_\nu^{-s}. \tag{3.167}$$

- For plane strain or plane stress problems use the damage criteria of Sect. 3.3.4.
- For small scale yielding, use Neuber plastic correction of an elastic analysis in order to obtain the local stress and plastic strain. Then make an integration of the unified damage law which can be either its simplest form deduced from the Gurson model in Sect. 3.4.5,

$$\dot{D} = (0.83 \cdot R_\nu)^{2.5} \dot{p} \quad \text{if} \quad p > 0.15, \quad D_c = 0.36, \tag{3.168}$$

its expression with saturated hardening,

$$\dot{D} = \left(\frac{\sigma_u^2 R_\nu}{2ES}\right)^s \dot{p} \quad \text{if} \quad p > \epsilon_{\text{pD}}, \tag{3.169}$$

or the original unified law,

$$\dot{D} = \left(\frac{\sigma_{\text{eq}}^2 R_\nu}{2ES(1-D)^2}\right)^s \dot{p} \quad \text{if} \quad p > \epsilon_{\text{pD}}. \tag{3.170}$$

- For large scale yielding, we advise using an elasto-plastic coupled with damage numerical analysis with the anisotropic unified damage law of Sect. 1.4.3,

$$\dot{D}_{ij} = \left(\frac{\overline{Y}}{S}\right)^s |\dot{\epsilon}^{\text{p}}|_{ij} \quad \text{if} \quad p > \epsilon_{\text{pD}}, \tag{3.171}$$

3.4 Numerical Failure Analysis

- without or with the effects of microdefects closure of Sect. 1.4.3,
- without or with initial conditions of plasticity and damage as discussed in Sect. 3.4.7.

- Finally, the more advanced ductile analysis within the scope of Chaps. 1 and 2 is an elasto-plastic coupled with damage analysis, including anisotropic damage, microdefects closure effects, non-natural initial conditions, and **plastic anisotropy** by use of the Hill criterion whose coupling with damage may be obtained as follows.

Damage anisotropy is modelled through the use of the second order damage variable $\boldsymbol{D}$, the associated tensor $\boldsymbol{H} = (\boldsymbol{1} - \boldsymbol{D})^{-1/2}$, and the effective stress

$$\tilde{\boldsymbol{\sigma}} = \left(\boldsymbol{H}\boldsymbol{\sigma}^{\mathrm{D}}\boldsymbol{H}\right)^{\mathrm{D}} + \frac{\sigma_{\mathrm{H}}}{1 - \eta D_{\mathrm{H}}}\boldsymbol{1} \,. \tag{3.172}$$

For ductile failures, modelling the elastic anisotropy is not of main importance as elasticity may often be neglected, but modelling the plastic anisotropy can be necessary as in the case of rolling sheets, for example. A possible approach simply consists of generalizing the use of the effective stress concept:

- Keep the elasticity law unchanged:

$$\boldsymbol{\epsilon}^{\mathrm{e}} = \frac{1+\nu}{E}\tilde{\boldsymbol{\sigma}} - \frac{\nu}{E}\mathrm{tr}\,\tilde{\boldsymbol{\sigma}}\,\boldsymbol{1} \tag{3.173}$$

- Use the Hill anisotropic yield criterion:

$$f = \sqrt{\tilde{\boldsymbol{\sigma}} : \underline{\boldsymbol{h}} : \tilde{\boldsymbol{\sigma}}} - R - \sigma_{\mathrm{y}} \tag{3.174}$$

where $\underline{\boldsymbol{h}}$ is the Hill fourth order tensor with minor and major symmetries, $h_{ijlk} = h_{ijkl} = h_{jikl} = h_{klij}$, and where the hardening R follows the same anisotropy

- Keep the damage evolution law:

$$\dot{\boldsymbol{D}} = \left(\frac{\overline{Y}}{S}\right)^s |\dot{\boldsymbol{\epsilon}}^{\mathrm{p}}| \qquad \text{if} \quad p > \epsilon_{\mathrm{pD}} \tag{3.175}$$

The only changes concerning the plasticity evolution laws are

$$\dot{\boldsymbol{\epsilon}}^{\mathrm{p}} = \dot{\lambda}\frac{\partial f}{\partial \boldsymbol{\sigma}} = \dot{\lambda}\frac{\partial f}{\partial \tilde{\boldsymbol{\sigma}}} : \frac{\partial \tilde{\boldsymbol{\sigma}}}{\partial \boldsymbol{\sigma}} \,, \tag{3.176}$$

which for incompressible plasticity ($\underline{\boldsymbol{h}}$ such as $\underline{\boldsymbol{h}} : \boldsymbol{\sigma} = \underline{\boldsymbol{h}} : \boldsymbol{\sigma}^{\mathrm{D}} = (\underline{\boldsymbol{h}} : \boldsymbol{\sigma})^{\mathrm{D}}$) gives

$$\dot{\boldsymbol{\epsilon}}^{\mathrm{p}} = \dot{\lambda}\frac{[\boldsymbol{H}\,(\underline{\boldsymbol{h}} : \tilde{\boldsymbol{\sigma}})\,\boldsymbol{H}]^{\mathrm{D}}}{\sqrt{\tilde{\boldsymbol{\sigma}} : \underline{\boldsymbol{h}} : \tilde{\boldsymbol{\sigma}}}} \,, \tag{3.177}$$

with the plastic multiplier $\dot{\lambda}$ calculated by means of the consistency condition $f = 0$, $\dot{f} = 0$ and where $R = R(r)$ and $r = \int_0^t \dot{r}\mathrm{d}t = \int_0^t \dot{\lambda}\mathrm{d}t$.

190   3 Ductile Failures

In non-proportional loading, plastic anisotropy may also result from the distorsion of the yield surface which features a smooth corner with an opposite flat end in the stresses space (Fig. 3.20). Such an "egg" effect may have an influence in metal forming, as in spring back prediction in deep drawing. Adequate (complex!) 3D models can be found in the literature (T. Kurtyka and M. Zyczkowski 1985, N.K. Gupta and A. Meyers 1994, M. François 2000, L. Vincent, S. Calloch and D. Marquis 2003 ...). To add the damage coupling, simply replace in the expressions for f the stress by the effective stress; the evolution law of the plastic strain coupled with damage is obtained by use of (3.176).

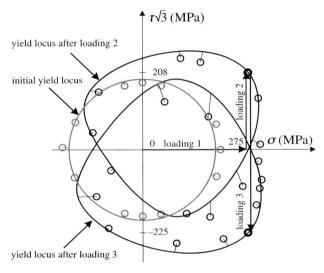

Fig. 3.20. Yield surface of a 2024-T4 aluminum alloy for non-proportional loadings (Lines: model after M. François (2000); small circles: experiments with an offset of $\delta\epsilon_p = 10^{-5}$ (M. Rousset 1985))

4

Low Cycle Fatigue

This second chapter on applications concerns the phenomenon of fatigue damage which occurs when materials are subjected to cyclic loadings. The classical way to describe fatigue consists of splitting the domain of the numbers of cycles to rupture into three parts corresponding to different strain behaviors and also different fields of applications: elasticity corresponds to relatively small stress amplitudes which induce large numbers of cycles to failure (larger than 10^5), this is "high cycle fatigue" treated in Chap. 6; **elasto-plasticity** corresponds to stresses above the yield stress which induce lower numbers of cycles to failure (smaller than 10^4), this is "**low cycle fatigue**;" elasto-visco-plasticity also corresponds to small number of cycles to failure (smaller than 10^4), but with time effects induced by creep, it is generally called "creep fatigue interaction" and is treated in Chap. 5.

From a physical point of view, the repeated variations of stress induce in metals alternate plastic strains which produce internal microstresses responsible of microdecohesions by slip band arrests. The microcracks initiated grow either inside the crystals or along the grains boundaries, depending on the materials and the loadings, up to coalescence corresponding to initiation of a mesocrack. **Plastic strain** and **stress** both participate in this phenomenon. This is the reason the unified damage law is able to model the low cycle fatigue and in 3D directly through the concepts of accumulated plastic strain and elastic energy.

Nevertheless, the main variable is the plastic strain range as it was pointed out independently in 1954 by S.S. Manson and L.F. Coffin in a famous empirical law called now the **Manson–Coffin** law. Many improvements followed (J.D. Morrow 1964, G.R. Halford 1975, A. Pineau 1980, D. Socie 1991) up to Continuum Damage Mechanics for which the difficulty to define a cycle is solved by a **time integration** of the damage law on the full time history of the loading.

The specific items of this chapter are the cyclic elasto-plastic stress concentration analysis, the effects of **loading history**, and the application to **multiaxial fatigue** and **elastomers**.

4.1 Engineering Considerations

Low cycle fatigue failures may occur when structures are subjected to **heavy cyclic loadings** which induce irreversible strains on small or large scale, giving rise to damage up to crack initiation and propagation. To avoid this phenomenon, a careful assessment of information on the development of damage is needed in order to quantify a safety margin. What is called here a cycle is most often the **service time** between the start and the stop of a plant operation or the repetition of a heavy loading. The number of cycles to rupture (N_R), corresponding to states of stress which induces plastic strains, is relatively small or "low":

- It can be on the order of 10 to 100 for aerospace rockets where some tests are necessary before launching or for metal forming by forging. The state of stress lies between the ultimate stress and the yield stress, $\sigma_u > \sigma > \sigma_y$.
- It can be on the order of 100 to 1000 for nuclear or thermal power plants, chemical plants, and many domestic apparatus as the butt hinges of polymeric boxes! The state of stress is somewhat higher than the yield stress, σ_y.
- It can be on the order of 1000 to 10000 for aircraft engines or car engines where, on some parts, the state of stress induces plastic strain on the order of the elastic strain, $\epsilon_p \approx \sigma_y/E$.
- $N_R > 10^5$ corresponds to high-cycle fatigue where the mesoscopic strains may be considered elastic, but not the microscopic ones. It is described in Chap. 6.
- For $10^4 < N_R < 10^5$, low- or high-cycle fatigue may be considered depending on the case and, moreover, on the degree of accuracy needed.

Low cycle fatigue is mainly governed by the dissipative strain at mesolevel and it is essentially encountered in **metals**, **polymers**, and **elastomers**.

4.2 Fast Calculation of Structural Failures

In low cycle fatigue, if the steady state of plastic strain is known at the critical point of a structure that is loaded periodically, there is no need to perform a complicated calculation to determine an approximative value of the number of cycles leading to a mesocrack initiation. Furthermore, if the plasticity is localized, there is a possibility to obtain the plastic strain range from a purely elastic calculation and a correction which is an extension of the Neuber method of Sect. 3.2.4. "Fast" also means "simple" here and the damage is considered isotropic and equally produced in tension and compression.

4.2.1 Uniaxial Behavior and Validation of the Damage Law[♂]

The basic low cycle fatigue characteristic of a material is its **Manson–Coffin curve** which gives the number of cycles to rupture of uniaxial specimens periodically loaded in tension-compression, at constant range of total strain $\Delta\epsilon$, and as a function of the range of plastic strain. An example is given in Fig. 4.1.

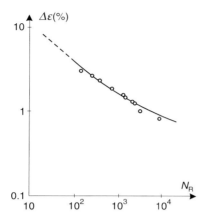

Fig. 4.1. Manson–Coffin curve of UDIMET 700 alloy at room temperature (J. Lemaitre and J.L. Chaboche 1985)

As in a log-log graph, the curve is close to a straight line. It can be represented by the empirical law of Manson–Coffin,

$$N_R = \left(\frac{C_{MC}}{\Delta\epsilon_p}\right)^{\gamma_{MC}}, \tag{4.1}$$

where C_{MC} and γ_{MC} are material parameters: C_{MC} depends on the temperature but γ_{MC} is close to 2 regardless of the material and the temperature.

Furthermore, an extrapolation of the curve to $N_R = 1$ corresponds to a test for which the range of plastic strain is close to two times the rupture plastic strain ϵ_{pR} in a monotonic tension test. Then

$$C_{MC} \approx 2\epsilon_{pR}. \tag{4.2}$$

When no fatigue tests are available for a given material, a rough estimation for an early design purpose is

$$N_R = \left(\frac{2\epsilon_{pR}}{\Delta\epsilon_p}\right)^2. \tag{4.3}$$

Let us recall that this Manson–Coffin curve, also called the "Wöhler curve in the range of low cycle fatigue" in Sect. 1.4.4, is one of the main data source for the identification of the unified damage law

194   4 Low Cycle Fatigue

$$\dot{D} = \left(\frac{Y}{S}\right)^s \dot{p} \quad \text{if} \quad p > p_D. \tag{4.4}$$

In the following, we assume that the damage parameters S, s, ϵ_{pD}, m, and D_c defined in Sect. 1.4 are known for each application. Nevertheless, we strongly advise checking each time it is possible with new tests results or informations that their values give good low cycle fatigue results (particularly for s and m). For more details, you should read Sects. 1.4.4 and 2.4 again. Both sections are devoted to the difficult problem of material parameters identification.

4.2.1.1 Case of Periodic Loading

Consider an uniaxial periodic loading of strain range $\Delta\epsilon$ at zero mean stress (Fig. 4.2) for the applied stress $\pm\sigma_M$.

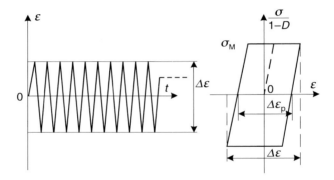

Fig. 4.2. Schematic low cycle fatigue

Beyond the damage threshold, the increment of damage per cycle ($\frac{\delta D}{\delta N}$) comes from a first integration of the uniaxial damage law, in which, for simplicity, the damaged material is considered as perfectly plastic at $\frac{\sigma}{1-D} \approx \sigma_M = const$:

$$\dot{D} = \left(\frac{\sigma^2}{2ES(1-D)^2}\right)^s |\dot{\epsilon}_p| \quad \text{if} \quad p > p_D$$
$$\Longrightarrow \quad \frac{\delta D}{\delta N} = \int_{1\,\text{cycle}} \dot{D} dt = \left(\frac{\sigma_M^2}{2ES}\right)^s 2\Delta\epsilon_p. \tag{4.5}$$

A second integration gives the number of cycles to rupture (N_R) corresponding to the critical value of the damage D_c for the initial condition:

$$N = N_D = \frac{p_D}{2\Delta\epsilon_p} \quad \rightarrow \quad D = 0, \tag{4.6}$$

with (see Sect. 1.4.1):

$$p_D = \epsilon_{pD} \left(\frac{\sigma_u - \sigma_y}{\frac{\sigma_{eq\,max} + \sigma_{eq\,min}}{2} - \sigma_y} \right)^m = \epsilon_{pD} \left(\frac{\sigma_u - \sigma_y}{\sigma_M - \sigma_y} \right)^m. \quad (4.7)$$

Then,

$$\boxed{N_R = N_D + \frac{D_c}{2\Delta\epsilon_p} \left(\frac{2ES}{\sigma_M^2} \right)^s.} \quad (4.8)$$

The number of cycles to rupture is a power function of the stress range and is a linear function of the plastic strain range. Just to compare with the Manson–Coffin law, consider σ_M as a power function of $\Delta\epsilon_p$ given by the cyclic tension-compression curve

$$\Delta\sigma = 2\sigma_M = K_c \Delta\epsilon_p^{1/M_c}. \quad (4.9)$$

Then,

$$N_R \approx const + \left(\frac{const}{\Delta\epsilon_p} \right)^{\frac{2s}{M_c} + 1}, \quad (4.10)$$

which shows that the value of the exponent s is on the order of $M_c/2$, at least for steels, to make $\frac{2s}{M_c} + 1 \approx 2$ as in the Manson–Coffin law.

Before using any of the unified damage law identified as explained in Sects. 1.4.4 and 2.4, we advise checking if the values of the materials in the formula (4.8) give the proper number of cycles to rupture in comparison with some available test data and eventually adjusting the most sensitive parameters (see Sect. 4.2.3).

4.2.1.2 Case of Non-Periodic Loading

When the range of stress varies as a function of the number of cycles, the problem of accumulation of damage arises. The **Palmgreen–Miner rule** (1924, 1945) consists of the simple summation of damages supposed to be proportional to the number of cycles for a given constant range. If n_i is the number of cycles of an applied strain range $\Delta\epsilon_i$ and N_{Ri} is the number of cycles to rupture corresponding to this range constant all along the process, the corresponding damage is supposed to be $n_i D_c / N_{Ri}$. Then for many loadings, the linear rule of summation is simply

$$\sum \frac{n_i D_c}{N_{Ri}} = D_c \quad \rightarrow \quad \boxed{\sum \frac{n_i}{N_{Ri}} = 1} \quad (4.11)$$

and $\sum n_i = N_R$ is the number of cycles to rupture corresponding to the entire sequence of $\Delta\epsilon_i$.

In fact, this rule is a property of any differential equation, linear or nonlinear, representing the damage evolution. But unfortunately, materials do not wish to obey mathematics! The accumulation of damage is most often nonlinear. Let us show it with the continuous damage law applied to two-level fatigue loadings.

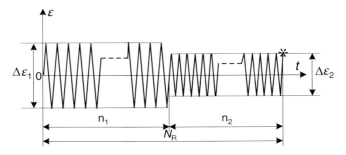

Fig. 4.3. Two level fatigue loading

Consider the loading represented in Fig. 4.3 and assume that the damage initiates (at N_{D1} cycles) during the first level ($n_1 > N_{D1}$). Using (4.5) together with $\sigma_{M1} = \frac{\Delta \sigma_1}{2}$ and $\Delta \epsilon_1 = const$, the damage $D(n_1)$ at the end of the first level is

$$D(n_1) = (n_1 - N_{D1}) \left(\frac{\sigma_{M1}^2}{2ES} \right)^s 2\Delta \epsilon_{p1}, \qquad (4.12)$$

or together with eq. (4.8),

$$D(n_1) = D_c \frac{n_1 - N_{D1}}{N_{R1} - N_{D1}}. \qquad (4.13)$$

The damage at the end of the second level n_2 is equal to the mesocrack initiation condition D_c. With $n_1 + n_2 = N_R$,

$$D_c(N_R) = D(n_1) + n_2 \left(\frac{\sigma_{M2}^2}{2ES} \right)^s 2\Delta \epsilon_{p2}, \qquad (4.14)$$

or

$$D_c(N_R) = D_c \frac{n_1 - N_{D1}}{N_{R1} - N_{D1}} + D_c \frac{n_2}{N_{R2} - N_{D2}}, \qquad (4.15)$$

or

$$\boxed{\frac{n_1}{N_{R1}} + \frac{n_2}{N_{R2}} \frac{1 - \dfrac{N_{D1}}{N_{R1}}}{1 - \dfrac{N_{D2}}{N_{R2}}} = 1.} \qquad (4.16)$$

This shows that

- $\dfrac{n_1}{N_{R1}} + \dfrac{n_2}{N_{R2}} < 1$ if the high level is first $\left(\dfrac{N_{D1}}{N_{R1}} > \dfrac{N_{D2}}{N_{R2}}\right)$
- $\dfrac{n_1}{N_{R1}} + \dfrac{n_2}{N_{R2}} > 1$ if the low level is first $\left(\dfrac{N_{D1}}{N_{R1}} < \dfrac{N_{D2}}{N_{R2}}\right)$
- The Miner rule $\dfrac{n_1}{N_{R1}} + \dfrac{n_2}{N_{R2}} = 1$ of linear accumulation is generally not recovered as it corresponds to the particular case $\dfrac{N_{D1}}{N_{R1}} = \dfrac{N_{D2}}{N_{R2}}$

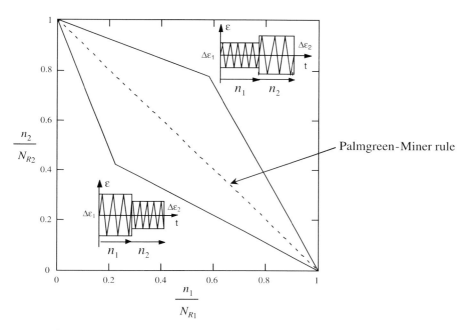

Fig. 4.4. Bilinear damage accumulation in two-level fatigue loadings

In fact, the rule obtained is a bilinear rule in the diagram n_2/N_{R2} vs n_1/N_{R1} because

(a) If $n_1 > N_{D1}$, the damage initiates during the first loading level. The number of cycles to rupture, $N_R = n_1 + n_2$ (with n_1 known), is given by (4.16).

(b) If $n_1 < N_{D1}$, the damage initiates during the second loading level. The number of cycles to rupture, N_R, is given by the damage equation, $\dfrac{N_R - N_D}{N_{R2} - N_{D2}} = 1$, where N_D is determined by the stored energy balance:

$$w_D = N_{D1}\Delta w_{s1} = N_{D2}\Delta w_{s2} = n_1 \Delta w_{s1} + (N_D - n_1)\Delta w_{s2} \quad (4.17)$$

or

$$N_D = n_1 + (N_{D1} - n_1)\frac{\Delta w_{s1}}{\Delta w_{s2}} \quad \text{with} \quad \frac{\Delta w_{s1}}{\Delta w_{s2}} = \frac{N_{D2}}{N_{D1}}. \qquad (4.18)$$

As $N_R = n_1 + n_2$ the final relationship is:

$$\boxed{\frac{n_1}{N_{R1}}\frac{N_{D2}}{N_{D1}}\frac{N_{R1}}{N_{R2}} + \frac{n_2}{N_{R2}} = 1.} \qquad (4.19)$$

An angular point corresponds to the case $N_{D1} = n_1$ of damage initiating at the exact time of the level transition. An example is given in Fig. 4.4 for $N_D/N_R = 0.225$ and 0.58.

4.2.2 Case of Proportional Loading ♂

The case of proportional loading is somewhat similar to the uniaxial case as all the tensors are colinear due to the fixed principal directions of the stress tensor (see Sect. 2.1.2). This is the case at all points of a structure if the applied loads vary proportionately with one parameter. If the loading is periodic, the number of cycles to rupture is obtained from two integrations of the damage law:

$$\dot{D} = \left(\frac{\sigma_{eq}^2 R_\nu}{2ES(1-D)^2}\right)^s \dot{p} \quad \text{if} \quad p > p_D. \qquad (4.20)$$

- *A first integration over one cycle:* the material is again considered perfectly plastic for each maximum stress $\frac{\sigma_{eq}}{1-D} \approx \sigma_{eq\,max} = const$; the triaxiality function R_ν is also constant due to the loading proportionality,

$$\frac{\delta D}{\delta N} = \left(\frac{\sigma_{eq\,max}^2 R_\nu}{2ES}\right)^s 2\Delta p \quad \text{if} \quad p > p_D, \qquad (4.21)$$

where Δp is the accumulated plastic strain increment over half of a cycle.
- *A second integration over the whole process* with the initial condition

$$N = N_D = \frac{p_D}{2\Delta p} \quad \to \quad D = 0, \qquad (4.22)$$

with

$$p_D = \epsilon_{pD}\left(\frac{\sigma_u - \sigma_y}{\sigma_{eq\,max} - \sigma_y}\right)^m. \qquad (4.23)$$

Then

$$\boxed{N_R = N_D + \frac{D_c}{2\Delta p}\left(\frac{2ES}{\sigma_{eq\,max}^2 R_\nu}\right)^s.} \qquad (4.24)$$

It is interesting to compare this number of cycles to rupture to the uniaxial case as

$$\frac{N_{\mathrm{R}}(\sigma_{ij}) - N_{\mathrm{D}}(\sigma_{ij})}{N_{\mathrm{R}}(\sigma) - N_{\mathrm{D}}(\sigma)} = \frac{\Delta\epsilon_{\mathrm{p}}}{\Delta p}\left(\frac{\sigma_{\mathrm{M}}^2}{\sigma_{\mathrm{eq\ max}}^2 R_\nu}\right)^s. \quad (4.25)$$

Using the damage equivalent stress $\sigma^\star = \sigma_{\mathrm{eq}} R_\nu^{1/2}$ gives the possibility to derive the number of cycles to rupture of a multiaxial state of stress defined by $\sigma_{\mathrm{eq\ max}}$, R_ν and the accumulated plastic strain range Δp from only one experimental point $N_{\mathrm{R}}(\Delta\epsilon_{\mathrm{p}}, \sigma_{\mathrm{M}})$ of the Manson–Coffin curve (knowing the exponent s) if the approximation $N_{\mathrm{D}}/N_{\mathrm{R}} = const$ is made:

$$\frac{N_{\mathrm{R}}(\sigma_{ij})}{N_{\mathrm{R}}(\sigma_{\mathrm{M}})} = \frac{\Delta\epsilon_{\mathrm{p}}}{\Delta p}\left(\frac{\sigma_{\mathrm{M}}^\star}{\sigma_{\mathrm{M}}}\right)^{-2s}. \quad (4.26)$$

4.2.3 Sensitivity Analysis♂♂

The general method is described in Sect. 2.4.6. It is applied here to the **number of cycles-to-crack-initiation** (N_{R}) found in Sect. 4.2.2 in order to obtain the relative influence of the loading and of the material parameters on N_{R}:

$$N_{\mathrm{R}} = N_{\mathrm{D}} + \frac{D_{\mathrm{c}}}{2\Delta p}\left(\frac{2ES}{\sigma_{\mathrm{eq\ max}}^2 R_\nu}\right)^s,$$
$$N_{\mathrm{D}} = \frac{\epsilon_{\mathrm{pD}}}{2\Delta p}\left(\frac{\sigma_{\mathrm{u}} - \sigma_{\mathrm{y}}}{\sigma_{\mathrm{eq\ max}} - \sigma_{\mathrm{y}}}\right)^m, \quad (4.27)$$

with $R_\nu = \frac{2}{3}(1+\nu) + 3(1-2\nu)T_X^2$.

The relative error $\frac{\delta N_{\mathrm{R}}}{N_{\mathrm{R}}}$ is determined by the logarithmic derivative of the above equation in the same manner as for ductile failure of Sect. 3.2.3:

$$\ln(N_{\mathrm{R}} - N_{\mathrm{D}}) = \ln D_{\mathrm{c}} - \ln 2 - \ln \Delta p$$
$$+ s\left(\ln 2 + \ln E + \ln S - 2\ln \sigma_{\mathrm{eq\ max}} - \ln R_\nu\right). \quad (4.28)$$

Here we take the absolute value $\delta x = |\mathrm{d}x|$ as the sign of the error is not known:

$$\frac{\delta(N_{\mathrm{R}} - N_{\mathrm{D}})}{N_{\mathrm{R}} - N_{\mathrm{D}}} = \frac{\delta N_{\mathrm{R}}}{N_{\mathrm{R}}}\frac{N_{\mathrm{R}}}{N_{\mathrm{R}} - N_{\mathrm{D}}} - \frac{\delta N_{\mathrm{D}}}{N_{\mathrm{D}}}\frac{N_{\mathrm{D}}}{N_{\mathrm{R}} - N_{\mathrm{D}}},$$
$$\frac{\delta N_{\mathrm{D}}}{N_{\mathrm{D}}} = \frac{\delta\epsilon_{\mathrm{pD}}}{\epsilon_{\mathrm{pD}}} + \frac{\delta\Delta p}{\Delta p} + \left|\ln\left(\frac{\sigma_{\mathrm{u}} - \sigma_{\mathrm{y}}}{\sigma_{\mathrm{eq\ max}} - \sigma_{\mathrm{y}}}\right)\right|\frac{\delta m}{m} + m\frac{\delta(\sigma_{\mathrm{u}} - \sigma_{\mathrm{y}})}{\sigma_{\mathrm{u}} - \sigma_{\mathrm{y}}}$$
$$+ m\frac{\delta(\sigma_{\mathrm{eq\ max}} - \sigma_{\mathrm{y}})}{\sigma_{\mathrm{eq\ max}} - \sigma_{\mathrm{y}}}, \quad (4.29)$$
$$\frac{\delta R_\nu}{R_\nu} = \frac{\delta\nu}{\nu}\frac{|6T_X^2 - \frac{2}{3}|\nu}{R_\nu} + \frac{\delta T_X}{T_X}\frac{6(1-2\nu)T_X^2}{R_\nu}.$$

The coefficients $S_{Ak}^{N_R}$ of the sensitivity matrix of Sect. 2.4.6 are defined by

$$\frac{\delta N_R}{N_R} = \sum_k S_{Ak}^{N_R} \frac{\delta \mathcal{A}_k}{\mathcal{A}_k}, \qquad (4.30)$$

$$\begin{aligned}
S_{T_X}^{N_R} &= \frac{N_R - N_D}{N_R} \frac{6s(1-2\nu)T_X^2}{R_\nu}, \\
S_{\sigma_{eq\,max}}^{N_R} &= \frac{N_R - N_D}{N_R} 2s + \frac{N_D}{N_R} m \frac{\sigma_{eq\,max}}{\sigma_{eq\,max} - \sigma_y}, \\
S_{\Delta p}^{N_R} &= \frac{N_R - N_D}{N_R} + \frac{N_D}{N_R} = 1, \\
S_E^{N_R} &= \frac{N_R - N_D}{N_R} s, \\
S_\nu^{N_R} &= \frac{N_R - N_D}{N_R} \frac{|6T_X^2 - \frac{2}{3}|\nu s}{R_\nu}, \\
S_{\sigma_y}^{N_R} &= \frac{N_D}{N_R} m \left[\frac{\sigma_y}{\sigma_u - \sigma_y} + \frac{\sigma_y}{\sigma_{eq\,max} - \sigma_y} \right], \\
S_{\epsilon_{pD}}^{N_R} &= \frac{N_D}{N_R}, \\
S_m^{N_R} &= \frac{N_D}{N_R} m \left| \ln\left(\frac{\sigma_u - \sigma_y}{\sigma_{eq\,max} - \sigma_y} \right) \right|, \\
S_{\sigma_u}^{N_R} &= \frac{N_D}{N_R} \frac{m\sigma_u}{\sigma_u - \sigma_y}, \\
S_S^{N_R} &= \frac{N_R - N_D}{N_R} s, \\
S_s^{N_R} &= \frac{N_R - N_D}{N_R} s \left| \ln\left(\frac{2ES}{\sigma_{eq\,max}^2 R_\nu} \right) \right|, \\
S_{D_c}^{N_R} &= \frac{N_R - N_D}{N_R}.
\end{aligned} \qquad (4.31)$$

The value of those coefficients are the height of the boxes in Fig. 4.5 for the following set of parameters representing an average of many materials: $T_X = 1$, $R_\nu = 2.07$, $\sigma_{eq\,max} = 400$ MPa, $\sigma_{eq\,max}/\sigma_y = 1.5$, $E = 200000$ MPa, $\nu = 0.3$, $\sigma_u/\sigma_y = 2$, $S = 2$ MPa, $s = 5$, $m = 2$, and $N_R/N_D = 2$.

The loading is represented by the triaxiality ratio, the stress, and the accumulated plastic strain. The influence of an error on the stress is the largest (and is larger than for ductile failures) but the influence of the accumulated plastic strain is small. The parameters which do not need a large accuracy are m, ϵ_{pD} and D_c. Altogether, do not expect an accuracy better than a factor 2 on the number of cycles to rupture.

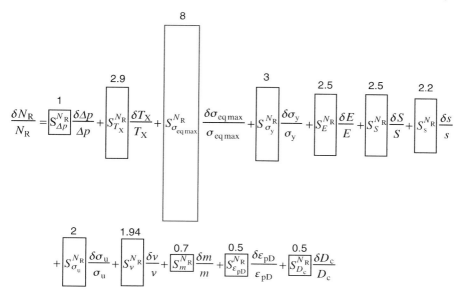

Fig. 4.5. Relative importance of each parameter in low cycle fatigue

4.2.4 Cyclic Elasto-Plastic Stress Concentration♂♂

As for the Neuber method described in Sect. 3.2.4, it is possible to determine the plastic strain range under the small scale yielding assumption from both

- A reference elastic calculation and
- The cyclic stress-strain curve of the material in tension-(compression).

In fact, as far as free edges are concerned and as long as plasticity remains very localized (this is often the case in fatigue), the strain energy density (SED) method gives better results in cyclic loading than the Neuber method. An illustrative example is shown in Sect. 4.4.2. This is why the cyclic Neuber method for unidimensional cases only is described here. Such a case is encountered on free edges of thin structures for which the state of stress is local tension-compression, with a stress triaxiality $T_X = 1/3$ and a stress triaxiality function $R_\nu = 1$.

4.2.4.1 Cyclic Neuber Method

The Neuber method in fatigue is written in terms of stress, strain, and plastic strain ranges over a cycle and is based on the local equality

$$\boxed{\Delta\sigma\Delta\epsilon = (\Delta\sigma\Delta\epsilon)_{\text{elas}} \quad \text{at the stress concentration point,}} \quad (4.32)$$

which is also written as

$$\Delta\sigma\Delta\epsilon = \frac{K_T^2}{E}(\Delta\sigma_n)^2, \quad (4.33)$$

with K_T as the stress concentration coefficient in elasticity and σ_n as the nominal stress (see Fig. 3.4).

A law of cyclic plasticity has to be considered in order to estimate the ranges $\Delta\sigma$, $\Delta\epsilon$, $\Delta\epsilon_p$, and $\Delta p = |\Delta\epsilon_p|$ (the accumulated plastic strain increment over half a cycle). The cyclic power law (4.9),

$$\Delta\epsilon_p = \left(\frac{\Delta\sigma}{K_c}\right)^{M_c}, \quad (4.34)$$

may be used but further explicit results will be obtained with the linear law

$$\Delta\epsilon = \frac{\Delta\sigma}{E} + \frac{\Delta\sigma - 2\sigma_y}{C_y}, \quad (4.35)$$

$$\Delta p = |\Delta\epsilon_p| = \frac{\Delta\sigma - 2\sigma_y}{C_y}, \quad (4.36)$$

which comes from the consideration of a linear kinematic hardening only where $X = C_y\epsilon_p$.

The equality

$$\left(\frac{\Delta\sigma}{E} + \frac{\Delta\sigma - 2\sigma_y}{C_y}\right)\Delta\sigma = \frac{K_T^2}{E}(\Delta\sigma_n)^2 \quad (4.37)$$

gives the platic stress range $\Delta\sigma$ when $K_T\Delta\sigma_n > 2\sigma_y$ for cyclic plasticity to occur. Equation (4.36) gives the accumulated plastic strain increment per cycle as

$$\frac{\delta p}{\delta N} = 2\Delta p = 2\frac{\Delta\sigma - 2\sigma_y}{C_y}. \quad (4.38)$$

If the loading is symmetric, $\sigma_{max} = -\sigma_{min} = \Delta\sigma/2$ gives the maximum and minimum stresses reached locally at the most loaded point.

If the loading is non-symmetric, the maximum stress σ_{max} has to be determined first by use of the monotonic Neuber method (see Sect. 3.2.4). Equation (4.37) gives $\Delta\sigma$ and the stress corresponding to the mimimun load is

$$\sigma_{min} = \sigma_{max} - \Delta\sigma. \quad (4.39)$$

We will consider next the SED method. In order to deal with non-symmetric loading we first need its expression for the monotonic case.

4.2.4.2 Monotonic SED Method

The strain energy density (SED) method assumes that at the stress concentration point, the strain energy density in confined plasticity is identical to the strain energy density at the same point, but calculated in elasticity.

For 1D monotonic loadings,

$$\int_0^\epsilon \sigma \mathrm{d}\epsilon = \int_0^{\epsilon_{\text{elas}}} \sigma_{\text{elas}} \mathrm{d}\epsilon_{\text{elas}} = \frac{K_T^2}{2E} \sigma_n^2 . \tag{4.40}$$

The local stress in plasticity, $\sigma > \sigma_y$, is estimated by solving this equation coupled with the constitutive law $\epsilon = \sigma/E + R^{-1}(\sigma - \sigma_y) = \sigma/E + g(\sigma)$, where $R(p)$ is the hardening rule.

For 3D monotonic loadings, the method solves

$$\int_0^{\epsilon_{ij}} \sigma_{ij} \mathrm{d}\epsilon_{ij} = \int_0^{\epsilon_{ij}^{\text{elas}}} \sigma_{ij}^{\text{elas}} \mathrm{d}\epsilon_{ij}^{\text{elas}} = \frac{1}{2} (\sigma_{ij} \epsilon_{ij})_{\text{elas}} , \tag{4.41}$$

locally written at the stress concentration point with the Hencky–Mises law $p = g(\sigma_{\text{eq}})$ as a complementary equation. It may be rewritten as

$$\frac{\sigma_{\text{eq}}^2 R_\nu}{2E} + \int_{\sigma_y}^{\sigma_{\text{eq}}} \sigma_{\text{eq}} g'(\sigma_{\text{eq}}) \mathrm{d}\sigma_{\text{eq}} = \frac{1}{2} (\sigma_{ij} \epsilon_{ij})_{\text{elas}} . \tag{4.42}$$

The term $(\sigma_{ij}\epsilon_{ij})_{\text{elas}}$ comes from the elastic reference computation, it is equal to $\sigma_{\text{eq}}^{\text{elas}\,2} R_\nu^{\text{elas}}/E$ with $\sigma_{\text{eq}}^{\text{elas}}$ and R_ν^{elas} as the von Mises stress and the triaxiality function calculated in elasticity, respectively.

The method gives σ_{eq} and p in plasticity. It needs the value of the triaxiality function R_ν. For general three-dimensional cases we use the proportional loading assumption $R_\nu = R_\nu^{\text{elas}}$. We use $R_\nu = 1$ for points located on free edges in plane stress. For points located on free edges in plane strain (see Sect. 2.1.4 for details concerning the calculation of the stress triaxiality in two-dimensional cases) we use $R_\nu = 1.27$.

Let us explicitly describe the three-dimensional method in two useful cases:

(a) *Monotonic SED method with linear isotropic hardening* $R = C_y p$. The von Mises stress in plasticity σ_{eq} is a solution of

$$\frac{\sigma_{\text{eq}}^2 R_\nu}{2E} + \frac{\langle \sigma_{\text{eq}}^2 - \sigma_y^2 \rangle}{2C_y} = \frac{1}{2} (\sigma_{ij} \epsilon_{ij})_{\text{elas}} , \tag{4.43}$$

which leads to the closed-form expression

4 Low Cycle Fatigue

$$\boxed{\sigma_{\text{eq}} = \sqrt{\frac{E(\sigma_{ij}\epsilon_{ij})_{\text{elas}} + \frac{E}{C_y}\sigma_y^2}{R_\nu + \frac{E}{C_y}}}} \quad (4.44)$$

and defines the auxiliary ratio

$$k_{\text{SED}} = \frac{\sigma_{\text{eq}}}{\sigma_{\text{eq}}^{\text{elas}}}. \quad (4.45)$$

The accumulated plastic strain is

$$p = \frac{\sigma_{\text{eq}} - \sigma_y}{C_y}. \quad (4.46)$$

(b) *Monotonic SED method with nonlinear isotropic hardening* $R = R_\infty^y(1-\exp(-b_y p))$. One has to solve the following equation numerically:

$$\frac{\sigma_{\text{eq}}^2 R_\nu}{E} + (R_\infty^y + \sigma_y)g(\sigma_{\text{eq}}) - \frac{1}{b_y}\langle\sigma_{\text{eq}} - \sigma_y\rangle = \frac{1}{2}(\sigma_{ij}\epsilon_{ij})_{\text{elas}}, \quad (4.47)$$

where $p = g(\sigma_{\text{eq}}) = -\frac{1}{b}\ln\left(\frac{R_\infty + \sigma_y - \sigma_{\text{eq}}}{R_\infty}\right)$.

The graphs of Fig. 4.6 show how the SED method differs from the Neuber method.

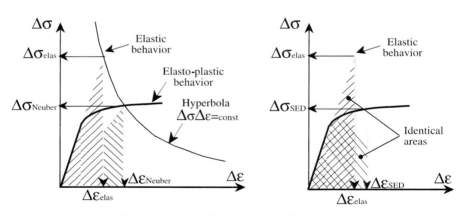

Fig. 4.6. Cyclic Neuber and SED methods

4.2.4.3 Cyclic SED Method

For locally-cyclic 1D state of stress, to obtain a closed-form solution, the SED method considers

$$\boxed{\int_0^{\Delta\epsilon} \Delta\sigma \, d\Delta\epsilon = \int_0^{\Delta\epsilon_{\text{elas}}} \Delta\sigma_{\text{elas}} \, d\Delta\epsilon_{\text{elas}} = \frac{K_T^2}{2E}(\Delta\sigma_n)^2} \qquad (4.48)$$

instead of (4.37) altogether with the cyclic constitutive law (4.34) or the linear law (4.35).

For three-dimensional loadings, the SED method gives the equivalent stress range $(\Delta\boldsymbol{\sigma})_{\text{eq}}$ as the solution of

$$\boxed{\int_0^{\Delta\epsilon_{ij}} \Delta\sigma_{ij} \, d\Delta\epsilon_{ij} = \left(\int_0^{\Delta\epsilon_{ij}} \Delta\sigma_{ij} \, d\Delta\epsilon_{ij}\right)_{\text{elas}}}, \qquad (4.49)$$

which has to be considered coupled with a law of cyclic plasticity $\Delta\epsilon_{ij} = \Delta\epsilon_{ij}(\Delta\sigma_{ij})$ and where $\Delta\sigma_{ij}$ and $\Delta\epsilon_{ij}$ stand for the stress and strain ranges.

In case of linear kinematic hardening, $\boldsymbol{X} = \frac{2}{3}C_y \boldsymbol{\epsilon}^p$, and with the hypothesis of a proportional loading, the cyclic plasticity law is

$$\begin{aligned}
\Delta\epsilon_{ij} &= \frac{1+\nu}{E}\Delta\sigma_{ij} - \frac{\nu}{E}\Delta\sigma_{kk}\delta_{ij} + \Delta\epsilon_{ij}^{\text{P}}, \\
\Delta\epsilon_{ij}^{\text{P}} &= \frac{3}{2C_y}\frac{\Delta\sigma_{ij}^{\text{D}}}{(\Delta\boldsymbol{\sigma})_{\text{eq}}} \langle(\Delta\boldsymbol{\sigma})_{\text{eq}} - 2\sigma_y\rangle,
\end{aligned} \qquad (4.50)$$

with the accumulated plastic strain increment over half a cycle equal to

$$\Delta p = \frac{\langle(\Delta\boldsymbol{\sigma})_{\text{eq}} - 2\sigma_y\rangle}{C_y}. \qquad (4.51)$$

The closed-form expression for the equivalent stress range is then

$$\boxed{(\Delta\boldsymbol{\sigma})_{\text{eq}} = \sqrt{\frac{E(\Delta\sigma_{ij}\Delta\epsilon_{ij})_{\text{elas}} + \dfrac{E}{C_y}4\sigma_y^2}{R_\nu + \dfrac{E}{C_y}}}}. \qquad (4.52)$$

To finish, the method defines the auxiliary ratio

$$k_{\text{SED}}^{\text{cyclic}} = \frac{(\Delta\boldsymbol{\sigma})_{\text{eq}}}{(\Delta\boldsymbol{\sigma})_{\text{eq}}^{\text{elas}}}. \qquad (4.53)$$

Compared to the monotonic case, the three-dimensional cyclic method for linear kinematic hardening is formally obtained by replacing σ_{eq} with $(\Delta\boldsymbol{\sigma})_{\text{eq}}$, $\sigma_{\text{eq}}^{\text{elas}}$ with $(\Delta\boldsymbol{\sigma})_{\text{eq}}^{\text{elas}}$, $(\sigma_{ij}\epsilon_{ij})_{\text{elas}}$ with $(\Delta\sigma_{ij}\Delta\epsilon_{ij})_{\text{elas}}$, and σ_y with $2\sigma_y$. Note that the same replacements within the Neuber monotonic expression for linear isotropic hardening gives the closed-form solution for the Neuber cyclic method with linear kinematic hardening.

4.2.4.4 Cyclic SED Method Applied to Fatigue Failures

Applying the cyclic method takes 3 steps:

1. **Perform the elastic reference computation**
   To simplify, consider the case of proportional loading for which all applied loads or displacements have fixed directions and are proportional to the scalar function of time $\alpha(t)$, periodical between $\alpha_{\min} < 0$ and $\alpha_{\max} > 0$. Due to elasticity's linearity, a single computation performed for the constant reference value $\alpha = \alpha^{\text{ref}}$ is needed. It gives the map of von Mises stress $\sigma_{\text{eq}}^{\text{ref}}(M)$, the stress triaxiality $T_X^{\text{ref}}(M)$, and $R_\nu = R_\nu^{\text{ref}}(M)$. Then, in any point of the structure,

$$\sigma_{\text{eq}}^{\text{elas}}(M,t) = \frac{|\alpha(t)|}{\alpha^{\text{ref}}} \sigma_{\text{eq}}^{\text{ref}}(M),$$

$$R_\nu(M,t) = R_\nu^{\text{ref}}(M),$$

$$\sigma_{\text{eq min}}^{\text{elas}}(M) = -\frac{\alpha_{\min}}{\alpha^{\text{ref}}} \sigma_{\text{eq}}^{\text{ref}}(M), \quad (4.54)$$

$$\sigma_{\text{eq max}}^{\text{elas}}(M) = \frac{\alpha_{\max}}{\alpha^{\text{ref}}} \sigma_{\text{eq}}^{\text{ref}}(M).$$

The equivalent stress range at any point of the elastic computation is then

$$(\Delta \sigma^{\text{elas}})_{\text{eq}} = \sigma_{\text{eq max}}^{\text{elas}} + \sigma_{\text{eq min}}^{\text{elas}} \quad (4.55)$$

and

$$(\Delta \sigma_{ij} \Delta \epsilon_{ij})_{\text{elas}} = \frac{(\Delta \sigma^{\text{elas}})_{\text{eq}}^2 R_\nu^{\text{ref}}}{E} = (\alpha_{\max} + |\alpha_{\min}|)^2 \frac{\sigma_{\text{eq}}^{\text{ref}\,2} R_\nu^{\text{ref}}}{E}. \quad (4.56)$$

2. **Apply the cyclic SED method**
   Determine the most loaded point as the point where the damage equivalent stress ($\sigma^\star = \sigma_{\text{eq}}^{\text{elas}} R_\nu^{1/2}$) is maximum. Use, for instance, Eq. (4.52) to determine at this point the equivalent stress range $(\Delta \sigma)_{\text{eq}}$ in localized plasticity. The accumulated plastic strain increment per cycle is (4.51):

$$\frac{\delta p}{\delta N} = \int_{1\,\text{cycle}} \dot{p}\,dt = 2\Delta p. \quad (4.57)$$

If the loading is non-symmetric, i.e., if $\alpha_{\min} \neq -\alpha_{\max}$, use the monotonic SED method (4.44) or (4.47) to determine the maximum von Mises stress $\sigma_{\text{eq max}}$. The von Mises stress for $\alpha = \alpha_{\min}$ is then

$$\sigma_{\text{eq min}} = (\Delta \sigma)_{\text{eq}} - \sigma_{\text{eq max}}. \quad (4.58)$$

If the loading is symmetric, simply take $\sigma_{\text{eq max}} = \sigma_{\text{eq min}} = \frac{(\Delta \sigma)_{\text{eq}}}{2}$.

3. **Calculate the damage and the number of cycles to rupture**
   Once $p = p_D$ is reached, the damage D increases twice per cycle, each time corresponding to yield in tension (plastic strain increment Δp_+, damage

increment ΔD_+) and compression (plastic strain increment Δp_-, damage increment ΔD_-).

For the fatigue loading between $\alpha_{\min}$ and $\alpha_{\max}$ to consider a linear kinematic hardening leads to a stabilized stress-strain cycle,

$$\Delta p_+ = \Delta p_- = \Delta p = \frac{1}{2}\frac{\delta p}{\delta N}, \qquad (4.59)$$

and a damage increment per cycle

$$\frac{\delta D}{\delta N} = \int_{1\,\text{cycle}} \dot{D} dt = \Delta D_+ + \Delta D_-. \qquad (4.60)$$

With sufficient accuracy for fatigue applications, neglect the coupling of the energy density release rate Y with damage and consider that it does not vary much between the applied load inducing the reach of the yield stress and the maximum load $\alpha_{\max}$ inducing the maximum von Mises stress $\sigma_{\text{eq max}}$, i.e.,

$$\Delta D_+ = \left(\frac{Y_{\max}}{S}\right)^s \Delta p_+ \quad \text{and} \quad Y_{\max} \approx \frac{\sigma^2_{\text{eq max}} R_\nu}{2E}. \qquad (4.61)$$

Close to the minimum load $\alpha_{\min}$ inducing the minimum von Mises stress $\sigma_{\text{eq min}}$,

$$\Delta D_- = \left(\frac{Y_{\min}}{S}\right)^s \Delta p_- \quad \text{and} \quad Y_{\min} \approx \frac{\sigma^2_{\text{eq min}} R_\nu}{2E}. \qquad (4.62)$$

The damage increment over one cycle of loading is then

$$\frac{\delta D}{\delta N} = \frac{(\sigma^{2s}_{\text{eq min}} + \sigma^{2s}_{\text{eq max}})R^s_\nu}{(2ES)^s}\Delta p \quad \text{if} \quad \begin{cases} \Delta\sigma_{\text{eq}} > 2\sigma_y, \\ p > p_\text{D} \end{cases} \qquad (4.63)$$

and $\frac{\delta D}{\delta N} = 0$ otherwise. Then the damage after N cycles is

$$D(N) = \frac{(\sigma^{2s}_{\text{eq min}} + \sigma^{2s}_{\text{eq max}})R^s_\nu \Delta p}{(2ES)^s}(N - N_\text{D}), \qquad (4.64)$$

where $N_\text{D} = \frac{p_\text{D}}{2\Delta p}$ is the number of cycles to reach $p = p_\text{D}$.

Finally,

$$\boxed{N_\text{R} = N_\text{D} + \frac{(2ES)^s D_\text{c}}{(\sigma^{2s}_{\text{eq min}} + \sigma^{2s}_{\text{eq max}})R^s_\nu \Delta p}} \qquad (4.65)$$

is the number of cycles-to-mesocrack-initiation.

For a complex history of loading, a numerical integration of the damage evolution law is necessary to obtain the mesocrack initiation condition of $D = D_\text{c}$.

4.2.5 Safety Margin and Crack Growth⚥

In fatigue, the sensible parameter on which to evaluate the security is the number of cycles. A safe design, for which the estimated number of cycles of loading for the whole life of a component is N_{service}, may be defined as

$$\boxed{N_{\text{service}} < \frac{N_R}{Saf}}, \qquad (4.66)$$

where N_R is the number of cycles to rupture determined by a calculation taking into account the estimated history of loading and Saf is the safety factor. $Saf = 20$, 10, or 5, depending on the "state of the art" of the application.

Once the mesocrack is initiated, the low cycle crack growth is generally pretty fast due to the high level of loading. So, there is often no need of a crack growth calculation. It is sufficient to check the condition of crack arrest or to check the whole ductile rupture by instability using the "R-curve" method described in Sect. 3.2.5.

4.3 Basic Engineering Examples

The same basic geometries and directions of loading as in Sect. 3.3 are examined here from the point of view of low cycle fatigue. The change is in the cyclic calculation of the stress concentration and the life duration expressed in terms of the number of cycles. Isotropic damage without the microdefects closure effect allows for closed-form solutions.

4.3.1 Plate or Members with Holes or Notches⚥

Consider, as in Sect. 3.3.1, any problem of a structure with a geometrical weakness to which corresponds a uniaxial stress concentration coefficient K_T, but with a cyclic loading on the nominal stress σ_n here:

$$\sigma_{\text{elas}} = K_T \sigma_n \quad \longrightarrow \quad \begin{cases} \sigma_{\text{elas}}^{\max} = K_T \sigma_n^{\max}, \\ \Delta\sigma_{\text{elas}} = K_T \Delta\sigma_n. \end{cases} \qquad (4.67)$$

Values of coefficients K_T for many cases are given in handbooks (see bibliography).

Plasticity occurs at least in the neighborhood of the extremal values of the stress and the cyclic SED uniaxial correction applies as the critical point is assumed to be at a free surface,

$$\begin{cases} \sigma_{\max} = k_{\text{SED}} \sigma_{\text{elas}} \\ \Delta\sigma = k_{\text{SED}}^{\text{cyclic}} \Delta\sigma_{\text{elas}} \end{cases} \longrightarrow \begin{cases} \sigma_{\max} = k_{\text{SED}} K_T \sigma_n^{\max}, \\ \sigma_{\min} = K_T \left(k_{\text{SED}} \sigma_n^{\max} - k_{\text{SED}}^{\text{cyclic}} \Delta\sigma_n \right), \end{cases} \qquad (4.68)$$

4.3 Basic Engineering Examples

with k_{SED} and $k_{\text{SED}}^{\text{cyclic}}$ given by the monotonic and cyclic SED methods of Sect. 4.2.4 (Eqs. (4.45) and (4.53)).

- If the loading is periodic at zero mean stress, equations of Sect. 4.2.1 apply: the number of cycles leading to a mesocrack initiation for a stress range $\Delta\sigma_n = 2\sigma_M$ is

$$N_R = \frac{\epsilon_{pD}}{2\Delta\epsilon_p}\left(\frac{\sigma_u - \sigma_y}{\sigma_M - \sigma_y}\right)^m + \frac{D_c}{2\Delta\epsilon_p}\left(\frac{2ES}{\sigma_M^2}\right)^s, \quad (4.69)$$

with σ_M and $\Delta\epsilon_p$ given by the cyclic SED (or the Neuber) method.
If 8 material parameters are too much, there is a possibility to derive a much simpler formula if one low cycle fatigue experimental result is known:
  - Assume that a maximum stress σ_{M0} and a plastic strain range $\Delta\epsilon_{p0}$ give a known number of cycles to rupture, N_{R0},
  - As the number of cycles N_D to reach the damage threshold is often small in comparison to N_R, it is neglected. Then

$$N_R \approx \frac{D_c}{2\Delta\epsilon_p}\left(\frac{2ES}{\sigma_M^2}\right)^s. \quad (4.70)$$

  - As the material parameters verify,

$$N_{R0} \approx \frac{D_c}{2\Delta\epsilon_{p0}}\left(\frac{2ES}{\sigma_{M0}^2}\right)^s. \quad (4.71)$$

Then, the number of cycles to failure reads:

$$\boxed{N_R \approx N_{R0}\frac{\Delta\epsilon_{p0}}{\Delta\epsilon_p}\left(\frac{\sigma_{M0}}{\sigma_M}\right)^{2s}.} \quad (4.72)$$

- If the loading is not periodic but if $\sigma_M(N)$ and $\Delta\epsilon_p(N)$ are known as functions of the number of cycles by the SED correction, the first integration of Sect. 4.2.1 applies:

$$\frac{\delta D}{\delta N} = \left(\frac{\sigma_M^2}{2ES}\right)^s 2\Delta\epsilon_p, \quad (4.73)$$

but the second integration is replaced by

$$D_c = \int_0^{N_R}\left(\frac{\sigma_M^2(N)}{2ES}\right)^s 2\Delta\epsilon_p(N)dN, \quad (4.74)$$

where the number of cycles to reach the damage threshold is neglected. This gives the number of cycles to mesocrack initiation as the implicit solution of a nonlinear equation, usually solved by use of mathematical software.

4.3.2 Pressurized Shallow Cylinders⚥

It is the same example as in Chap. 3 except that the pressure varies periodically between 0 and P_M (Fig. 4.7)

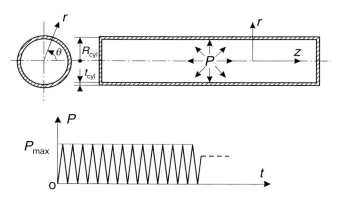

Fig. 4.7. Cyclic pressurized cylindrical cylinder

Let us recall from the stress analysis of the pressurized cylinder of Sect. 3.3.2 that

$$\sigma_{eq} = \frac{\sqrt{3}}{2}\frac{PR_{cyl}}{t_{cyl}}, \qquad T_X = \frac{\sigma_H}{\sigma_{eq}} = \frac{1}{\sqrt{3}}, \quad \text{and} \quad R_\nu = \frac{5-4\nu}{3}, \qquad (4.75)$$

with P as the internal pressure, R_{cyl} as the cylinder radius and t_{cyl} as its thickness.

The difference with the monotonic loading is in the plastic constitutive equation which must take into account the effect of cyclic loading by the kinematic hardening. There are two possibilities (Fig. 4.8):

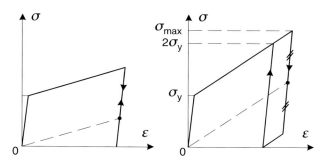

Fig. 4.8. Elastic (left) and plastic (right) shakedown in cyclic loading

1. The material has a small kinematic hardening for the stress considered; there is adaptation or elastic shakedown after the very first cycle which makes all the subsequent cycles elastic. The accumulated plastic strain rate remains zero and there is no damage and no failure by low cycle fatigue. There can be failure by high-cycle fatigue, in that case use the two-scale damage model (applications in Chap. 6). This condition of adaptation arises if

$$\sigma_{\text{eq max}} \leq 2\sigma_y \quad \text{or} \quad P_{\max} \leq \frac{4}{\sqrt{3}}\sigma_y \frac{t_{\text{cyl}}}{R_{\text{cyl}}}. \tag{4.76}$$

2. The material has a kinematic hardening high enough for the stress considered, $\sigma_{\text{eq max}} > 2\sigma_y$, so that hysteresis loops exist, leading to a stabilized cycle of plastic shakedown. Then $\dot{p} \neq 0$ and the phenomenon of low cycle fatigue occurs.

In this case, the damage law,

$$\dot{D} = \left(\frac{\sigma_{\text{eq}}^2 R_\nu}{2ES(1-D)^2}\right)^s \dot{p} \quad \text{if} \quad p > p_D, \tag{4.77}$$

has to be integrated once over a cycle to give $\frac{\delta D}{\delta N}$ and once over the whole process.

Considering linear kinematic hardening and neglecting the variation of the damage over one cycle,

$$\frac{\delta D}{\delta N} = 2\int_{2\sigma_y}^{\sigma_{\text{eq max}}} \left(\frac{\sigma_{\text{eq}}^2 R_\nu}{2ES(1-D)^2}\right)^s \frac{\mathrm{d}\sigma_{\text{eq}}}{C_y(1-D)} \tag{4.78}$$

or

$$\frac{\delta D}{\delta N} = \frac{2}{C_y(2s+1)}\left(\frac{R_\nu}{2ES(1-D)^2}\right)^s \frac{\left[\sigma_{\text{eq max}}^{2s+1} - (2\sigma_y)^{2s+1}\right]}{1-D}. \tag{4.79}$$

For the second integration,

$$\int_0^{D_c}(1-D)^{2s+1}\mathrm{d}D = \frac{2}{C_y(2s+1)}\left(\frac{R_\nu}{2ES}\right)^s\left[\sigma_{\text{eq max}}^{2s+1} - (2\sigma_y)^{2s+1}\right]\int_{N_D}^{N_R}\mathrm{d}N \tag{4.80}$$

or

$$\boxed{N_R = N_D + \frac{2s+1}{2s+2}\frac{1-(1-D_c)^{2s+2}}{\frac{2}{C_y}\left(\frac{R_\nu}{2ES}\right)^s\left[\sigma_{\text{eq max}}^{2s+1} - (2\sigma_y)^{2s+1}\right]}, \quad N_D = \frac{p_D}{2\Delta p},} \tag{4.81}$$

with

$$p_D = \epsilon_{pD} \left(\frac{\sigma_u - \sigma_y}{\frac{\sigma_{eq\,Max}}{2} - \sigma_y} \right)^m, \quad \Delta p = \frac{\sigma_{eq\,max} - 2\sigma_y}{C_y},$$

$$\sigma_{eq\,max} = \frac{\sqrt{3}}{2} \frac{P_{max} R_{cyl}}{t_{cyl}}, \quad \text{and} \quad R_\nu = \frac{5 - 4\nu}{3}. \quad (4.82)$$

For design purposes, it is interesting to see that the ratio R_{cyl}/t_{cyl} acts on the number of cycles to rupture at the power $(2s + 1)$.

4.3.3 Cyclic Bending of Beams♂

Cyclic bending is the fatigue process by which kids or older people break a metallic wire with an increase of the temperature which may burn the fingers! This is the same problem of a beam loaded in circular bending of Sect. 3.3.3, but here the applied curvature $1/\rho$ is periodic with a mean value equal to 0 (Fig. 4.9). Please refer to Sect. 5.3.4 for the consideration of the coupling with temperature.

The critical point where the mesocrack will initiate is of course on the upper or lower part of the beam where the plastic strain varies between $\epsilon_{p\,max}$ and $\epsilon_{p\,min} = -\epsilon_{p\,max}$, a value related to the radius of curvature ρ by the simple Bernoulli theory of beams in which the elastic strain is neglected:

$$\frac{l}{\rho} = \frac{l\epsilon_{p\,max}}{\frac{h}{2}} \quad \text{or} \quad \epsilon_{p\,max} = \frac{h}{2\rho}. \quad (4.83)$$

Then the problem reduces to a simple case of tension-compression plastic-strain imposed between $-h/2\rho$ and $h/2\rho$, where the redistribution of stress due to plasticity and damage is neglected.

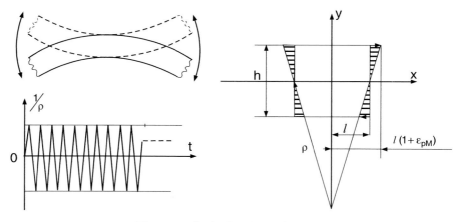

Fig. 4.9. Cyclic bending of a beam

Back to Sect. 4.2.1 and neglecting also the number of cycles to reach the damage threshold, we have

$$N_R = \frac{D_c}{2\Delta\epsilon_p}\left(\frac{2ES}{\sigma_M^2}\right)^s, \qquad (4.84)$$

with $\Delta p = \epsilon_{p\,max} - \epsilon_{p\,min} = \dfrac{h}{\rho}$.

The term σ_M is the stress corresponding to $\epsilon_{p\,max}$ which may be determined from the cyclic stress-strain curve of the material. Taking

$$\Delta\epsilon_p = \left(\frac{2\sigma_M}{K_c}\right)^{M_c}, \qquad (4.85)$$

this shows that the number of cycles to a mesocrack initiation is proportional to the radius of curvature at the power $(1 + \frac{2s}{M_c})$, which is on the order of 2 for steel (as $s \approx M_c/2$), as follows:

$$\boxed{N_R = \frac{D_c}{2}\left(\frac{8ES}{K_c^2}\right)^s\left(\frac{\rho}{h}\right)^{1+\frac{2s}{M_c}}.} \qquad (4.86)$$

4.4 Numerical Failure Analysis

Practical applications concern structures made of different mechanical components and different materials submitted to complex loadings, fatigue loadings here. The full computations of whole structures are difficult to perform as a fully-coupled damage analysis is computer-time consuming and as, compared to monotonic applications, the time increments are multiplied by a factor of 10 to 10000 cycles of fatigue.

The two ways to handle Continuum Damage Mechanics in low cycle fatigue are:

- **The fully coupled analysis** which needs big computers and modern computation methods such as sub-structuring and parallelism. It is of course the best quality one can expect for structures computation. It is also the easiest way for an FE code user to make the calculation and analyze the results. The use of a large time increment method (P. Ladevèze 1989) eventually coupled with parallelism computing allows the user to save time by calculating the stabilized cycle straightforwardly.
  The damage maps may be drawn at any time increment. The most loaded point is easily determined as the location of the maximum damage. The time or number of cycles to rupture corresponds to the reach of the critical damage D_c or the phenomenon of strain localization (see Sect. 1.6).

- **The post-processing analysis** which needs a reference computation performed either in elasticity or elasto-plasticity, and the application of a post-processor solving the damage constitutive equations cycle by cycle, with a time step integration for each cycle. This assumes of course no damage coupling with the strain behavior and in most cases yields a lower bound on the number of cycles to rupture that is a safe result regarding the security. Furthermore, for periodic loadings there is a possibility to apply the jump-in-cycles procedure of Sect. 2.1.5 to save computer time.

4.4.1 Effects of Loading History♂♂
(R. Desmorat and J. Besson 2003)

Due to the nonlinearities occurring in the damage growth, the accumulation of the damages created by different amplitudes of loading is not simple. Their order of occurrence has an effect that is sometimes beneficial, sometimes not. Let us illustrate this using some examples. The material is the SOLDUR 355 steel of Sect. 3.4.2 but with kinematic hardening here. Anisotropic damage and microdefects closure effect are considered together with (see Sect. 1.5.3):

- $E = 230000$ MPa, $\nu = 0.3$ for elasticity
- $\sigma_y = 375$ MPa as yield stress
- $X_\infty^y = 120$ MPa, $\gamma_y = 25$, $C_y = \gamma_y X_\infty^y = 3000$ MPa for hardening
- $\epsilon_{pD} = 2.5 \cdot 10^{-2}$, $\sigma_u = 474$ MPa, $m = 2.5$ for the damage threshold
- $h_a = 0$, $S = 0.43$ MPa, $s = 4$, $\eta = 2.8$, $D_c = 0.5$ for damage

The calculations are performed with the ZeBuLon computer code using the full elasto-plastic damage coupling with the implicit scheme of Sect. 2.2.5. The damage threshold is determined by (1.117) of Sect. 1.4.1 for multilevel fatigue applications. It is

$$p_D = \epsilon_{pD} \left(\frac{\sigma_u - \sigma_y}{\sigma_{M1} - \sigma_y} \right)^m \tag{4.87}$$

when damage initiates during the first level of stress range σ_{M1}. It is

$$p_D = \left[\epsilon_{pD}^{1/m} \frac{\sigma_u - \sigma_y}{\sigma_{M2} - \sigma_y} + p_1^{1/m} \frac{\sigma_{M2} - \sigma_{M1}}{\sigma_{M2} - \sigma_y} \right]^m \tag{4.88}$$

when damage initiates during the second level of stress range σ_{M2} after n_1 cycles, at the first level up to an accumulated plastic strain $p(n_1) = p_1$.

The mesh is quite simple: a single volume element! Each cycle is discretized with a minimum of 40 time steps automatically calculated for optimized convergence.

Considering several levels of 1D loadings, it has been shown in Sect. 4.2.1 that the deviation from the linear accumulation rule of Palmgreen–Miner is due to the damage threshold. In fact, a numerical analysis avoiding any approximation gives the same effect, but in a more complete manner. A set of two-level loadings is considered here, made of n_1 cycles of a first symmetric cyclic loading at constant strain range $\Delta\epsilon_1$, followed by n_2 cycles of a second symmetric loading at constant $\Delta\epsilon_2$ up to mesocrack initiation in the range of 10 to 10^4 cycles.

First, it is interesting to see the variation in the number of cycles leading to the damage threshold N_D as a function of the number of cycles to rupture N_R. Figure 4.10 shows that the ratio N_D/N_R increases largely with N_R, as shown by many experiments (A. Pineau 1980).

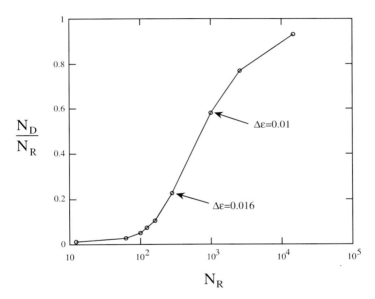

Fig. 4.10. Evolution of the number of cycles-to-damage-threshold in low cycle fatigue

Second, after the damage threshold, the evolution of the damage is not too far to be linear as a function of the number of cycles (Fig. 4.11). This justifies some early fatigue theories where the damage was set as $D = N/N_R$, but here it should be $D = \langle N - N_D \rangle / (N_R - N_D)$.

Last, the accumulation diagram corresponding to the two-level fatigue loading of Fig. 4.3 is computed for two different strain ranges. Figure 4.12 shows the deviation with respect to the Palmgreen–Miner linear rule represented here by the straight line $[(1,0),(0,1)]$. It is somewhat more precise than the diagram of Fig. 4.4 drawn for the same values of N_D/N_R. Additional simu-

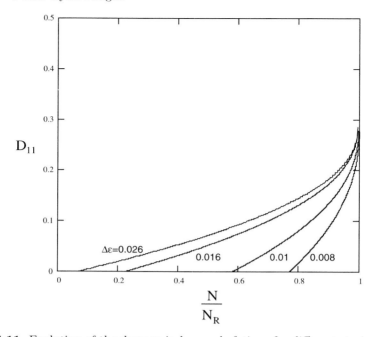

Fig. 4.11. Evolution of the damage in low cycle fatigue for different strain amplitudes

lations and experiments show that the deviation is more and more important as the difference between the two loading amplitudes increases.

4.4.2 Multiaxial and Multilevel Fatigue Loadings♂♂

As an example, consider the biaxial testing steel specimen of Fig. 4.13 which exhibits stress concentrations, localized plasticity, and damage at notches. The sample is 120 mm long and 4.5 mm thick, which allows for the plane stress assumption. A multilevel fatigue loading is applied and the failure conditions are determined after a single elastic reference computation by use of the cyclic SED method of Sect. 4.2.4, followed by the time integration of the unified damage law. The edges are in pure tension with a stress triaxiality ratio of 1/3 which corresponds to $R_\nu = 1$.

The experiment is performed on the triaxial testing machine of Fig. 4.14. Two in-phase proportional loads $F_1(t) = F_2(t) = F(t)$ are applied on the lateral sides of the specimen. The total loading consists of thirteen blocks i of cyclic $F^{(i)}(t)$ varying between a zero minimum load and a constant maximum load $F_{\max}^{(i)}$ (Fig. 4.15). The first block is made of 38000 cycles at $F_{\max}^{(1)} = 35$ kN, then the load is increased by 5 kN every 100 cycles up to $F_{\max}^{(13)} = 95$ kN. Failure, characterized by a mesoscopic crack of 0.5 mm in the edge, occurs at this last level after $N_{\text{Rexp}}^{(13)} = 3050$ cycles. This corresponds to a total number of cycles to rupture of 42150.

The material is a 2-1/4 CrMo steel at room temperature. For strains smaller than 4%, the monotonic plastic behavior is approximated by a linear hardening $R = C_y p$; for cyclic loading with strains smaller than 4%, it is approximated by a linear kinematic hardening $\boldsymbol{X} = \frac{2}{3} C_y \boldsymbol{\epsilon}^p$. The full set of material parameters defined in Sects. 1.4 and 1.5 is:

- $E = 200000$ MPa, $\nu = 0.3$ for elasticity
- $\sigma_y = 180$ MPa as yield stress
- $C_y = 6000$ MPa for hardening
- $\sigma_u = R_\infty^y + \sigma_y = 450$ MPa as ultimate stress
- $\epsilon_{pD} = 0.12$, $m = 2$, $\sigma_f^\infty = 140$ MPa, $S = 2.8$ MPa, $s = 2$, $D_c = 0.2$ for damage with a damage threshold in fatigue given by

$$p_D = \epsilon_{pD} \left(\frac{\sigma_u - \sigma_f^\infty}{\frac{\Delta \sigma}{2} - \sigma_f^\infty} \right)^m . \tag{4.89}$$

The procedure of the numerical prediction is as follows:

1. **Perform the elastic reference computation for a load $F_1 = F_2 = F_{\text{ref}}$.**
   Define for each level $\alpha_{\min} = F_{\min}/F_{\text{ref}} = 0$ and $\alpha_{\max} = F_{\max}/F_{\text{ref}}$ (with $\alpha^{\text{ref}} = 1$). The von Mises stress at the edge concentration point is $\sigma_{\text{eq}}^{\text{ref}} = 129.2$ MPa for $F_{\text{ref}} = 10$ kN. At this point, $R_\nu = 1$ and

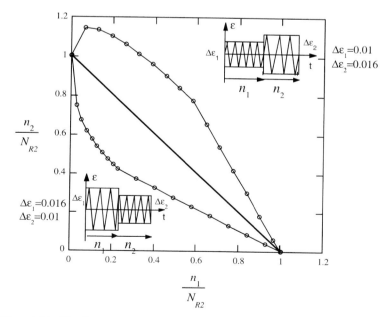

Fig. 4.12. Nonlinear damage accumulation in two-level fatigue loadings

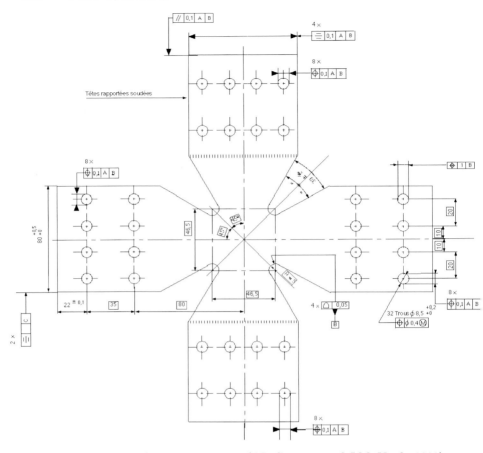

Fig. 4.13. Biaxial testing specimen (J.P. Sermage and J.M. Virely 1998)

$$E\left(\sigma\epsilon\right)_{\text{elas}} = E\left(\Delta\sigma\Delta\epsilon\right)_{\text{elas}} = \left(\alpha_{\max}\sigma_{\text{eq}}^{\text{ref}}\right)^2 . \qquad (4.90)$$

Check that for each level $(\Delta\sigma)_{\text{elas}} > 2\sigma_{\text{y}}$ so that plasticity (at least) must occur locally.

2. **Apply the Strain Energy Density method for each level to take into account the plasticity.**

Calculate the maximum stress from the monotonic SED method:

$$\sigma_{\max} = \sqrt{\dfrac{E\left(\sigma\epsilon\right)_{\text{elas}} + \dfrac{E}{C_{\text{y}}}\sigma_{\text{y}}^2}{1 + \dfrac{E}{C_{\text{y}}}}} . \qquad (4.91)$$

Fig. 4.14. Triaxial testing machine ASTREE of LMT-Cachan

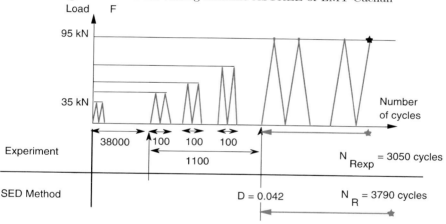

Fig. 4.15. Multilevel fatigue loading

Calculate the stress range $\Delta\sigma$, the plastic strain range $\Delta\epsilon_\mathrm{p}$, and the plastic strain increment over one cycle of the considered level $\dfrac{\delta p}{\delta N}$:

$$\Delta\sigma = \sqrt{\dfrac{E\,(\Delta\sigma\Delta\epsilon)_\mathrm{elas} + 4\dfrac{E}{C_\mathrm{y}}\sigma_\mathrm{y}^2}{1 + \dfrac{E}{C_\mathrm{y}}}}, \qquad (4.92)$$

$$\dfrac{\delta p}{\delta N} = 2\Delta\epsilon_\mathrm{p} = 2\dfrac{\Delta\sigma - 2\sigma_\mathrm{y}}{C_\mathrm{y}}. \qquad (4.93)$$

At the beginning of the last level, the accumulated plastic strain is

$$p = 38000 \frac{\delta p}{\delta N}^{(1)} + 100 \frac{\delta p}{\delta N}^{(2)} + \ldots + 100 \frac{\delta p}{\delta N}^{(12)} = 47.06. \quad (4.94)$$

3. **Calculate the damage and the number of cycles to rupture.**
   (a) *Determine the number of cycles at damage initiation (N_D)*
   The damage threshold for the first level $F_{\max}^{(1)} = 35$ kN is $p_D^{(1)} = 6.7$.
   Then
   $$N_D = \frac{p_D^{(1)}}{2\Delta\epsilon_p^{(1)}} = 6650 \text{ cycles}. \quad (4.95)$$
   Check that $N_D < 38000$ cycles so that the damage initiates at this level.
   (b) *Calculate the damage increment for each level*
   According to (4.63),
   $$\begin{aligned}\frac{\delta D}{\delta N} &= \frac{(\sigma_{\text{eq min}}^{2s} + \sigma_{\text{eq max}}^{2s})R_\nu^s}{2(2ES)^s} \frac{\delta p}{\delta N} \\ &= \frac{\sigma_{\max}^{2s} + (\Delta\sigma - \sigma_{\max})^{2s}}{2(2ES)^s} \frac{\delta p}{\delta N}.\end{aligned} \quad (4.96)$$

   (c) *Determine the number of cycles to rupture ($N_R^{(13)}$)*
   The damage increments over each block are summed up to $D = D_c$. At the beginning of the last level,
   $$D = (38000 - N_D)\frac{\delta D}{\delta N}^{(1)} + 100\frac{\delta D}{\delta N}^{(2)} + \ldots + 100\frac{\delta D}{\delta N}^{(12)} = 0.042 \quad (4.97)$$

   and a number of cycles at crack initiation $N_R^{(13)}$ is estimated as
   $$N_R^{(13)} = \frac{D_c - 0.042}{\frac{\delta D}{\delta N}^{(13)}} = 3790 \text{ cycles}, \quad (4.98)$$

   compared to $N_{R\exp}^{(13)} = 3050$ and which corresponds to a total number of cycles $N_R = 39100 + 3790 = 42890$ instead of the 42150 cycles obtained experimentally. The accumulated plastic strain at failure is then
   $$p_R = 47.06 + 3790\frac{\delta p}{\delta N}^{(13)} = 112.6. \quad (4.99)$$

From the first level to the last, the accumulated plastic strain per cycle ($\frac{\delta p}{\delta N}$) varies from $0.1\ 10^{-2}$ to $1.73\ 10^{-2}$ whereas the damage per cycle ($\frac{\delta D}{\delta N}$) varies from $8.94\ 10^{-7}$ to $4.17\ 10^{-5}$.

4.4.3 Damage and Fatigue of Elastomers♂♂♂
(S. Cantournet and R. Desmorat, 2002)

Elastomers, at least for fatigue purposes, may be modeled by **hyperelasticity with internal friction** coupled with **damage**. No plasticity! A dissipative phenomenon occurs, however, due to internal sliding of the macromolecular chains on themselves and the black carbon filler particles. Internal viscosity is also an additional dissipative mechanism not taken into account in this section.

The unified damage law governed by the accumulated plastic strain rate used up to now is useless. Fortunately, its generalization to any dissipative phenomenon described in Sect. 1.4.5 applies. It just needs to be formulated within the finite strains framework (see Sect. 3.4.1), replacing ϵ with the Green–Lagrange deformation $\boldsymbol{E}$, $\boldsymbol{\sigma}$ with the second Piola–Kirchhoff stress tensor $\boldsymbol{S}$, and with internal variables:

- The internal inelastic strain $\boldsymbol{E}^\pi$ (instead of ϵ^π). It is associated with the opposite of a stress denoted by $\boldsymbol{S}^\pi$.
- The internal sliding variable $\boldsymbol{\alpha}$ associated with the residual microstress tensor $\boldsymbol{X}$.
- The isotropic damage variable D, associated with the opposite of the energy density release rate Y.

4.4.3.1 Hyperelasticity with Internal Friction Coupled with Damage

The state potential is (for the isothermal case),

$$\rho_0 \psi = (1-D)\left[w_1(\boldsymbol{E}) + w_2(\boldsymbol{E}-\boldsymbol{E}^\pi)\right] + \frac{1}{2} C_x \boldsymbol{\alpha} : \boldsymbol{\alpha}, \qquad (4.100)$$

with ρ_0 as the density of the undeformed material and where

- w_1 is a hyperelastic energy density such as the Mooney, Hart–Smith or Lambert–Diani–Rey densities. For incompressible materials, $J = \det \boldsymbol{F} = 1$ and

$$w_1 = moon_1(I_1 - 3) + moon_2(I_2 - 3) \qquad \text{Mooney,}$$
$$w_1 = h_1 \int \exp\left(h_3(I_1 - 3)^2\right) dI_1 + 3h_2 \ln \frac{I_2}{3} \qquad \text{Hart–Smith,}$$
$$w_1 = h_1 \int \exp\left(h_3(I_1 - 3)^{a_1}\right) dI_1 + 3h_2 \int \frac{dI_2}{I_2^{a_2}} \qquad \text{Lambert–Diani–Rey.}$$

$$(4.101)$$

Here, $I_1 = \operatorname{tr} \boldsymbol{C}$ and $I_2 = \frac{1}{2}\left[(\operatorname{tr} \boldsymbol{C})^2 - \operatorname{tr} \boldsymbol{C}^2\right]$ are the first two invariants of the dilatation tensor $\boldsymbol{C} = \boldsymbol{F}^\mathrm{T}\boldsymbol{F}$; and $moon_i$ and h_i are the hyperelasticity material parameters. For quasi-incompressible materials, the third invariant $I_3 = J^2$ does not remain strictly equal to unity. A compressibility term that is a function of I_3 is often introduced. For example, the Mooney density for compressible or quasi-incompressible elastomers is written as

$$w_1 = moon_1(I_1 - I_3 - 2) + moon_2(I_2 - 2I_3 - 1) + \frac{1}{2}K_v(I_3 - 1)^2, \quad (4.102)$$

where K_v is the compressibility modulus.
- w_2 is the second order term of the Mooney–Rivlin development with inelastic strain $\boldsymbol{E}^\pi$,

$$w_2 = 4C_{20}\left[\mathrm{tr}(\boldsymbol{E} - \boldsymbol{E}^\pi)\right]^2 = C_{20}\left(I_1 - 2\mathrm{tr}\boldsymbol{E}^\pi - 3\right)^2. \quad (4.103)$$

For incompressible materials, the internal pressure P is calculated from the boundary conditions using a pressure-displacement FE formulation. The state laws then become

$$\boxed{\begin{aligned}
\boldsymbol{S} &= \rho_0 \left.\frac{\partial \psi}{\partial \boldsymbol{E}}\right|_{J=1} = (1-D)\frac{\partial(w_1 + w_2)}{\partial \boldsymbol{E}} - P\boldsymbol{C}^{-1}, \\
\boldsymbol{S}^\pi &= -\rho_0 \frac{\partial \psi}{\partial \boldsymbol{E}^\pi} = (1-D)\frac{\partial w_2}{\partial \boldsymbol{E}}, \\
\boldsymbol{X} &= \rho_0 \frac{\partial \psi}{\partial \boldsymbol{\alpha}} = C_x \boldsymbol{\alpha}, \\
Y &= -\rho_0 \frac{\partial \psi}{\partial D} = w_1 + w_2.
\end{aligned}} \quad (4.104)$$

This defines the effective stresses $\tilde{\boldsymbol{S}} = \dfrac{\boldsymbol{S}}{1-D}$ and $\tilde{\boldsymbol{S}}^\pi = \dfrac{\boldsymbol{S}^\pi}{1-D}$.

For compressible or quasi-incompressible materials, the first law of (4.104) is simply

$$\boldsymbol{S} = \rho_0 \frac{\partial \psi}{\partial \boldsymbol{E}} = (1-D)\frac{\partial(w_1 + w_2)}{\partial \boldsymbol{E}}. \quad (4.105)$$

The reversibility criterion (such as $f < 0 \Rightarrow$ hyperelasticity) is:

$$\boxed{f = \left\|\frac{\boldsymbol{S}^\pi}{1-D} - \boldsymbol{X}\right\| - \sigma_s = \sqrt{\left(\tilde{\boldsymbol{S}}^\pi - \boldsymbol{X}\right):\left(\tilde{\boldsymbol{S}}^\pi - \boldsymbol{X}\right)} - \sigma_s,} \quad (4.106)$$

with σ_s the reversibility limit in the $\boldsymbol{S}^\pi$ stress plane.

The dissipation potential is

$$F = f + \frac{\gamma}{2C_x}\boldsymbol{X}:\boldsymbol{X} + \frac{S}{(s+1)(1-D)}\left(\frac{Y}{S}\right)^{s+1}, \quad (4.107)$$

giving rise to the evolutions laws through the normality rule,

$$\dot{\boldsymbol{E}}^\pi = \dot{\mu}\frac{\partial F}{\partial \boldsymbol{S}^\pi}, \quad \dot{\boldsymbol{\alpha}} = -\dot{\mu}\frac{\partial F}{\partial \boldsymbol{X}}, \quad \text{and} \quad \dot{D} = \dot{\mu}\frac{\partial F}{\partial Y}, \quad (4.108)$$

with $\dot{\mu}$ as the internal sliding multiplier determined from the consistency condition $f = 0$, $\dot{f} = 0$. They lead to

$$\boxed{\begin{aligned}\dot{\boldsymbol{E}}^\pi &= \frac{\dot{\mu}}{1-D}\frac{\tilde{\boldsymbol{S}}^\pi - \boldsymbol{X}}{\|\tilde{\boldsymbol{S}}^\pi - \boldsymbol{X}\|}, \\ \dot{\boldsymbol{X}} &= \left[C_x \dot{\boldsymbol{E}}^\pi - \gamma \boldsymbol{X}\dot{\pi}\right](1-D),\end{aligned}}$$
(4.109)

and the damage evolution law for elastomeric materials valid for monotonic loading as well as fatigue loading:

$$\boxed{\begin{aligned}\dot{D} &= \left(\frac{Y}{S}\right)^s \dot{\pi} \quad \text{if} \quad \pi > \pi_D \\ D &= D_c \quad \longrightarrow \quad \text{mesocrack initiation,}\end{aligned}}$$
(4.110)

where $\pi = \int_0^t (1-D)\dot{\mu}\,dt = \int_0^t \|\dot{\boldsymbol{E}}^\pi\|\,dt$ is, as expected, the **cumulative measure of the internal sliding**.

Compared to hyperelasticity, 8 additional material parameters are introduced:

- The reversibility threshold σ_s
- C_{20}, C_x, and γ for internal friction: the parameters C_{20} and C_x model the hysteresis loops when the parameter γ models the stress softening
- π_D, S, s, and D_c for damage evolution and mesocrack initiation

4.4.3.2 Hysteresis and Cyclic Softening of Filled Rubber

With neither damage nor viscosity, the model represents the hysteresis and the stress softening of filled elastomers. Figure 4.16 shows the cyclic tensile curves of styrene butadiene rubber (SBR) filled with silica particles.

Note that in terms of hyperelasticity, considering Hart–Smith density allows for a good modelling of the up-turn at very large strains. For fatigue applications, take the model which best represents the hysteresis loops in the range of deformation under consideration.

4.4.3.3 Fatigue Curve of Filled Natural Rubber

The model of hyperelasticity with internal friction coupled with damage allows us to compute the number of cycles to rupture in unidimensional fatigue. The numerical procedure is detailed in the next section for structures but applies, of course, to cyclic tension.

Consider here a filled natural rubber (NR) tested both in cyclic tension and fatigue. The material parameters are

- $moon_1 = 1.848$ MPa, $moon_2 = 0.264$ MPa, $K_v = 1188$ MPa
- $\sigma_s = 1$ MPa, $C_{20} = 1$ MPa, $C_x = 10$ MPa, $\gamma = 5$
- $\pi_D = 0$, $S = 5.6$ MPa, $s = 5$, $D_c = 0.2$

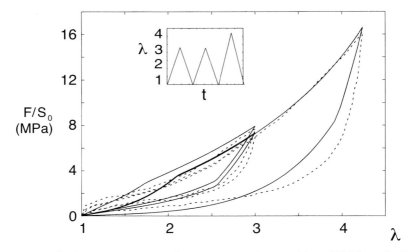

Fig. 4.16. Cyclic tensile curve of styrene butadiene rubber (SBR) with Hart–Smith density. The dotted line represents experimental results from A. Lapra (2001) ($h_1 = 1.6$ MPa, $h_2 = 0.28$ MPa, $h_3 = 5.29\ 10^{-4}$, $\sigma_s = 1.16$ MPa, $C_{20} = 0.14$ MPa, $C_x = 0.93$ MPa, $\gamma = 0.57$, $D = 0$).

The fatigue curve of the maximum applied elongation $\lambda_{\max}$ (larger principal component of the transformation gradient $\boldsymbol{F}$) versus the number of cycles to rupture N_R is calculated and compared to experiments with success in Fig. 4.17. The measured elongation to rupture in tension, $\lambda_R = 7.2$, is also reported on the diagram which shows the ability of the single generalized damage law $\dot{D} = (Y/S)^s \dot{\pi}$ to model monotonic as well as fatigue damages.

4.4.3.4 Fatigue of a Metal/Elastomer Joint

As an example, consider the case of the double lap joint of Fig. 4.18 submitted to fatigue loading. Half of the joint is meshed in three dimensions with 27-node quadratic elements. The bottom ends are fixed and the same displacement $\vec{U}(t) = U(t)\vec{e}_3$ is applied cyclically on the top end between $U = 0$ and $U = U_{\max} = 45$ mm at a loading rate of 100 mm/min. Considering the damage pattern, the most damaged zones are the corners points of the rubber parts where a crack initiates (after a number of cycles $N_R^{\text{exp}} = 30$), with a damage propagation mainly along the interfaces.

The metallic parts are made of steel ($E = 200000$ MPa, $\nu = 0.3$) which remain elastic during the test. The elastomer is made of filled natural rubber.

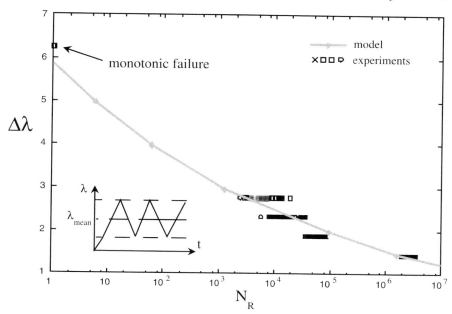

Fig. 4.17. Experimental and calculated fatigue curves for filled natural rubber (NR) (mean elongation $\lambda_{\mathrm{mean}} = 2.53$)

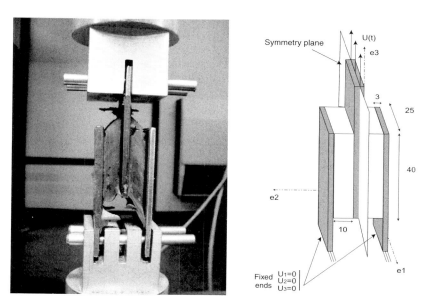

Fig. 4.18. Metal/elastomer joint (dimensions in mm)

226   4 Low Cycle Fatigue

The uncoupled procedure to estimate both the fatigue damage and the number of cycles to rupture is the one described in Sect. 2.1. It is detailed here for the damage law considered:

Step 1. **Perform the nonlinear structure FEA with no damage up to a stabilized cycle.** This constitutes the reference computation. The result of the analysis is the history of the stresses $S_{ij}(t)$ or $\boldsymbol{S}(t)$, the strains $E_{ij}(t)$ or $\boldsymbol{E}(t)$, and the internal variables $\boldsymbol{S}^{\pi}(t)$, $\boldsymbol{E}^{\pi}(t)$, $\boldsymbol{X}(t)$, and $\boldsymbol{\alpha}(t)$. It is also the global hysteretic response of the structure force F vs displacement U.

Step 2. **Determine the most loaded point where $Y = w_1 + w_2$ is maximum** in order to calculate the damage increment (step 3), the damage history (step 4), and the number of cycles to rupture (step 5) at this point. If damage maps are needed, perform the steps 3 to 5 at the Gauss points of the damaged zone considered.

Step 3. **Calculate the damage increment over one cycle.** The calculation is made over the stabilized cycle as a post-processing of the reference computation. This is usual in fatigue as it makes the analysis much faster with rather high accuracy,

$$\frac{\delta D}{\delta N} = \int_{1\,\text{cycle}} \left(\frac{Y(t)}{S}\right)^s \dot{\pi}(t)\mathrm{d}t$$
$$= \int_{1\,\text{cycle}} \left(\frac{w_1(\boldsymbol{E}(t)) + w_2(\boldsymbol{E}(t) - \boldsymbol{E}^{\pi}(t))}{S}\right)^s \|\dot{\boldsymbol{E}}^{\pi}(t)\|\mathrm{d}t \,. \tag{4.111}$$

Step 4. **Calculate the damage history at the considered point(s).** The derivative $\dfrac{\delta D}{\delta N}$ is constant for a given maximum applied displacement $U_{\max}$. Then,

$$D(N) = \frac{\delta D}{\delta N} \cdot (N - N_{\mathrm{D}}), \tag{4.112}$$

with N_{D} as the number of cycles at damage initiation. Here, $N_{\mathrm{D}} = 0$ as $\pi_{\mathrm{D}} = 0$.

Damage maps may then be drawn. The computed damage pattern is in accordance with the experiment: it exhibits a damage larger in the rubber corner points and is localized along the interfaces (Fig. 4.19).

Step 5. **Calculate the number of cycles to crack initiation corresponding to $D = D_{\mathrm{c}}$:**

$$N_{\mathrm{R}} = N_{\mathrm{D}} + \frac{D_{\mathrm{c}}}{\dfrac{\delta D}{\delta N}} \quad \text{at the most loaded point.} \tag{4.113}$$

These calculations have been made for the metal/elastomer joint. The number of cycles at crack initiation corresponds to the reach of $D = D_{\mathrm{c}}$ at the corner point A. The result is $N_{\mathrm{R}} = 39$ cycles, compared to $N_{\mathrm{R}}^{\exp} = 30$ which corresponds to an accuracy of about 30%.

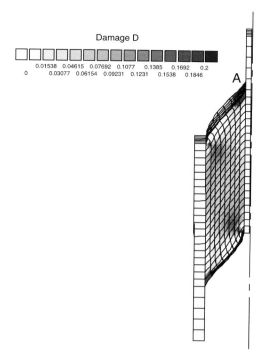

Fig. 4.19. Damage pattern at crack initiation in the metal/elastomer joint (mesh of half of the joint)

4.4.4 Predeformed and Predamaged Initial Conditions♂♂

As in Sect. 3.4.7 for ductile failures, it is interesting to know how the number of cycles to rupture can be modified in low cycle fatigue by changes in the initial conditions of the main variables. Forming processes, accidents, or over loadings may induce, for a "forming" or an "accident" stress σ_{eq}^0, initial fields of:

- The accumulated plastic strain p_0
- The corresponding value of the strain hardening variables R_0 and $\boldsymbol{X}_0$
- The stored energy $w_{\text{s}}^0 \approx A(\sigma_{\text{eq}}^0 - \sigma_{\text{f}}^\infty)p_0^{1/m}$,
- The damage D_0 or $\boldsymbol{D}_0$
- The internal residual stresses

They have to be introduced as initial conditions in the structure calculation, at each Gauss point if a finite element analysis is used, to determine the stresses $\sigma_{\text{eq}}(t)$ and $\sigma_{\text{H}}(t)$, as well as the plastic strain $p(t)$ or Δp. They also have to be introduced in the damage law.

- The initial plastic strain (p_0) or the stored energy density (w_{s}^0) modifies the number of cycles corresponding to the damage initiation (N_{D})

through the change in damage threshold written in terms of accumulated plastic strain. For a fatigue loading with a periodic, signed von Mises stress between $-\sigma_{\text{eq min}}$ and $\sigma_{\text{eq max}}$ applied after the forming process or the accident, the stored energy threshold reads

$$w_s = A(\sigma_u - \sigma_f^\infty)\epsilon_{\text{pD}}^{1/m}$$
$$= A(\sigma_{\text{eq}}^0 - \sigma_f^\infty)p_0^{1/m} + A\left(\frac{\sigma_{\text{eq min}} + \sigma_{\text{eq max}}}{2} - \sigma_f^\infty\right)(p_D^{1/m} - p_0^{1/m}) \quad (4.114)$$

and corresponds to the accumulated plastic strain threshold $p = p_D^{\text{new}}$,

$$p_D^{\text{new}} = \left[\epsilon_{\text{pD}}^{1/m} \frac{\sigma_u - \sigma_f^\infty}{\frac{1}{2}(\sigma_{\text{eq min}} + \sigma_{\text{eq max}}) - \sigma_f^\infty} \right. $$
$$\left. + p_0^{1/m} \frac{\frac{1}{2}(\sigma_{\text{eq min}} + \sigma_{\text{eq max}}) - \sigma_{\text{eq}}^0}{\frac{1}{2}(\sigma_{\text{eq min}} + \sigma_{\text{eq max}}) - \sigma_f^\infty}\right]^m, \quad (4.115)$$

becoming $p_D^{\text{new}} = p_0$ if $p_0 \geq p_D^0 = \epsilon_{\text{pD}}\left(\dfrac{\sigma_u - \sigma_f^\infty}{\sigma_{\text{eq}}^0 - \sigma_f^\infty}\right)^{1/m}$.

- The initial damage corresponding to $p_0 \geq p_D$ acts in the integration of the damage rate from D_0 to D_c.

For example, in the case of periodic, proportional loading of Sect. 4.2.2, the formula (4.24) is replaced by

$$N_R = \frac{p_D^{\text{new}} - p_0}{2\Delta p} + \frac{D_c}{2\Delta p}\left(\frac{2ES}{\sigma_{\text{eq max}}^2 R_\nu}\right)^s \quad \text{if} \quad p_0 < p_D^0$$
$$N_R = \frac{D_c - D_0}{2\Delta p}\left(\frac{2ES}{\sigma_{\text{eq max}}^2 R_\nu}\right)^s \quad \text{if} \quad p_0 \geq p_D^0. \quad (4.116)$$

4.4.5 Hierarchic Approach up to Non-Proportional Effects♂,♂♂♂

As for ductile failure predictions in Sect. 3.4.8, the different models of low cycle fatigue may be classified in an increasing order of accuracy corresponding (unfortunately) to an increasing order of difficulty.

- The Manson–Coffin law is certainly the first model to consider for an early design with a periodical loading or if only the plastic strain range is known. The number of cycles-to-crack initiation (N_R) at the critical point where the state of stress is one dimensional is simply

$$N_R = \left(\frac{C_{\text{MC}}}{\Delta \epsilon_p}\right)^{\gamma_{\text{MC}}} \quad \text{or} \quad N_R \approx \left(\frac{2\epsilon_{\text{pR}}}{\Delta \epsilon_p}\right)^2. \quad (4.117)$$

- If the loading is periodic by blocks, the linear Palmgreen–Miner rule may give an order of magnitude of the result but without the possibility to know the sign of the error:

$$\sum \frac{n_i}{N_{Ri}} = 1, \qquad N_R = \sum n_i. \qquad (4.118)$$

- When the local loading is 3D, the Manson–Coffin law does not apply but if it is proportional and periodic the integrated unified damage law may be used. For a symmetric periodic loading,

$$N_R = N_D + \frac{D_c}{2\Delta p}\left(\frac{ES}{\sigma_{eq\,max}^2 R_\nu}\right)^s,$$

$$\text{with} \quad N_D = \frac{\epsilon_{pD}}{2\Delta p}\left(\frac{\sigma_u - \sigma_f^\infty}{\sigma_{eq\,max} - \sigma_f^\infty}\right)^m. \qquad (4.119)$$

The equations require knowledge of the value of 5 material damage parameters plus E, ν, σ_f^∞, and σ_u. They can be reduced to one by using the equivalence between 1D and 3D through the damage equivalent stress $\sigma^\star$ if one point of the Manson–Coffin curve is known:

$$\frac{N_R(\sigma_{ij})}{N_R(\sigma_M)} \approx \frac{\Delta\epsilon_p}{\Delta p}\left(\frac{\sigma_M^\star}{\sigma_M}\right)^{2s}. \qquad (4.120)$$

- The last possibility is the integration in time of the unified damage law if the loading is not periodic but is a complex function of time, either 1D or 3D, either for isotropic or anisotropic damage, and either deterministic or random by the Monte Carlo method:

$$\dot{D} = \left(\frac{Y}{S}\right)^s |\dot{\epsilon}^p| \qquad \text{if} \quad w_s > w_D. \qquad (4.121)$$

In case of periodic loadings, the stored energy damage initiation criterion is equivalent to

$$\dot{D} \neq 0 \quad \text{if} \quad p > p_D = \epsilon_{pD}\left(\frac{\sigma_u - \sigma_f^\infty}{\frac{(\Delta\sigma)_{eq}}{2} - \sigma_f^\infty}\right)^m. \qquad (4.122)$$

Remark – *Note that the extension of the Neuber and the Strain Energy Density methods is a big help in determining $\Delta\epsilon_p$ or Δp from an elastic calculation.*

In matters related to plasticity, classical isotropic and kinematic hardening laws have been considered so far. The Armstrong–Frederick nonlinear law $\dot{\boldsymbol{X}} = \frac{2}{3}C\dot{\boldsymbol{\epsilon}}^p - \gamma \boldsymbol{X}\dot{p}$ represents the Baushinger effect of the apparent yield stress decrease in cyclic loadings and also the cyclic stress softening of non-symmetric strain-controlled loadings. But it is not suitable for more complex hardening mechanisms such as non-proportionality effects or ratcheting, which is the continuous increase of the plastic strain at constant stress range (even when no damage occurs). Other kinematic hardening laws must then be considered. They are usually written for undamaged materials and need to be coupled with damage by the effective stress concept.

- A better modelling of hardening (for a wider range of plastic strains or for uniaxial ratcheting) is obtained by use of several kinematic hardening variables $\boldsymbol{X}_i$ related to nonlinear laws (Z. Mroz 1967), coupled here with either isotropic or anisotropic damage (no summation on subscripts i):

$$\boldsymbol{X} = \sum_i \boldsymbol{X}_i,$$

$$\dot{\boldsymbol{X}}_i = \frac{2}{3} C_i \dot{\boldsymbol{\epsilon}}^{\mathrm{p}} - \gamma_i \boldsymbol{X}_i \dot{r}, \quad (4.123)$$

$$\dot{\boldsymbol{\epsilon}}^{\mathrm{p}} = (\boldsymbol{H} \dot{\boldsymbol{e}}^{\mathrm{p}} \boldsymbol{H})^{\mathrm{D}} \quad \text{with} \quad \dot{\boldsymbol{e}}^{\mathrm{p}} = \boldsymbol{n}^{\mathrm{X}} \dot{r} = \frac{3}{2} \frac{\tilde{\boldsymbol{\sigma}}^{\mathrm{D}} - \boldsymbol{X}}{(\tilde{\boldsymbol{\sigma}} - \boldsymbol{X})_{\mathrm{eq}}} \dot{r},$$

where $\dot{\boldsymbol{\epsilon}}^{\mathrm{p}} = \boldsymbol{n}^{\mathrm{X}} \dot{p}$ is the plastic strain rate in plasticity ($\boldsymbol{D} = \boldsymbol{0}$) and $\dot{\boldsymbol{\epsilon}}^{\mathrm{p}} = \boldsymbol{n}^{\mathrm{X}} \dot{r}$ is the effective plastic strain rate in plasticity coupled with damage.

- Another way is to keep only one kinematic variable but to introduce in its evolution law a term taking into account the rotation of the stress vector to model the ratcheting and non-proportional effects. Many nonlinear laws have been proposed but are most often not coupled with damage (H. Burlet and G. Cailletaud 1987, A. Benallal and D. Marquis 1988, E. Tanaka 1994, N. Ohno and J.D. Wang 1994, T. Kurtyka 1996, S. Calloch 1996, M. François 2000). For instance, the Burlet–Cailletaud model states that

$$\dot{\boldsymbol{X}} = \frac{2}{3} C \dot{\boldsymbol{\epsilon}}^{\mathrm{p}} - \gamma (\boldsymbol{X} : \boldsymbol{n}^{\mathrm{X}}) \, \boldsymbol{n}^{\mathrm{X}} \dot{p} \quad (4.124)$$

and the Ohno–Wang model uses

$$\dot{\boldsymbol{X}} = \frac{2}{3} C \dot{\boldsymbol{\epsilon}}^{\mathrm{p}} - \gamma (\boldsymbol{X} : \boldsymbol{X})^\alpha \langle \boldsymbol{X} : \dot{\boldsymbol{\epsilon}}^{\mathrm{p}} \rangle \boldsymbol{X}. \quad (4.125)$$

In order to introduce the damage coupling, one way is to define a non-associated potential F_{X} such as the nonlinear kinematic hardening law derived from $F = f + F_{\mathrm{X}}$. Replacing the stress $\boldsymbol{\sigma}$ by the effective stress $\tilde{\boldsymbol{\sigma}}$ in F before taking the derivatives gives the nonlinear kinematic hardening law fully coupled with damage. For instance, the potential

$$F_{\mathrm{X}} = \frac{3\gamma}{4C} (\boldsymbol{X} : \boldsymbol{X})^{\alpha_1} \langle \boldsymbol{X} : \boldsymbol{n}^{\mathrm{X}} \rangle^{\alpha_2} \quad \text{and} \quad \boldsymbol{n}^{\mathrm{X}} = \frac{3}{2} \frac{\underline{\boldsymbol{E}} : \boldsymbol{\epsilon}^{\mathrm{eD}} - \frac{2}{3} C \boldsymbol{\alpha}}{(\underline{\boldsymbol{E}} : \boldsymbol{\epsilon}^{\mathrm{eD}} - \frac{2}{3} C \boldsymbol{\alpha})_{\mathrm{eq}}}$$
$$(4.126)$$

gives

$$\dot{X} = \frac{2}{3}C\dot{e}^{\mathrm{p}} - \gamma\alpha_2(X:X)^{\alpha_1}\langle X:n^{\mathrm{X}}\rangle^{\alpha_2-1}n^{\mathrm{X}}\dot{r} \\ - \gamma\alpha_1(X:X)^{\alpha_1-1}\langle X:\dot{e}^{\mathrm{p}}\rangle^{\alpha_2}X, \quad (4.127)$$

with the first back stress term proportional to n^{X} as in the Burlet–Cailletaud law and the second one proportional to X as in the Ohno–Wang law.

- Finally, the treatment for plastic anisotropy is similar to the one for the ductile failure case (see Sect. 3.4.8). Use as yield function:

$$f = \sqrt{(\tilde{\sigma} - X) : \underline{h} : (\tilde{\sigma} - X)} - R - \sigma_{\mathrm{y}} \quad \text{where} \quad \underline{h} \text{ is the Hill tensor} \quad (4.128)$$

and take

$$F = f + \frac{3}{4}\,X : \underline{C}^{-1} : \underline{\Gamma} : X \quad (4.129)$$

as the dissipation potential with the state law $X = \frac{2}{3}\underline{C}:\alpha$ and with $\underline{C}$ and $\underline{\Gamma}$ as material-dependent fourth order tensors. The nonlinear anisotropic kinematic hardening law with no damage ($\underline{C}^{-1} : \underline{\Gamma}$ assumed to be symmetric) becomes

$$\dot{X} = \frac{2}{3}\underline{C} : \dot{e}^{\mathrm{p}} - \underline{\Gamma} : X\dot{p}. \quad (4.130)$$

The coupling of the kinematic hardening evolution law with damage is gained using the thermodynamics framework as

$$\dot{\epsilon}^{\mathrm{p}} = \dot{r}\frac{\partial F}{\partial\sigma} \quad \text{and} \quad \dot{\alpha} = -\dot{r}\frac{\partial F}{\partial X} \quad (4.131)$$

lead to

$$\dot{X} = \frac{2}{3}\underline{C} : \dot{e}^{\mathrm{p}} - \underline{\Gamma} : X\dot{r}, \quad (4.132)$$

where $\dot{e}^{\mathrm{p}} = \dot{r}\dfrac{\underline{h}:(\tilde{\sigma} - X)}{\sqrt{(\tilde{\sigma}-X):\underline{h}:(\tilde{\sigma}-X)}}$ and $\dot{\epsilon}^{\mathrm{p}} = (H\,\dot{e}^{\mathrm{p}}\,H)^{\mathrm{D}}$.

5

Creep, Creep-Fatigue, and Dynamic Failures

Generally, an elevation of temperature in most materials induces a decrease of their strength and a sensitivity to the strain rate. Then, **creep** and **creep-fatigue** failures are associated with **temperature** and **time**. They are associated with temperature because these phenomena are thermally activated from mid to elevated temperatures that are above about one third of the absolute melting temperature for metals. The time association arises because the temperature induces viscous effects depending explicitely upon time.

From a physical point of view, creep damage in **metals** is essentially the nucleation and growth of **intergranular** microcracks up to crystals triple points where the coalescence of microcracks induce a mesocrack. Under fatigue loading, a net of **intragranular** microcracks develops and interacts with the creep microcracks net. From the mechanical point of view, this interaction is nonlinear and so is the cumulation of damages resulting of different stress amplitudes. In **polymers**, damage occurs as a result of rupture of molecular bonds in zones of defects or impurities that induce a strain rate effect with temperature, particularly the transition state or rubber state.

Modelling creep effects was done during the 20th century, beginning with E.N. Andrade (1910) and F.H. Norton (1929). The introduction of a damage variable was in fact first made for creep by L.M. Kachanov (1958), Y.N. Rabotnov (1968), and used then by J. Lemaitre (1971), J. Hult (1972), F.A. Leckie, and D. Hayhurst (1974). Its extension to creep-fatigue damage was followed by J.L. Chaboche and J. Lemaitre (1974), M. Chranowski (1975) to become an engineering tool to design mechanical components against creep or creep-fatigue failures (J. Skrzypek and A. Ganczarski 2002).

This chapter has the same organization as others with a **visco-plastic stress** concentration analysis, an expression of the **yield stress** for **adiabatic** deformations strain rate and temperature-dependent, and examples of **creep-fatigue**, **dynamic** analysis for **crash** problems and **penetration** of projectiles.

5.1 Engineering Considerations

All engineering components submitted to loadings at elevated temperature may be subjected to creep damage effects in quasi-static loadings and to creep-fatigue damage effects in cyclic loadings. This is the case for all metallic materials and polymers. These effects can be classified with respect to the lifetime:

- Short range effects in rockets components where the lifetime is of a few minutes or in the coating of the steel ladles in the metallurgy of steel elaboration where the lifetime is only on the order of 200 to 500 service hours.
- Middle range effects in gas turbines of airplanes or car engines where the lifetime is on the order of 5000 to 10000 hours.
- Large range effects in the chemistry industry and thermal or nuclear power plants where the lifetime is of the order on 10 to 50 years.
- Extremely large range duration as those considered in the conservation of radioactive wastes – up to 100, 1000, or even 10000 years!

The analysis grows more difficult as the time range increases at least for three reasons: as the time increases there are additional physical effects like diffusion, corrosion, and aging which are not usually taken into account in the mechanical constitutive equations; there is a time scale of some months or years above which there is almost no possibility to perform tests and ensure good identification of the material parameters; last but not least, the probability of the occurrence of unpredicted events increases with time. Furthermore, be careful at the two extrema of the plastic strain rates where viscous effects must be taken into account even at room temperature:

- $\dot{\epsilon}_p > 1\text{--}10^4$ s^{-1} is the range of rates occurring in shocks, crash, or loadings by accident. It is usually called **dynamic plasticity** because it happens at room temperature but it is formally labelled as **visco-plasticity** as the strain rate modifies the material response, especially if the reference identification has been performed at usual strain rates of about 10^{-4} s^{-1}. For example, the usual engineering yield stress may increase 10–50% at high strain rates of about 10^4 s^{-1}.
- $\dot{\epsilon}_p < 10^{-10}$ s^{-1} at the other extreme is the range which may exist in steels at room temperature loaded at small stresses, below the engineering yield stress. The resulting **slow creep** or **relaxation** may change the stress concentrations, induce leakage in pressurized vessels, or decrease the tension in the cables of pre-stressed oncrete by relaxation of the steel cables associated with the creep of the concrete.

5.2 Fast Calculation of Structural Failures

Despite the fact that temperature and time add two more variables, it is possible to perform approximate calculations of the damage to help in early

5.2.1 Uniaxial Behavior and Validation of the Damage Law ♂

5.2.1.1 Creep Damage

Pure creep is the time evolution of the plastic strain under constant stress. Generally, the damage begins to grow at the minimum strain rate after the primary creep and increases the strain rate up to rupture (Fig. 5.1).

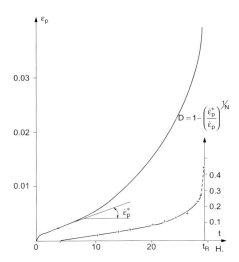

Fig. 5.1. Damage during a pure creep test on superalloy IN 100 at 1000°C (J. Lemaitre and J.L. Chaboche 1985)

The first law proposed in 1958 to describe the creep damage is the unidimensional **Kachanov law** without any threshold:

$$\dot{D} = \left[\frac{\sigma}{A_D(1-D)}\right]^{r_D}, \qquad (5.1)$$

where A_D and r_D are material parameters depending upon the temperature.

The time-to-rupture t_R at constant stress and constant temperature is the solution of this differential equation for $D = D_c$, with the initial condition $t = 0 \rightarrow D = 0$,

$$t_R = \frac{1 - (1 - D_c)^{r_D+1}}{r_D + 1} \left(\frac{\sigma}{A_D}\right)^{-r_D}. \qquad (5.2)$$

This helps us draw from experiments the isochronous curves which give the time-to-rupture function of the constant stress applied (Fig. 5.2).

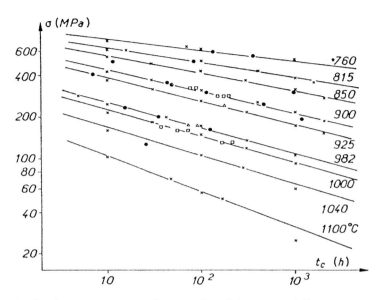

Fig. 5.2. Isochronous curves of superalloy IN 100 at different temperatures (J. Lemaitre and J.L. Chaboche 1985)

This kind of tests may be helpful in the validation of the unified damage law for creep damage if the plastic strain to rupture ϵ_{pR} and the plastic strain at the inflexion point corresponding to the threshold ϵ_{pD} are recorded.

The integration of the differential law for $\sigma = const$ is particularly easy:

$$\dot{D} = \left(\frac{\sigma^2}{2ES(1-D)^2}\right)^s |\dot{\epsilon}_p| \quad \text{if} \quad \epsilon_p > \epsilon_{pD} \tag{5.3}$$

or

$$\int_0^{D_c} (1-D)^{2s} dD = \left(\frac{\sigma^2}{2ES}\right)^s \int_{\epsilon_{pD}}^{\epsilon_{pR}} d\epsilon_p, \tag{5.4}$$

leading to

$$\epsilon_{pR} - \epsilon_{pD} = \left(\frac{\sigma^2}{2ES}\right)^{-s} \frac{1}{2s+1}\left[1-(1-D_c)^{2s+1}\right]^{-1}. \tag{5.5}$$

The slope of the best straight line fitting the points $\log(\epsilon_{pR}-\epsilon_{pD})$ vs $\log \sigma$ for different creep tests "must" give $-2s$, close to the value identified by the procedures of Sects. 1.4.4 and 2.4.

5.2.1.2 Creep-Fatigue Damage

If the loading is a sequence of constant stresses with hold times and unloadings or if it is a cyclic loading, the phenomenon of fatigue occurs as a superimposed effect. It is called **creep-fatigue interaction** because the earliest models were a combination of two terms: one for creep, and one for fatigue without or with couplings to obtain a model of nonlinear interaction.

The simple Taira rule of linear interaction (1952) applies to isothermal uniaxial cyclic loading with $0 \leq \sigma \leq \sigma_{\max}$ and with a hold time of Δt. If N_R is the number of cycles to rupture corresponding to a time $t_R \approx \Delta t \cdot N_R$; if N_{RF} is the number of cycles to rupture in pure fatigue ($\Delta t \approx 0$) for the same stress range; and if t_{RC} is the time-to-rupture in pure creep for the same maximum stress $\sigma = \sigma_{\max}$ for all t, the Taira rule reads:

$$\boxed{\frac{t_R}{t_{RC}} + \frac{N_R}{N_{RF}} = 1} \quad (5.6)$$

Easy to apply, it may give an order of magnitude of the time or of the number of cycles to rupture if nothing else is available.

An interesting point, the unified damage law developed in Chap. 1,

$$\dot{D} = \left(\frac{\sigma^2}{2ES(1-D)^2}\right)^s |\dot{\epsilon}_p| \quad \text{if} \quad p > \epsilon_{pD}, \quad (5.7)$$

can represent both creep damage and fatigue damage as the damage rate is governed by the plastic strain with no difference for creep and fatigue as, from the point of view of strains, a fatigue loading is nothing more than cyclic visco-plasticity.

Numerically, it is possible to show that the interaction between creep damage and fatigue damage is nonlinear if a damage threshold and kinematic hardening are taken into consideration (see Sect. 5.4.2). A small creep reduces considerably the fatigue life, a physical property which is not contained in the Taira rule.

5.2.2 Case of Proportional Loading♂

The case of proportional loading allows us to solve a 3D problem by manipulation of scalars only as all the tensors are colinear (see Sect. 2.1.2). As already stated, many engineering problems involve one load or multiple loads all varying proportionately to one parameter to ensure the local property of proportional loading.

In **monotonous creep**, if the history of stress is known as a function of time $\boldsymbol{\sigma}(t)$ at the critical point, one only has to calculate the von Mises equivalent stress $\sigma_{eq}(t)$ and the constant triaxiality ratio (σ_H/σ_{eq}). In the unified damage law, the accumulated plastic strain rate is replaced by the pure viscous Norton law in which the yield stress and the hardening are neglected here (but not the coupling with the damage),

$$\dot{p} = \left[\frac{\sigma_{\text{eq}}}{K_N^0(1-D)}\right]^{N_0}. \tag{5.8}$$

One has to solve a differential equation fully similar to Kachanov law (5.1),

$$\dot{D} = \left[\frac{\sigma_{\text{eq}}^2 R_\nu}{2ES(1-D)^2}\right]^s \left[\frac{\sigma_{\text{eq}}}{K_N^0(1-D)}\right]^{N_0} \quad \text{if} \quad p > p_D \tag{5.9}$$

or

$$\int_0^D (1-D)^{2s+N_0} \mathrm{d}D = \int_{t_D}^t \left[\frac{\sigma_{\text{eq}}^2(t) R_\nu}{2ES}\right]^s \left[\frac{\sigma_{\text{eq}}(t)}{K_N^0}\right]^{N_0} \mathrm{d}t, \tag{5.10}$$

where the time t_D to reach the damage threshold p_D is given by

$$p_D = \int_0^{t_D} \left[\frac{\sigma_{\text{eq}}}{K_N^0}\right]^{N_0} \mathrm{d}t. \tag{5.11}$$

Then,

$$D = 1 - \left[1 - (2s+N_0+1)\int_{t_D}^t \frac{\sigma_{\text{eq}(t)}^{2s+N_0} R_\nu^s}{(2ES)^s (K_N^0)^{N_0}} \mathrm{d}t\right]^{\frac{1}{2s+N_0+1}} \tag{5.12}$$

and the time-to-rupture t_R is given by

$$\boxed{\int_{t_D}^{t_R} \frac{\sigma_{\text{eq}(t)}^{2s+N_0} R_\nu^s}{(2ES)^s (K_N^0)^{N_0}} \mathrm{d}t = \frac{1-(1-D_c)^{2s+N_0+1}}{2s+N_0+1}.} \tag{5.13}$$

For example, for a **constant state of stress** $p_D = \epsilon_{pD}$ and

$$t_{\text{RC}} = t_{\text{DC}} + \frac{1-(1-D_c)^{2s+N_0+1}}{2s+N_0+1} \frac{(2ES)^s (K_N^0)^{N_0}}{\sigma_{\text{eq}}^{2s+N_0} R_\nu^s},$$

$$t_{\text{DC}} = \epsilon_{pD} \left(\frac{K_N^0}{\sigma_{\text{eq}}}\right)^{N_0}. \tag{5.14}$$

The difference with an uniaxial case characterized by the same equivalent stress $\sigma = \sigma_{\text{eq}}$ lies only in the triaxiality function R_ν and gives an easy way to solve any 3D problem (in proportional loading!),

$$\boxed{\frac{t_{\text{RC}}(\sigma_{ij}) - t_{\text{DC}}(\sigma_{ij})}{t_{\text{RC}}(\sigma) - t_{\text{DC}}(\sigma)} = R_\nu^{-s},} \tag{5.15}$$

with $t_{\text{DC}}(\sigma_{ij}) = t_{\text{DC}}(\sigma) = \epsilon_{pD}\left(\frac{\sigma}{K_N^0}\right)^{-N_0}.$

The case of **creep-fatigue** (non-monotonous creep) is also treated by the above equations ((5.12) and (5.13)) if the same assumptions apply.

The integral is only a bit more difficult to calculate. For example for the periodic repeated loading of Fig. 5.3, the damage increment over one cycle $\frac{\delta D}{\delta N} = \int_{1\text{ cycle}} \dot{D} dt$ (assume D is constant over one cycle in the calculation of the integral) is given by the integral involving the stress rate during loadings and unloadings where $|\dot{\sigma}_{eq}| = \dot{\sigma}_{eq}^f$ considered constant,

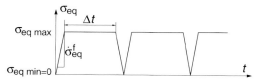

Fig. 5.3. Creep-fatigue periodic loading

$$dD = \left[\frac{\sigma_{eq}^2(t) R_\nu}{2ES(1-D)^2}\right]^s \left[\frac{\sigma_{eq}(t)}{K_N^0(1-D)}\right]^{N_0} dt \quad \text{with} \quad dt = \frac{d\sigma_{eq}}{\dot{\sigma}_{eq}^f}, \quad (5.16)$$

$$\frac{\delta D}{\delta N} = \left[\frac{R_\nu}{2ES(1-D)^2}\right]^s \left[\frac{1}{K_N^0(1-D)}\right]^{N_0}$$
$$\times \left[2\int_0^{\sigma_{eq\,max}} \sigma_{eq}^{2s+N_0} \frac{d\sigma_{eq}}{\dot{\sigma}_{eq}^f} + \int_0^{\Delta t} \sigma_{eq\,max}^{2s+N_0} dt\right]. \quad (5.17)$$

A second integration over the number of cycles gives the condition to rupture,

$$\int_0^{D_c} (1-D)^{2s+N_0} \delta D = \left[\frac{R_\nu}{2ES}\right]^s \left[\frac{1}{K_N^0}\right]^{N_0}$$
$$\times \left[\frac{2}{2s+N_0+1} \frac{\sigma_{eq\,max}^{2s+N_0+1}}{\dot{\sigma}_{eq}^f} + \sigma_{eq\,max}^{2s+N_0} \Delta t\right] (N_R - N_D), \quad (5.18)$$

where N_D is the number of cycles to reach the damage threshold (see Sect. 1.4.1):

$$N_D \approx \frac{p_D}{\Delta t}\left(\frac{K_N^0}{\sigma_{eq\,max}}\right)^{N_0} \quad \text{with} \quad p_D = \epsilon_{pD}\left(\frac{2\sigma_u}{\sigma_{eq\,max}}\right)^m, \quad (5.19)$$

$$\boxed{N_R = N_D + \frac{1-(1-D_c)^{2s+N_0+1}}{2s+N_0+1} \frac{(2ES)^s (K_N^0)^{N_0}}{\sigma_{eq\,max}^{2s+N_0} R_\nu^s} \times \left[\frac{2}{2s+N_0+1} \frac{\sigma_{eq\,max}}{\dot{\sigma}_{eq}^f} + \Delta t\right]^{-1}.} \quad (5.20)$$

In the last brackets, the first term corresponds to fatigue damage and the second to creep damage. The last equation involves temperature-dependent material parameters which may become time-dependent in the integral through the eventual history of temperature.

Here again, the number of cycles to rupture of a **3D problem** is easily deduced from the known number of cycles to rupture of an **uniaxial case** characterized by the stress $\sigma(t)$ which varies between 0 and $\sigma_M = \sigma_{\text{eq max}}$,

$$\boxed{\frac{N_R(\sigma_{ij}) - N_D(\sigma_{ij})}{N_R(\sigma_M) - N_D(\sigma_M)} = R_\nu^{-s}.} \tag{5.21}$$

Equations (5.14) and (5.20) allow us to determine the **3D creep-fatigue interaction diagram** N_R/N_{RF} function of t_R/t_{RC} for a given stress triaxiality where:

- N_{RF} is the number of cycles to rupture in pure fatigue (hold time $\Delta t = 0$),

$$N_{\text{RF}} = N_{\text{DF}} + \frac{\left[1 - (1 - D_c)^{2s+N_0+1}\right](2ES)^s \left(K_N^0\right)^{N_0} \dot{\sigma}_{\text{eq}}^f}{2\sigma_{\text{eq max}}^{2s+N_0+1} R_\nu^s} \tag{5.22}$$

- t_{RC} is the time to rupture in pure creep,

$$t_{\text{RC}} = t_{\text{DC}} + \frac{1 - (1 - D_c)^{2s+N_0+1}}{2s + N_0 + 1} \frac{(2ES)^s \left(K_N^0\right)^{N_0}}{\sigma_{\text{eq max}}^{2s+N_0} R_\nu^s} \tag{5.23}$$

- N_R is the number of cycles to rupture in creep-fatigue
- $t_R = N_R \Delta t$ is the time spent in creep

One has

$$(N_{\text{RF}} - N_{\text{DF}}) \frac{2}{2s + N_0 + 1} \frac{\sigma_{\text{eq max}}}{\dot{\sigma}_{\text{eq}}^f} = t_{\text{RC}} - t_{\text{DC}} \tag{5.24}$$

and

$$(N_R - N_D) \left[\frac{2}{2s + N_0 + 1} \frac{\sigma_{\text{eq max}}}{\dot{\sigma}_{\text{eq}}^f} + \Delta t\right] = t_{\text{RC}} - t_{\text{DC}}, \tag{5.25}$$

which lead to

$$(N_R - N_D) \frac{t_{\text{RC}} - t_{\text{DC}}}{N_{\text{RF}} - N_{\text{DF}}} + (N_R - N_D) \Delta t \frac{t_{\text{RC}} - t_{\text{DC}}}{t_{\text{RC}} - t_{\text{DC}}} = t_{\text{RC}} - t_{\text{DC}} \tag{5.26}$$

and, therefore, the 3D nonlinear creep-fatigue interaction through the values of the thresholds N_D, N_{DF}, t_D, t_{DC}:

$$\boxed{\frac{N_R - N_D}{N_{\text{RF}} - N_{\text{DF}}} + \frac{t_R - t_D}{t_{\text{RC}} - t_{\text{DC}}} = 1,} \tag{5.27}$$

where the stress triaxiality and the maximum von Mises stress $\sigma_{\text{eq max}}$ are the same for fatigue, creep and creep-fatigue. Note that a damage evolution with no threshold ($p_D = 0$) implies $t_{DC} = t_D = 0$ and $N_{DF} = N_D = 0$. In that case, eq. (5.27) recovers the Taira rule (5.6) of linear creep-fatigue interaction.

If the reference creep time to rupture $t_{RC}(\sigma_M)$ and fatigue number of cycles to rupture $N_{RF}(\sigma_M)$ are known from an uniaxial loading ($R_\nu = 1$) at the same $\sigma_{\text{eq max}} = \sigma_M$ as the 3D loading, one can write the 3D creep-fatigue interaction as

$$\boxed{\frac{N_R(\sigma_{ij}) - N_D(\sigma_{ij})}{N_{RF}(\sigma_M) - N_{DF}(\sigma_M)} + \frac{t_R(\sigma_{ij}) - t_D(\sigma_{ij})}{t_{RC}(\sigma_M) - t_{DC}(\sigma_M)} = R_\nu^{-s}.} \quad (5.28)$$

5.2.3 Sensitivity Analysis♂♂

As in Chap. 3 (Sect. 3.2.3) for the plastic strain to rupture and in Chap. 4 (Sect. 4.2.3) for the number of cycles to rupture, it is possible to perform a sensitivity analysis for the **time to rupture in creep** in accordance with the general method of Sect. 2.4.6. Start with the closed-form solution in proportional loading (5.14), where

$$t_R = t_D + \frac{1 - (1 - D_c)^{2s+N_0+1}}{2s + N_0 + 1} \left(\frac{2ES}{\sigma_{\text{eq}}^2 R_\nu}\right)^s \left(\frac{K_N^0}{\sigma_{\text{eq}}}\right)^{N_0}, \quad (5.29)$$

with

$$t_D = \epsilon_{pD} \left(\frac{K_N^0}{\sigma_{\text{eq}}}\right)^{N_0} \quad (5.30)$$

and

$$R_\nu = \frac{2}{3}(1 + \nu) + 3(1 - 2\nu)T_X^2. \quad (5.31)$$

Taking the logarithm gives

$$\ln(t_R - t_D) = \ln\left(1 - (1 - D_c)^{2s+N_0+1}\right) - \ln(2s + N_0 + 1)$$
$$+ s \ln\left(\frac{2ES}{\sigma_{\text{eq}}^2 R_\nu}\right) + N_0 \ln\left(\frac{K_N^0}{\sigma_{\text{eq}}}\right), \quad (5.32)$$

with

$$\ln\left(1 - (1 - D_c)^{2s+N_0+1}\right) = \ln\left(1 - \exp\left[(2s + N_0 + 1)\ln(1 - D_c)\right]\right) \quad (5.33)$$

and the derivative

$$\text{d}\ln\left(1 - (1 - D_c)^{2s+N_0+1}\right) = \frac{(2s + N_0 + 1)(1 - D_c)^{2s+N_0}}{1 - (1 - D_c)^{2s+N_0+1}} \text{d}D_c$$
$$- \frac{(1 - D_c)^{2s+N_0+1} \ln(1 - D_c)}{1 - (1 - D_c)^{2s+N_0+1}} \text{d}N_0 \quad (5.34)$$
$$- \frac{2(1 - D_c)^{2s+N_0+1} \ln(1 - D_c)}{1 - (1 - D_c)^{2s+N_0+1}} \text{d}s.$$

Taking the derivative with the notation $\delta x = |\mathrm{d}x|$ to ensure upper bounds on the errors gives

$$\frac{\delta(t_\mathrm{R} - t_\mathrm{D})}{t_\mathrm{R} - t_\mathrm{D}} = \frac{\delta t_\mathrm{R}}{t_\mathrm{R}} \frac{t_\mathrm{R}}{t_\mathrm{R} - t_\mathrm{D}} - \frac{\delta t_\mathrm{D}}{t_\mathrm{D}} \frac{t_\mathrm{D}}{t_\mathrm{R} - t_\mathrm{D}},$$

$$\frac{\delta t_\mathrm{D}}{t_\mathrm{D}} = \frac{\delta \epsilon_\mathrm{pD}}{\epsilon_\mathrm{pD}} + N_0 \frac{\delta K_N^0}{K_N^0} + N_0 \frac{\delta \sigma_\mathrm{eq}}{\sigma_\mathrm{eq}} + N_0 \left|\ln \frac{K_N^0}{\sigma_\mathrm{eq}}\right| \frac{\delta N_0}{N_0}, \qquad (5.35)$$

$$\frac{\delta R_\nu}{R_\nu} = \frac{|6T_\mathrm{X}^2 - \frac{2}{3}|\nu}{R_\nu} \frac{\delta \nu}{\nu} + \frac{6(1-2\nu)T_\mathrm{X}^2}{R_\nu} \frac{\delta T_\mathrm{X}}{T_\mathrm{X}}.$$

The sensitivity on the rupture time is finally $\dfrac{\delta t_\mathrm{R}}{t_\mathrm{R}} = \sum_k S_{\mathcal{A}_k}^{t_\mathrm{R}} \dfrac{\delta \mathcal{A}_k}{\mathcal{A}_k}$, where each $S_{\mathcal{A}_k}^{t_\mathrm{R}}$ is the sensitivity coefficient of the parameters $\mathcal{A}_k$ with respect to the time to rupture t_R,

$$S_{T_\mathrm{X}}^{t_\mathrm{R}} = \frac{t_\mathrm{R} - t_\mathrm{D}}{t_\mathrm{R}} \frac{6s(1-2\nu)T_\mathrm{X}^2}{R_\nu},$$

$$S_{\sigma_\mathrm{eq}}^{t_\mathrm{R}} = N_0 + \frac{t_\mathrm{R} - t_\mathrm{D}}{t_\mathrm{R}} 2s,$$

$$S_E^{t_\mathrm{R}} = \frac{t_\mathrm{R} - t_\mathrm{D}}{t_\mathrm{R}} s,$$

$$S_\nu^{t_\mathrm{R}} = \frac{t_\mathrm{R} - t_\mathrm{D}}{t_\mathrm{R}} \frac{|6T_\mathrm{X}^2 - \frac{2}{3}|\nu s}{R_\nu},$$

$$S_{K_N^0}^{t_\mathrm{R}} = N_0,$$

$$S_{N_0}^{t_\mathrm{R}} = N_0 \left|\ln \frac{\sigma_\mathrm{eq}}{K_N^0}\right| + \frac{t_\mathrm{R} - t_\mathrm{D}}{t_\mathrm{R}} \left(\frac{1}{2s + N_0 + 1} + \frac{(1-D_\mathrm{c})^{2s+N_0+1} \ln(1-D_\mathrm{c})}{1 - (1-D_\mathrm{c})^{2s+N_0+1}}\right),$$

$$S_{\epsilon_\mathrm{pD}}^{t_\mathrm{R}} = \frac{t_\mathrm{D}}{t_\mathrm{R}},$$

$$S_S^{t_\mathrm{R}} = \frac{t_\mathrm{R} - t_\mathrm{D}}{t_\mathrm{R}} s,$$

$$S_s^{t_\mathrm{R}} = \frac{t_\mathrm{R} - t_\mathrm{D}}{t_\mathrm{R}} s \left|\ln \frac{\sigma_\mathrm{eq}^2 R_\nu}{2ES} + \frac{2}{2s + N_0 + 1} + \frac{2(1-D_\mathrm{c})^{2s+N_0+1} \ln(1-D_\mathrm{c})}{1 - (1-D_\mathrm{c})^{2s+N_0+1}}\right|,$$

$$S_{D_\mathrm{c}}^{t_\mathrm{R}} = \frac{t_\mathrm{R} - t_\mathrm{D}}{t_\mathrm{R}} \frac{(2s + N_0 + 1)(1-D_\mathrm{c})^{2s+N_0} D_\mathrm{c}}{1 - (1-D_\mathrm{c})^{2s+N_0+1}}.$$

To give relative values of these coefficients, an example of average values of the parameters representative of many materials is visualized by the height of the boxes in the scheme of Fig. 5.4. This is for $\sigma_\mathrm{eq} = 100$ MPa, $T_\mathrm{X} = 1$, $R_\nu = 2.07$, and $t_\mathrm{D}/t_\mathrm{R} = 1/2$ as quantities defining the loading and $E = 140000$ MPa, $\nu = 0.3$, $K_N^0 = 700$ MPa·s$^{1/N_0}$, $N_0 = 8$, $\epsilon_\mathrm{pD} = 0.2$, $S = 0.15$ MPa, $s = 2$, and $D_\mathrm{c} = 0.2$ as values of the material parameters at a mid-level temperature.

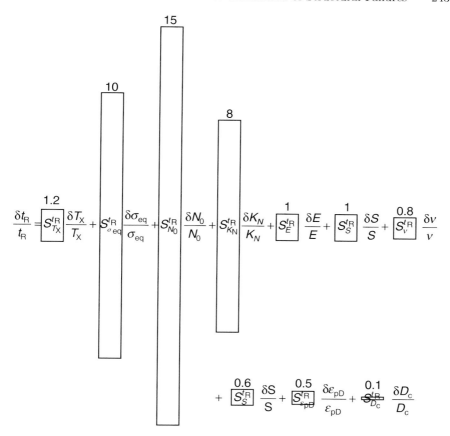

Fig. 5.4. Relative importance of each parameter in creep failures

Here again, the parameters which have the largest influence on the accuracy of the time to failure are those defining the loading and particularly the stress acting on both the plastic strain and the damage rates. Note the extreme importance of the Norton exponent N_0. The damage exponent s, the damage threshold ϵ_{pD} and the critical damage D_c are found to be of less importance.

For creep-fatigue, the calculation is very similar as only one logarithmic function is added, giving rise to two more terms related to the loading: the loading rate $\dot{\sigma}_{eq}$ and the hold time Δt, and to three sensitivity terms $S_{\sigma_{eq\,max}}$, S_s, and S_{N_0} for $\frac{\delta \sigma_{eq\,max}}{\sigma_{eq\,max}}$, $\frac{\delta s}{s}$, and $\frac{\delta N_0}{N_0}$. The number of cycles to rupture is given by (5.20) and the sensitivity to the hold time Δt or the fatigue stress rate $\dot{\sigma}_{eq}^f$ are

$$S_{\Delta t}^{N_R} = \frac{N_R - N_D}{N_R} \frac{\Delta t}{\frac{2}{2s + N_0 + 1} \frac{\sigma_{eq\,max}}{\dot{\sigma}_{eq}^f} + \Delta t} + \frac{N_D}{N_R}$$

$$S_{\dot{\sigma}_{eq}^f}^{N_R} = \frac{N_R - N_D}{N_R} \frac{\frac{2}{2s + N_0 + 1} \frac{\sigma_{eq\,max}}{\dot{\sigma}_{eq}^f}}{\frac{2}{2s + N_0 + 1} \frac{\sigma_{eq\,max}}{\dot{\sigma}_{eq}^f} + \Delta t}.$$
(5.36)

5.2.4 Elasto-Visco-Plastic Stress Concentration ♂♂

Representing time-dependent viscosity effects such as creep and relaxation by post-processing an elastic computation is a difficult task as the reference calculation is time-independent.

In the case of **localized visco-plasticity (small scale yielding)**, such a reference computation coupled with the Neuber method allows us to determine the history of the von Mises stress and the accumulated plastic strain.

In case of **fully (visco-)plastified** mechanical components (**large scale yielding**), a time-dependent reference computation is needed. In order to derive a fast method, the idea is to perform an elasto-visco-plastic computation only on a Representative Volume Element with the nominal strain $\epsilon_n(t)$ or stress $\sigma_n(t)$ as uniaxial loading, and to use the corresponding solution coupled with the knowledge of the notch elastic stress concentration coefficient K_T.

5.2.4.1 Small Scale Yielding – the Neuber Method

In the case of visco-plasticity confined in a small stress concentration area, the main part of the structure remains elastic and is submitted to a relaxation state on the yielding part. Apply the Neuber method coupled with Norton law to yield

$$\frac{\sigma_{eq}^2 R_\nu}{E} + \sigma_{eq} p = (\sigma_{ij}\epsilon_{ij})_{elas} \quad \text{and} \quad \dot{p} = \left\langle \frac{\sigma_{eq} - R - X_{eq} - \sigma_y}{K_N} \right\rangle^N. \quad (5.37)$$

For a given triaxiality ratio, this gives a relationship between the von Mises stress and the plastic strain

$$\sigma_{eq} = \frac{\sqrt{E^2 p^2 + 4ER_\nu (\sigma_{ij}\epsilon_{ij})_{elas}} - Ep}{2R_\nu}, \quad (5.38)$$

and leads to the law $p(t)$ and then $\sigma_{eq}(t)$ at the stress concentration point as the solution of the following first order differential equation:

$$\dot{p} = \left\langle \frac{\frac{1}{2R_\nu}\left(\sqrt{E^2 p^2 + 4ER_\nu (\sigma_{ij}\epsilon_{ij})_{elas}} - Ep\right) - R(p) - X_{eq}(p) - \sigma_y}{K_N} \right\rangle^N.$$
(5.39)

This leads to a decreasing von Mises stress over a long period of time and corresponds to a relaxation deformation process.

5.2.4.2 Large Scale Yielding (M. Chaudonneret 1978, 1985)

For the case of a structure undergoing a full visco-plastic state of deformation, the complete nonlinear computation is difficult to avoid! Nevertheless this can be done for the engineering cases of notches by following an extended Neuber procedure based on an uniaxial visco-plastic reference calculation made only on a volume element, on which the nominal stress $\sigma_n(t)$ or strain $\epsilon_n(t)$ is applied. The loading may be creep, fatigue, or creep-fatigue.

The information concerning the geometry of the notch is introduced in the rate form of the Neuber heuristic $\sigma\epsilon = \sigma_{elas}\epsilon_{elas}$ of Chap. 3 through the use of the elastic stress concentration coefficient K_T (see Fig. 3.4):

$$\sigma\dot{\epsilon} + \dot{\sigma}\epsilon = K_T^2(\sigma_n\dot{\epsilon}_n + \dot{\sigma}_n\epsilon_n), \qquad (5.40)$$

where the couple (σ, ϵ) is determined from the stress-strain field at the most loaded point of the notch and where σ_n and ϵ_n is the nominal stress and strain, respectively. Coupled with the strain partition,

$$\epsilon = \epsilon^e + \epsilon^p = \frac{\sigma}{E} + \epsilon^p \quad \text{and} \quad \epsilon_n = \frac{\sigma_n}{E} + \epsilon_n^p, \qquad (5.41)$$

one gets:

$$\dot{\sigma} = \frac{1}{2\sigma + E\epsilon^p} \left\{ K_T^2 \left[(2\sigma_n + E\epsilon_n^p)\dot{\sigma}_n + E\sigma_n\dot{\epsilon}_n^p \right] - E\sigma\dot{\epsilon}^p \right\}. \qquad (5.42)$$

This expression associated with a viscosity law such as the Norton law,

$$\dot{\epsilon}^p = \left\langle \frac{|\sigma - X| - R - \sigma_y}{K_N} \right\rangle^N \operatorname{sgn}(\sigma - X) \quad \text{and} \quad \dot{p} = |\dot{\epsilon}^p|, \qquad (5.43)$$

coupled with the consideration of both nonlinear isotropic hardening $R = R(p)$ and kinematic hardening $\dot{X} = C\dot{\epsilon}^p - \gamma X|\dot{\epsilon}^p|$ allows us to determine the stress and strain histories, $\sigma(t)$ and $\epsilon(t)$. This just needs the use of a step-by-step differential scheme such as the Euler or better Runge–Kutta schemes.

If the reference computation is made in elasticity, the method gives back the expression (5.37) for small scale yielding in which $R_\nu = 1$, $\sigma_{eq} = \sigma$, and $p = \epsilon_p$.

5.2.4.3 Mesocrack Initiation

Calculate the damage by performing the time integration of the unified damage law,

$$\dot{D} = \left(\frac{\sigma^2}{2ES} \right)^s |\dot{\epsilon}_p| \quad \text{if} \quad \epsilon_p > p_D, \qquad (5.44)$$

$$D(t) = \int_{t_\text{D}}^{t} \left(\frac{\sigma^2(t)}{2ES}\right)^s |\dot{\epsilon}_\text{p}(t)| \text{d}t, \tag{5.45}$$

with t_D as the time at damage initiation given by

$$p_\text{D} = \int_0^{t_\text{D}} \left\langle \frac{|\sigma(t) - X(t)| - R(t) - \sigma_\text{y}}{K_N} \right\rangle^N \text{d}t. \tag{5.46}$$

A mesocrack initiates when $D = D_\text{c}$ for a time $t = t_\text{R}$.

Remark – *In the previous equations, $\sigma(t)$ replaces the effective stress $\tilde{\sigma}$ of the fully coupled analysis. This means that the Neuber method is in fact*

$$\tilde{\sigma}\epsilon = K_\text{T}^2 \sigma_\text{n} \epsilon_\text{n} \tag{5.47}$$

and gives an estimation of the effective stress directly.

5.2.4.4 Large Scale Yielding – Approximate Method

When the material is largely visco-plastic, i.e., when elasticity is negligible, a simpler, but 3D, form than the previous Neuber method may be derived. The Neuber fundamental hypothesis becomes

$$\tilde{\sigma}\epsilon_\text{p} = K_\text{T}^2 \sigma_\text{n} \epsilon_\text{n} \qquad \text{for the uniaxial case,} \tag{5.48}$$

or

$$\tilde{\sigma}_\text{eq} p = K_\text{T}^2 \sigma_\text{eq n} p_\text{n} \qquad \text{for the 3D case,} \tag{5.49}$$

where $\tilde{\sigma}$ and $\tilde{\sigma}_\text{eq}$ are the effective stress and effective von Mises stress, respectively, at the most loaded point.

The Neuber heuristic (5.49) considered altogether with the pure viscous Norton law,

$$\dot{p} = \left(\frac{\tilde{\sigma}_\text{eq}}{K_N^0}\right)^{N_0} \quad \text{and} \quad \dot{p}_\text{n} = \left(\frac{\sigma_\text{eq}}{K_N^0}\right)^{N_0}, \tag{5.50}$$

leads to

$$K_N^0 \dot{p}^{1/N_0} p = K_\text{T}^2 K_N^0 \dot{p}_\text{n}^{1/N_0} p_\text{n} \tag{5.51}$$

and

$$\boxed{p = K_\text{T}^{\frac{2N_0}{N_0+1}} p_\text{n} \quad \text{and} \quad \dot{p} = K_\text{T}^{\frac{2N_0}{N_0+1}} \dot{p}_\text{n},} \tag{5.52}$$

where $K_\text{T}^{\frac{2N_0}{N_0+1}}$ acts as a creep strain concentration coefficient.

The effective stress is then

$$\tilde{\sigma}_\text{eq} = K_N^0 \dot{p}^{1/N_0} = K_\text{T}^{\frac{2}{N_0+1}} K_N^0 \dot{p}_\text{n}^{1/N_0} = K_\text{T}^{\frac{2}{N_0+1}} \sigma_\text{eq n}. \tag{5.53}$$

The damage initiated at the stress concentration point after a service time t_D is a solution of

$$p_D = \int_0^{t_D} \dot{p}(t) dt = K_T^{\frac{2N_0}{N_0+1}} \int_0^{t_D} \dot{p}_n(t) dt, \qquad (5.54)$$

i.e., when $p_n(t_D) = K_T^{-\frac{2N_0}{N_0+1}} p_D$.

Failure occurs when $D(t = t_R)$ reaches locally D_c with

$$D = \int_{t_D}^{t} \left(\frac{\tilde{\sigma}_{eq}^2 R_\nu}{2ES}\right)^s \left(\frac{\tilde{\sigma}_{eq}}{K_N^0}\right)^{N_0} dt = \frac{K_T^{\frac{4s+2N_0}{N_0+1}}}{(2ES)^s (K_N^0)^{N_0}} \int_{t_D}^{t} \sigma_{eq\,n}^{2s+N_0}(t) R_\nu^s(t) dt \qquad (5.55)$$

where $K_T^{\frac{4s+2N_0}{N_0+1}}$ acts as a damage rate concentration coefficient.

An example is given in Sect. 5.3.2.

5.2.5 Safety Margin and Crack Growth

In creep, the security is governed by the duration without any accident so that the essential parameter is the time t. A safe design of a component must prove that the service time is below the time t_R of a crack initiation somewhere in the structure. Here, we introduce again a safety factor Saf which defines the safety margin

$$t_{\text{service}} < \frac{t_R}{Saf}. \qquad (5.56)$$

If creep fatigue is involved, the engineering parameter can be either the number of cycles N or the accumulated time t and the safety margins are defined as

$$N_{\text{service}} < \frac{N_R}{Saf} \quad \text{or} \quad t_{\text{service}} < \frac{t_R}{Saf}. \qquad (5.57)$$

When dealing with service durations larger than several years, much care must be taken as the identification of the parameters is most often based on short range tests. In situ control of components in service and updated calculations are advised each time new information arrives. Anyway, the safety factor must be large enough so that $Saf = 5, 10, 20$, or more depending on the expected duration of the component. A more advanced concept consists of evaluating the rupture probability while taking into account the scatter of the material parameters and the uncertainties upon the loading.

There is no simple way to determine the crack growth rate but the numerical simulation by a structure calculation with elements removal in elasto-visco-plasticity coupled with damage is possible. The viscosity, playing a regularization role, ensures the convergence of the numerical algorithms (see Sect. 1.6.2).

5.3 Basic Engineering Examples

Here we describe some simple results that can help us quickly estimate creep or creep-fatigue effects on failure of mechanical components. The accuracy is poor as the temperature increases the nonlinearities regarding to stress-strain and time relations. If an accuracy of a few percent may be expected in predictions in elasticity, it increases to 10–20% in plasticity and up to 50–100% in visco-plasticity. Only fatigue is worse: a factor of 10 (1000% !) may happen in high cycle fatigue!

5.3.1 Strain Rate and Temperature-Dependent Yield Stress♂

What is called **"dynamic plasticity"** is in fact visco-plasticity as the high strain rates play a role even at room temperature. Furthermore, the short times involved induce almost adiabatic plasticity for which the effect of plastic dissipation increases the temperature. Here we derive an approximate formula to obtain the yield stress as well as function of the strain rate and the temperature, from its value at room temperature or an other reference temperature.

The effect of the temperature T is introduced by considering the visco-plastic strain as a thermally-activated phenomenon governed by the Arrhenius law which makes the unidimensional plastic strain rate dependent on the temperature, as in

$$\dot{\epsilon}_\mathrm{p} \approx g_{\sigma,\epsilon_\mathrm{p}}(\sigma, \epsilon_\mathrm{p}) \exp\left(-\frac{H}{T}\right), \qquad (5.58)$$

where the parameter H is related to the energy of activation of the process.

We then introduce the pure viscous Norton law at a reference temperature T_0,

$$\dot{\epsilon}_\mathrm{p} = \left(\frac{\sigma}{K_N^0}\right)^{N_0} g_{\epsilon_\mathrm{p}}(\epsilon_\mathrm{p}) \exp\left(\frac{H}{T_0} - \frac{H}{T}\right). \qquad (5.59)$$

First, consider the engineering yield stress $\sigma_{\mathrm{y}02}(\dot{\epsilon}_{\mathrm{p}0}, T_0)$ corresponding to a plastic strain $\epsilon_{\mathrm{p}0} = 0.2\ 10^{-2}$, a reference plastic strain rate $\dot{\epsilon}_{\mathrm{p}0}$, and the reference temperature T_0:

$$\sigma_{\mathrm{y}02}(\dot{\epsilon}_{\mathrm{p}0}, T_0) = K_N^0 \left(\frac{\dot{\epsilon}_{\mathrm{p}0}}{g_{\epsilon_\mathrm{p}}(\epsilon_{\mathrm{p}0})}\right)^{1/N_0}. \qquad (5.60)$$

Second, consider the engineering yield stress $\sigma_{\mathrm{y}02}(\dot{\epsilon}_\mathrm{p}, T)$ corresponding to a larger plastic strain rate $\dot{\epsilon}_\mathrm{p}$ and a larger temperature T,

$$\sigma_{\mathrm{y}02}(\dot{\epsilon}_\mathrm{p}, T) = K_N^0 \left[\frac{\dot{\epsilon}_\mathrm{p}}{g_{\epsilon_\mathrm{p}}(\epsilon_{\mathrm{p}0})} \exp\left\{-\left(\frac{H}{T_0} - \frac{H}{T}\right)\right\}\right]^{1/N_0}. \qquad (5.61)$$

Finally, for the 3D case, $\dot{\epsilon}_p$ becomes $\dot{p}$ and the engineering yield stress (both strain-rate- and temperature-dependent) is approximately given by

$$\sigma_{y02}(\dot{p}, T) \approx \sigma_{y02}(\dot{\epsilon}_{p0}, T_0) \left(\frac{\dot{p}}{\dot{\epsilon}_{p0}}\right)^{1/N_0} \exp\left\{-\frac{H}{N_0}\left(\frac{1}{T_0} - \frac{1}{T}\right)\right\}, \quad (5.62)$$

an expression which needs the knowledge of:

- The Norton parameters K_N^0 and N_0 at the reference temperature T_0 not too far from the temperature considered because, unfortunately, they are not exactly constant with the temperature,
- The yield stress σ_{y02} at a reference strain rate $\dot{\epsilon}_{p0}$ and temperature T_0, and
- The parameter H "adjusted" on the yield stress at least at one other temperature.

To finish, expression (5.62) can also be used at room temperature, particularly in dynamics problems such as shocks, crash, or perforation, considering $\dot{\epsilon}_{p0} \approx \dot{\epsilon}_0$ so that

$$\sigma_{y02}(\dot{p}) = \sigma_{y02}(\dot{\epsilon}_0) \left(\frac{\dot{p}}{\dot{\epsilon}_0}\right)^{1/N_0}. \quad (5.63)$$

At room temperature, the exponent N_0 is large ($N_0 \approx 20$ to 100) and the only possibility to identify its value for each material is to perform relaxation tests at large values of initial strain rate. For steels, the plastic strain rate may vary over several decades in a few hours (Fig. 5.5).

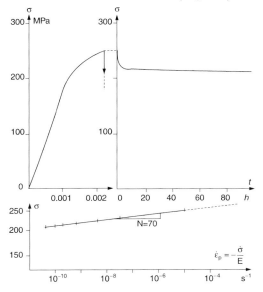

Fig. 5.5. Relaxation tests to identify pure viscous Norton parameters for 316 stainless steel at room temperature (J. Lemaitre and J.L. Chaboche 1985)

5.3.2 Plates or Members with Holes or Notches

This is the same problem as in Sects. 3.3.1 and 4.3.1 where a hole or a notch increases the nominal stress σ_n on its edge due to the elastic concentration coefficient K_T. Consider the case of a fully visco-plastic component where the damage may initiate and reach the critical localized damage at the stress concentration point.

Use the approximate Neuber method of Sect. 5.2.4 which states for a local uniaxial state of stress σ and $\epsilon \approx \epsilon_p$ that

$$\tilde{\sigma}\epsilon = K_T^2 \sigma_n \epsilon_{pn}, \tag{5.64}$$

and consider the pure viscous Norton law:

$$\dot{\epsilon}_p = \left(\frac{\tilde{\sigma}}{K_N^0}\right)^{N_0} \quad \text{and} \quad \dot{\epsilon}_{pn} = \left(\frac{\sigma_n}{K_N^0}\right)^{N_0}, \tag{5.65}$$

with $\tilde{\sigma} = \sigma/(1-D)$ as the effective stress at the stress concentration point.

Determine then the history of the strain $\epsilon_p(t)$ and the effective stress $\tilde{\sigma}(t)$. The local damage is given by (5.55) and a mesocrack initiates at the time t_R when D reaches the critical value D_c.

5.3.2.1 Creep Mesocrack Initiation

Consider first the case of creep: a constant far field σ_∞ leads to a constant nominal stress σ_n (see Fig. 3.4). The full procedure takes 3 steps:

1. **Perform the reference calculation**
   The solution $\epsilon_{pn}(t)$ for creep at $\sigma_n = const$ is

$$\epsilon_{pn}(t) = \left(\frac{\sigma_n}{K_N^0}\right)^{N_0} t. \tag{5.66}$$

2. **Apply the approximate Neuber method**
   The Neuber heuristic and the Norton law lead to

$$\epsilon_p(t) = K_T^{\frac{2N_0}{N_0+1}} \epsilon_{pn}(t) \tag{5.67}$$

   and the effective stress is

$$\tilde{\sigma} = K_T^{\frac{2}{N_0+1}} K_N^0 \dot{\epsilon}_{pn}^{1/N_0} = K_T^{\frac{2}{N_0+1}} \sigma_n = const. \tag{5.68}$$

3. **Calculate the damage evolution**
   The damage initiates at $t = t_D$ when the plastic strain reaches the damage threshold ϵ_{pD} locally,

$$t_D = K_T^{-\frac{2N_0}{N_0+1}} \epsilon_{pD} \left(\frac{K_N^0}{\sigma_n}\right)^{N_0}. \tag{5.69}$$

For a stress concentration coefficient of $K_T = 3$, this means that the damage initiates at the stress concentration point for a time t_D that is 3 to 9 times shorter that the time t_{Dn} of damage occurrence in the entire cross section as $t_{Dn} = \epsilon_{pD} \left(\frac{K_N^0}{\sigma_n} \right)^{N_0}$. The damage history is

$$D(t) = K_T^{\frac{4s+2N_0}{N_0+1}} \frac{\sigma_n^{2s+N_0}}{(2ES)^s (K_N^0)^{N_0}} (t - t_D). \tag{5.70}$$

Finally, failure occurs when $D = D_c$ for a creep time t_R represented by

$$\boxed{t_R = K_T^{-\frac{2N_0}{N_0+1}} \epsilon_{pD} \left(\frac{K_N^0}{\sigma_n} \right)^{N_0} + K_T^{-\frac{4s+2N_0}{N_0+1}} \frac{(2ES)^s (K_N^0)^{N_0} D_c}{\sigma_n^{2s+N_0}}.} \tag{5.71}$$

The second term (tertiary creep) of the time to rupture varies as the inverse of the stress to the power $(2s + N_0)$ which can be on the order of 5 to 20 that of a very large stress sensitivity. Then be careful when defining a safety factor.

5.3.2.2 Creep-Fatigue

The far field $\sigma_\infty(t)$ varies periodically between 0 and $\sigma_{\infty\,\max}$, with a hold time Δt at $\sigma_{\infty\,\max}$. The nominal stress varies between 0 and $\sigma_{n\,\max}$ in the same manner. This is creep-fatigue for which the resolution is similar to the one detailed in Sect. 5.2.2. The approximate Neuber method allows us to again calculate the history of the accumulated plastic strain $p(t)$ and the damage $D(t)$ at the stress concentration point.

The three-step procedure is the same as for the creep case:

1. **Perform the reference calculation**
   One has to calculate the plastic strain increment per cycle,

$$\frac{\delta p_n}{\delta N} = \int_{1\,\text{cycle}} |\dot{\epsilon}_{pn}| dt. \tag{5.72}$$

During the monotonous part of the loading, consider an applied load linear in time with $\sigma_n = \dot{\sigma}_n^f t$, where

$$\dot{\epsilon}_{pn} = \left(\frac{\sigma_n}{K_N^0} \right)^{N_0} = \left(\frac{\dot{\sigma}_n^f}{K_N^0} \right)^{N_0} t^{N_0}$$

$$\implies \Delta \epsilon_{pn} = \frac{1}{(N_0+1)(K_N^0)^{N_0}} \frac{\sigma_{n\,\max}^{N_0+1}}{\dot{\sigma}_n^f}. \tag{5.73}$$

During the creep part,

$$\Delta\epsilon_{\text{pn}} = \left(\frac{\sigma_{\text{n max}}}{K_N^0}\right)^{N_0} \Delta t. \tag{5.74}$$

Then,

$$\boxed{\frac{\delta p_{\text{n}}}{\delta N} = \left(\frac{\sigma_{\text{n max}}}{K_N^0}\right)^{N_0} \left[\frac{2}{N_0+1}\frac{\sigma_{\text{n max}}}{\dot{\sigma}_{\text{n}}^f} + \Delta t\right].} \tag{5.75}$$

2. **Apply the approximate Neuber method**

   The Neuber heuristic $\tilde{\sigma}_{\text{eq}} p = K_T^2 \sigma_n \epsilon_n$ and the previous Norton law give the local plastic strain as

$$p(t) = K_T^{\frac{2N_0}{N_0+1}} p_n(t), \tag{5.76}$$

the local plastic increment per cycle as

$$\frac{\delta p}{\delta N} = K_T^{\frac{2N_0}{N_0+1}} \frac{\delta p_n}{\delta N}, \tag{5.77}$$

and the effective stress at the stress concentration point as

$$\tilde{\sigma}(t) = K_T^{\frac{2}{N_0+1}} K_N^0 \dot{p}_n^{1/N_0}(t) = K_T^{\frac{2}{N_0+1}} \sigma_n(t). \tag{5.78}$$

3. **Calculate the damage evolution**

   Damage initiates after a number of cycles $N = N_D$ when $p = \frac{\delta p}{\delta N} N$ reaches the damage threshold p_D. The corresponding number of cycles is then

$$N_D = \frac{p_D}{\frac{\delta p}{\delta N}} = K_T^{-\frac{2N_0}{N_0+1}} \frac{p_D}{\frac{\delta p_n}{\delta N}}, \tag{5.79}$$

with p_D given by (5.19) of Sect. 5.2.2,

$$p_D = \epsilon_{\text{pD}} \left(\frac{2\sigma_u}{\sigma_{\text{eq max}}}\right)^m = K_T^{-\frac{2m}{N_0+1}} \epsilon_{\text{pD}} \left(\frac{2\sigma_u}{\sigma_{\text{n max}}}\right)^m. \tag{5.80}$$

The damage history is given by the time integration of the damage law,

$$\dot{D} = \left(\frac{\tilde{\sigma}^2}{2ES}\right)^s \left(\frac{\tilde{\sigma}}{K_N^0}\right)^{N_0} \quad \text{if} \quad p > p_D, \tag{5.81}$$

first over one cycle, $\frac{\delta D}{\delta N} = \int_{1 \text{ cycle}} \dot{D} dt$, then over the whole loading. During the monotonous loading part with $\dot{\sigma}_n = \dot{\sigma}_n^f = const$, as in Sect. 5.2.2,

$$\dot{D} = \frac{\tilde{\sigma}^{2s+N_0}}{(2ES)^s (K_N^0)^{N_0}} = K_T^{\frac{4s+2N_0}{N_0+1}} \frac{\dot{\sigma}_n^{f\ 2s+N_0}}{(2ES)^s (K_N^0)^{N_0}} t^{2s+N_0} \qquad (5.82)$$

gives a damage growth of

$$\Delta D = \frac{K_T^{\frac{4s+2N_0}{N_0+1}}}{(2s+N_0+1)(2ES)^s (K_N^0)^{N_0}} \frac{\sigma_{n\ \max}^{2s+N_0+1}}{\dot{\sigma}_n^{f\ 2s+N_0+1}}. \qquad (5.83)$$

During the creep part of each cycle,

$$\dot{D} = \frac{K_T^{\frac{4s+2N_0}{N_0+1}}}{(2ES)^s (K_N^0)^{N_0}} \sigma_{n\ \max}^{2s+N_0} \implies \Delta D = \frac{K_T^{\frac{4s+2N_0}{N_0+1}}}{(2ES)^s (K_N^0)^{N_0}} \sigma_{n\ \max}^{2s+N_0} \Delta t. \qquad (5.84)$$

Then,

$$\frac{\delta D}{\delta N} = \frac{K_T^{\frac{4s+2N_0}{N_0+1}}}{(2ES)^s (K_N^0)^{N_0}} \sigma_{n\ \max}^{2s+N_0} \left[\frac{2}{2s+N_0+1} \frac{\sigma_{n\ \max}}{\dot{\sigma}_n} + \Delta t \right] \qquad (5.85)$$

and the number of creep-fatigue cycles to rupture corresponds finally to $D(N_R) = \frac{\delta D}{\delta N} \cdot (N_R - N_D) = D_c$ so that

$$\boxed{\begin{aligned} N_R &= N_D + K_T^{-\frac{4s+2N_0}{N_0+1}} \frac{(2ES)^s (K_N^0)^{N_0} D_c}{\sigma_{n\ \max}^{2s+N_0}} \\ &\quad \times \left[\frac{2}{2s+N_0+1} \frac{\sigma_{n\ \max}}{\dot{\sigma}_n} + \Delta t \right]^{-1} \end{aligned}}$$

with

$$\boxed{N_D = K_T^{-\frac{2(N_0+m)}{N_0+1}} \epsilon_{pD} \frac{(2\sigma_u)^m (K_N^0)^{N_0}}{\sigma_{n\ \max}^{N_0+m}} \left[\frac{2}{N_0+1} \frac{\sigma_{n\ \max}}{\dot{\sigma}_n^f} + \Delta t \right]^{-1}}.$$

(5.86)

Even for such a complex case, the number of cycles to rupture by mesocrack initiation is fully determined from the knowledge of the nominal strain $\sigma_n(t)$ and the elastic stress concentration coefficient K_T. Again, the sensitivity to the stress is high and a small hold time has already a strong negative effect on the component lifetime.

5.3.3 Pressurized Shallow Cylinder⌀

From the analysis of the Sect. 3.3.2 (Fig. 3.6), the stresses are expressed as a function of the applied pressure P, the radius R_{cyl}, and the thickness t_{cyl} of the cylinder as

$$\sigma_{eq}(t) = \frac{\sqrt{3}}{2} \frac{P(t) R_{cyl}}{t_{cyl}}, \qquad T_X = \frac{\sigma_H}{\sigma_{eq}} = \frac{1}{\sqrt{3}}, \quad \text{and} \quad R_\nu = \frac{5 - 4\nu}{3}. \quad (5.87)$$

The problem is to determine the time of a mesocrack initiation, t_R, when the pressure is kept constant (creep) or when it varies periodically with periods of hold time Δt at $P = P_{max}$ (creep-fatigue). We neglect here the time-to-damage initiation.

For creep, the integration of the damage law with a pure viscous Norton law is straightforward:

$$\dot{D} = \left[\frac{\sigma_{eq}^2 R_\nu}{2ES(1-D)^2} \right]^s \left[\frac{\sigma_{eq}}{K_N^0 (1-D)} \right]^{N_0} \quad (5.88)$$

or

$$\int_0^{D_c} (1-D)^{2s+N_0} dD = \left[\frac{\sigma_{eq}^2 R_\nu}{2ES} \right]^s \left[\frac{\sigma_{eq}}{K_N^0} \right]^{N_0} t_R, \quad (5.89)$$

and finally:

$$\boxed{t_R = \frac{1 - (1 - D_c)^{2s+N_0+1}}{2s + N_0 + 1} \frac{(2ES)^s K_N^{0 \; N_0}}{\left(\dfrac{5-4\nu}{3} \right)^s \left(\dfrac{\sqrt{3}}{2} \dfrac{P_{max} R_{cyl}}{t_{cyl}} \right)^{2s+N_0}}.} \quad (5.90)$$

What is important is that t_R varies as the pressure P to the power $2s + N_0$, which can be on the order of 10 to 15 for metals. This means that an increase of the pressure of 5% may reduce the time to an explosion by a factor of about 2 or more!

For creep-fatigue induced by a pressure maintained constant at $P = P_{max}$ for N periods of time Δt, the integration of the damage law is the one of Sect. 5.2.2 (Eqs. (5.16) to (5.20)), with of course the particular values (5.87) for the state of stress and stress triaxiality and a constant pressure rate $\dot{P}$ or $-\dot{P}$ during the loading or unloading parts of the cycle.

The number of cycles to rupture is

$$\boxed{N_R = \frac{(1 - (1 - D_c)^{2s+N_0+1})(2ES)^s K_N^{0 \; N_0}}{(2s + N_0 + 1) \left(\dfrac{5-4\nu}{3} \right)^s \left(\dfrac{\sqrt{3}}{2} \dfrac{P_{max} R_{cyl}}{t_{cyl}} \right)^{2s+N_0}} \times \left[\frac{2}{2s + N_0 + 1} \frac{P_{max}}{\dot{P}} + \Delta t \right]^{-1}.} \quad (5.91)$$

In practical applications, the creep term is almost always much larger than the fatigue term, i.e.,

$$\Delta t \gg \frac{2}{2s + N_0 + 1} \frac{P_{\max}}{\dot{P}} . \tag{5.92}$$

Again, this means that the fatigue damage may sometimes be neglected, but **never the creep damage**.

5.3.4 Adiabatic Dynamics Post-Buckling in Bending♂♂

This is the same problem as the plastic post-buckling of beams treated in Sect. 3.3.3 except that it corresponds to a crash or a shock problem where dynamic plasticity, that is visco-plasticity, occurs. Then two phenomena must be considered:

- The strain rate effect which requires us to consider a visco-plastic constitutive equation.
- The large increase in temperature due to a plastic dissipation in a short time which necessitates the consideration of the thermomechanical coupling.

The problem is made as simple as possible in order to obtain the energy absorbed in the material in a closed-form useful for qualitative discussion about the influence of each parameter.

- The geometrical description is in Fig. 3.7 of Chap. 3 where the Bernoulli hypothesis allows us to write

$$\epsilon = \epsilon_\mathrm{M} \frac{2y}{h} . \tag{5.93}$$

- The energy absorbed by the full plastic hinge of the beam $l = \pi h/2$ is

$$W = 2bl \int_0^{h/2} \int_0^{t_\mathrm{M}} \dot{w} \, \mathrm{d}t \mathrm{d}y , \tag{5.94}$$

where t_M is the time to achieve the process and $\dot{w}$ is the plastic power density

$$\dot{w} = \sigma \dot{\epsilon}_\mathrm{p} . \tag{5.95}$$

- The visco-plastic constitutive equation is the pure viscous Norton law ritten with the damage effective stress $\tilde{\sigma} = \sigma/(1-D)$, where the coupling with the temperature is a power function approximation of the Arrhenius term $\exp\left(\frac{H}{T}\right)$:

$$\dot{\epsilon}_\mathrm{p} = \left[\frac{\tilde{\sigma}}{K_N^0}\right]^{N_0} \left(\frac{T}{T_0}\right)^{H_N} , \tag{5.96}$$

where T_0 is the reference temperature at which the material parameters K_N^0 and N_0 are identified. H_N is the temperature exponent on the order of 5 to 10 for steels in the range of 100 to 500°C.

- The damage constitutive equation is the simplified expression (assuming the strain hardening saturated) obtained in Sect. 3.2.1 with the damage threshold ϵ_{pD} taken to be equal to zero. The coupling of the damage with the temperature is neglected as at least part of it is already taken into account through the plastic strain ϵ_p, so that

$$D = D_c \frac{\epsilon_p}{\epsilon_{pR}}. \tag{5.97}$$

- The temperature derives from the adiabatic condition written as the heat equation,

$$\rho C_h \dot{T} = \dot{w}, \tag{5.98}$$

where ρ is the density, $\rho = 7800$ kg/m^3 for steels, C_h is the specific heat, and $C_h \approx 500$ J/kg °C for steels. Then

$$\dot{T} = \frac{\dot{w}}{\rho C_h} \quad \text{or} \quad T = T_0 + \frac{w}{\rho C_h}. \tag{5.99}$$

- Some other hypotheses can make the person in charge of the calculation happy:
  - Elastic strain neglected so that $\epsilon_p = \epsilon$
  - Process at constant strain rate, $\dot{\epsilon} = const = \dot{\epsilon}^\star \rightarrow \epsilon = \dot{\epsilon}^\star t$

The calculation is now easy:

$$\dot{w} = K_N^0 \left(1 - D_c \frac{\epsilon_p}{\epsilon_{pR}}\right) \left(\frac{T}{T_0}\right)^{-\frac{H_N}{N_0}} \dot{\epsilon}_p^{\frac{N_0+1}{N_0}}, \tag{5.100}$$

$$\int_0^w \left(1 + \frac{w}{\rho C_h T_0}\right)^{\frac{H_N}{N_0}} dw = \int_0^t K_N^0 \left(1 - D_c \frac{\dot{\epsilon}^\star}{\epsilon_{pR}} t\right) \dot{\epsilon}^{\star \frac{N_0+1}{N_0}} dt, \tag{5.101}$$

$$\frac{\rho C_h T_0 N_0}{H_N + N_0}\left[\left(1 + \frac{w}{\rho C_h T_0}\right)^{\frac{H_N + N_0}{N_0}} - 1\right] = K_N^0 \left(t - D_c \frac{\dot{\epsilon}^\star}{\epsilon_{pR}} \frac{t^2}{2}\right) \dot{\epsilon}^{\star \frac{N_0+1}{N_0}}, \tag{5.102}$$

or

$$w = \rho C_h T_0 \left\{ \left[1 + \frac{(H_N + N_0)K_N^0}{\rho C_h T_0 N_0}\left(1 - D_c \frac{\dot{\epsilon}^\star}{2\epsilon_{pR}} t\right) \dot{\epsilon}^{\star \frac{N_0+1}{N_0}} t \right]^{\frac{N_0}{H_N + N_0}} - 1\right\}. \tag{5.103}$$

Taking the total time t_M as the kinematic measure of the process,

$$\dot{\epsilon}^\star = \dot{\epsilon}^\star(y) = \frac{2\epsilon_M}{t_M} \frac{y}{h}. \tag{5.104}$$

The energy absorbed, $W = 2bl \int_0^{h/2} w(y, t_M) dy$, is

$$W = 2blh\rho C_\text{h} T_0 \left\{ \int_0^{\frac{1}{2}} \left[1 + \frac{(H_N + N_0) K_N^0 (2\epsilon_\text{M})^{\frac{N_0+1}{N_0}}}{\rho C_\text{h} T_0 N_0 t_\text{M}^{1/N_0}} \right. \right.$$
$$\left. \left. \times \left(1 - D_\text{c} \frac{\epsilon_\text{M}}{\epsilon_\text{pR}} u \right) u^{\frac{N_0+1}{N_0}} \right]^{\frac{N_0}{H_N+N_0}} \mathrm{d}u - \frac{1}{2} \right\},$$

(5.105)

with $\epsilon_\text{M} = \frac{\pi h - \pi h/2}{\pi h/2} = 1$ and $l = \pi h/2$ for a full bending.

Without any coupling nor damage,

$$\dot{w}_0 = K_N^0 \dot{\epsilon}_\text{p}^{\frac{N_0+1}{N_0}}$$

(5.106)

and

$$W_0 = blh \frac{N_0 K_N^0}{2N_0 + 1} \epsilon_\text{M}^{\frac{N_0+1}{N_0}} t_\text{M}^{-1/N_0}.$$

(5.107)

In both cases, the energy absorbed increases as the process is faster but the increase is weak as N_0 is always large and t_M is always short, thus keeping the variations of the term t_M^{-1/N_0} small.

Some applications show that the coupling with the damage is more important than the coupling with the temperature. In (5.105), the damage coupling is governed by the term $D_\text{c} \epsilon_\text{M}/\epsilon_\text{pR}$ while the temperature coupling is mainly governed by the dimensionless ratio $K_N^0/\rho C_\text{h} T_0 t_\text{M}^{1/N_0}$. Altogether they lead to a ratio W/W_0 on the order of 0.7 to 0.9 for metals.

5.4 Numerical Failure Analysis

At high temperatures, many materials become visco-plastic and the main local loadings encountered in mechanical components are creep, creep-fatigue, and relaxation. Viscosity enhances the stress redistribution, with a strong effect on damage and failure.

A structure computation made in elasto-visco-plasticity coupled with damage gives the damage map history and the crack initiation conditions. If the damage constitutive equations are not available in an FE code, implement them by use of the numerical schemes of Chap. 2. For non-isothermal loadings, it is important to have a complete set of material parameters for the temperature range of the application in mind. Identify the parameters at a few constant temperatures (see for instance Sects. 1.4.4, 1.5.1.1 and 2.4). Use the least square method to fit the temperature dependency with non-linear functions of T. Don't be afraid to use nonlinear functions as smooth functions are better than linear interpolation. For example, for a material parameter $\mathcal{A}$, a general possibility is

$$\mathcal{A} = a_1 + a_2 T + a_3 \exp(a_4 T).$$

(5.108)

The use of the anisotropic damage law is again advised as it needs only one additional material parameter which in practice may be taken as $\eta = 3$. The consideration of the quasi-unilateral conditions of microdefects closure is not necessary if the loading remains positive as is often the case in creep and creep-fatigue.

A fully-coupled analysis is very costly and for cyclic loading, one may use an elasto-visco-plasticity computation followed by a damage post-processing (uncoupled analysis). An example is given in Sect. 5.4.3.

One of the very first thing to do is to test the numerical implementation of the constitutive laws on a simple reference calculation (and also the quality of the mesh, the time increments, and the convergence of the computation). The academic problem of the hollow sphere under external pressure is especially designed for this purpose (see Sect. 5.4.1). It uses simple assumptions for both viscosity and damage.

5.4.1 Hollow Sphere under External Pressure
(A. Benallal, R. Billardon and L. Moret–Bailly 1991)

This example has been designed to check the numerical procedures introduced in an FE code. It is built on the possibility to exhibit a closed-form solution in elasto-visco-plasticity fully coupled with damage when the material behavior is modelled by the following constitutive equations of perfect visco-plasticity (no hardening) with a linear Norton law:

$$\boldsymbol{\epsilon} = \boldsymbol{\epsilon}^e + \boldsymbol{\epsilon}^p,$$
$$\boldsymbol{\epsilon}^e = \frac{1+\nu}{E(1-D)}\boldsymbol{\sigma} - \frac{\nu}{E(1-D)} \operatorname{tr}\boldsymbol{\sigma}\,\mathbf{1},$$
$$\dot{\boldsymbol{\epsilon}}^p = \frac{3}{2}\frac{\boldsymbol{\sigma}^D}{\sigma_{eq}}\dot{p}, \tag{5.109}$$

with

$$\dot{p} = \left\langle \frac{\frac{\sigma_{eq}}{1-D} - \sigma_y}{K_N^y} \right\rangle^{N_y} \quad \text{with} \quad N_y = 1.$$

The damage evolution law is written as

$$\dot{D} = \frac{Y^s}{S_0}\dot{p} \tag{5.110}$$

so that $S_0 = S^s$ recovers the unified damage law $\dot{D} = (Y/S)^s \dot{p}$. With $s = 0$ and $\sigma_y = 0$, it corresponds to the simple law

$$\dot{D} = \frac{\sigma_{eq}}{S_0 K_N^0 (1-D)} \tag{5.111}$$

which does not take into account the triaxiality effect. But when considered altogether with $\nu = 0.5$, it allows for a closed-form solution of the hollow sphere problem.

Consider an external pressure loading $P(t)$ applied on the sphere of Fig. 5.6. The solution at time t, at a radius point r, is of the form

$$D = 1 - g(t) + \frac{f(t)}{S_0 K_N^0 r^3}$$

$$\sigma_{eq} = (1 - D)\left\{\sigma_y(1 - \exp(-\mu t)) + \frac{\dot{f}(t)}{r^3}\right\}, \quad (5.112)$$

where $\mu = E/K_N^0$ and

$$g(t) = 1 - \frac{\sigma_y}{S_0 K_N^0}\left[t - \frac{1}{\mu} + \frac{1}{\mu}\exp(-\mu t)\right]. \quad (5.113)$$

- Up to mesocrack initiation (which occurs at the inner radius when $D(a) = 1$),

$$f(t) = \frac{1}{\alpha}\left[-\beta g(t) + \sqrt{\beta^2 g^2(t) + 2\alpha\left(\Pi(t) - S_0 K_N^0(g^2(t) - 1)\ln\frac{b}{a}\right)}\right], \quad (5.114)$$

where $\Pi(t) = \int_0^t P(\tau)d\tau$,

$$\alpha = \frac{\left[\frac{1}{b^6} - \frac{1}{a^6}\right]}{3 S_0 K_N^0}, \quad \text{and} \quad \beta = \frac{2}{3}\left[\frac{1}{a^3} - \frac{1}{b^3}\right]. \quad (5.115)$$

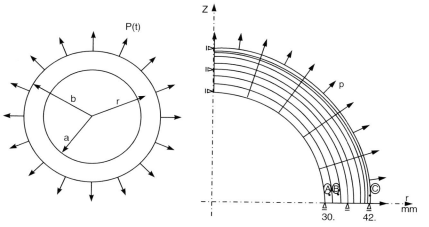

Fig. 5.6. Hollow sphere under external pressure and FE meshing

- Beyond mesocrack initiation, the time evolutions of the radius $c(t)$ of the spherical damaged zone and of the function $f(t)$ satisfying the coupled differential equations are

$$P(t) = 2\sigma_y g(t) \ln \frac{b}{c(t)} + \left(\frac{2\sigma_y}{3S_0 K_N^0} f(t) - \frac{2}{3} \dot{f}(t) g(t) \right) \left[\frac{1}{b^3} - \frac{1}{c^3(t)} \right]$$
$$+ \frac{\dot{f}(t) f(t)}{3S_0 K_N^0} \left[\frac{1}{b^6} - \frac{1}{c^6(t)} \right]$$
$$\dot{c}(t) = \frac{\sigma_y c^{4/3}(t)}{3f(t)} + \frac{c(t) \dot{f}(t)}{3f(t)} .$$

(5.116)

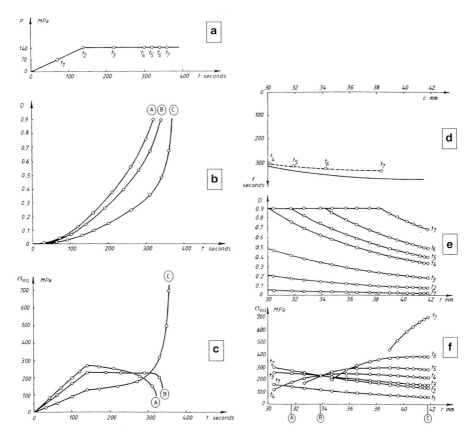

Fig. 5.7. a Loading history. **b** Damage vs time at Gauss points A, B and C. **c** von Mises stress vs time at Gauss points A, B and C. **d** Damage initiation front $c(t)$ (FE solution: open circles; closed-form solution: solid line). **e** Damage vs radius for different times. **f** von Mises stress vs radius for different times

This problem is used to check an FE implementation in ABAQUS (the UMAT VISCOENDO of LMT-Cachan). The mesh made of forty 8-node quadrilateral elements is depicted in Fig. 5.6 where the boundary conditions are indicated. Due to axial symmetry, one quarter of the sphere is modelled. The pressure history is given in Fig. 5.7a.

5.4.1.1 Validation of the FE implementation

Figures 5.7b and 5.7c illustrate the evolutions with time of the damage $D(t)$ and von Mises stress $\sigma_{eq}(t)$ at the three Gauss points A, B, and C marked in Fig. 5.6. The agreement between the analytical solution (continuous line) and the FE analysis (open circles) validates the numerical implementation.

The curves of Fig. 5.7d for times t_4 to t_7 give an idea of the propagation of the damage initiation front radius $c(t)$. The slight deviation between the two solutions observed is due to the fact that in the closed form solution, D_c is taken to be equal to 1 instead of 0.9 for FE computations, and hence leads to a larger time for initiation.

Figure 5.7e shows the damage redistribution along the radius obtained with ABAQUS or different times t_1 to t_7 (see Fig. 5.7a). Since the numerical critical value for damage is taken as $D_c = 0.9$, t_4 appears as the time to the initiation of a mesocrack at an inner radius of $a = 30$ mm.

5.4.1.2 Interpretation of the Results

Figure 5.7f shows the role of the coupling of the damage model with elasto-visco-plastic equations. Since the fully-coupled approach has been used in this example, both the effect of the damage gradient within the structure and the effect of the damage front advance (modelled by the removal of completely broken elements) on the stress field can be observed: first, at the beginning of the loading (between times t_2 and t_4), the redistribution of the equivalent stress σ_{eq} is only due to the damage evolution (which is quicker at the inner radius); second, beyond the time to mesocrack initiation t_4, although the pressure load is constant, very large increases in the von Mises stress correspond to the propagation of the front which becomes the effective inner radius. In other words there is a growth of the inner cavity.

Besides, it can be noticed in Figs. 5.7c and 5.7f that the point B plays the role of the "skeletal" point of the "Reference Stress Method" (D.L. Mariott and F.A. Leckie 1970), where the equivalent stress remains constant during the pure creep loading (for t greater than t_2) despite the redistribution of the stresses within the structure.

5.4.2 Effect of Loading History: Creep-Fatigue

Let us study again creep-fatigue interaction but with now the full set of elasto-visco-plasticity coupled with damage constitutive equations of Table 1.3. The loading is described in Fig. 5.3. The material is a stainless alloy IN 100 at 827°C and the material parameters are

- For elasticity: $E = 170000$ MPa
- For viscosity (Norton law): $K_N = 450$ MPa·s$^{1/N}$ and $N = 7.5$
- For isotropic hardening modelling cyclic softening: $R_\infty = -15$ MPa, $b = 100$ with an initial yield stress $\sigma_y = 30$ MPa
- For kinematic hardening: $C = 138000$ MPa, $\gamma = 1200$ (then $X_\infty = C/\gamma = 115$ MPa)
- For damage: $S = 0.2$ MPa, $s = 1$, $\epsilon_{pD} = 0.005$, and $D_c = 0.3$

The damage threshold p_D is a constant equal to ϵ_{pD}.

A numerical resolution of the constitutive laws is necessary to plot the creep-fatigue interaction diagram of N_R/N_{RF} vs t_R/t_{RC} for a given maximum stress σ_M, a fatigue stress rate $\dot\sigma^f$, and different hold times Δt (with N_R as the number of creep-fatigue cycles and $t_R = N_R \cdot \Delta t$ as the time spent in creep). The computation is performed here on a single element in ABAQUS by using VISCOENDO UMAT.

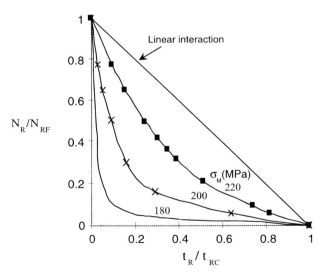

Fig. 5.8. Computed creep-fatigue interaction diagram for the stainless alloy IN 100 at 827°C (J.P. Sermage 1998)

Using the above, we obtain a strong nonlinear creep-fatigue interaction (Fig. 5.8): the larger the maximum stress the stronger the nonlinearity, as observed experimentally for this material. The material response obtained is very sensitive to creep as a small hold time reduces the lifetime substantially: for a maximum stress of $\sigma_M = 200$ MPa (corresponding to $N_{RF} = 1720$ cycles, $t_{RC} = 18 \cdot 10^4$s), 10% of creep decreases by half the lifetime.

Another important result is that this nonlinearity is mainly governed by the kinematic hardening. Setting $R = const = R_\infty$ does not change the results much.

5.4.3 Creep-Fatigue and Thermomechanical Loadings^{◊◊◊}
(J.P. Sermage, J. Lemaitre and R. Desmorat 2000)

Both coupled and uncoupled analyses are compared here to experimental results concerning the Maltese cross shape specimen of Fig. 4.13, loaded by two forces in its plane, F_1 and F_2. The specimen is heated with a controlled eddy current heating system which ensures a "uniform" temperature field at $\pm 5°C$. The real temperature is measured by thermocouples and introduced as a given temperature field in the structure calculation.

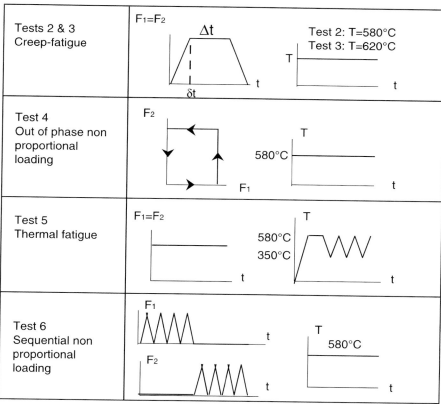

Fig. 5.9. Tests performed

264   5 Creep, Creep-Fatigue, and Dynamic Failures

Six tests are performed with forces histories $F_1(t)$, $F_2(t)$, and temperature histories given in Fig. 5.9. The first test performed at room temperature is the multilevel fatigue test of Chap. 4. The tests 2 to 6 are:

- Two **in-phase creep-fatigue interaction tests** with $F_1 = F_2$ between 0 and $F_{\max} = 28$ kN at constant temperature T=580°C (test 2) and T=620°C (test 3). The loads $F_{1\,\max} = F_{2\,\max} = 28$ kN have a hold time of $\Delta t = 40$ s. The period of the loading is $\Delta t + 2\delta t = 50$ s.
- One **multilevel out-of-phase non-proportional loading test** at T=580°C (test 4). The maximum applied loads are $F_{1\,\max} = F_{2\,\max} = 28$ kN for the first 1000 cycles, $F_{1\,\max} = F_{2\,\max} = 35$ kN for the next 1000 cycles, and then $F_{1\,\max} = F_{2\,\max} = 38$ kN up to failure.
- One **thermal fatigue test** at constant loads $F_1 = F_2 = F_{\max} = 32$ kN (after an initial load from 0 to $F_{\max}$ in 50 s) and variable temperature (test 5). The temperature varies between the maximum 580°C and the minimum 300°C within a period of 600 s.
- One **sequential non-proportional loading test** (test 6). The specimen is submitted to a non-proportional loading in sequences of 4 cycles of periodic $F_1(t)$ between 0 and $F_{\max} = 35$ kN at $F_2 = 0$ followed by 4 cycles of periodic $F_2(t)$ between 0 and $F_{\max} = 35$ kN at $F_1 = 0$; the duration of a loading-unloading is 20 s which corresponds to 160 s per full cycle.

These histories of loading are the inputs:

- For a fully coupled FE analysis using ABAQUS and VISCOENDO UMAT.
- For an uncoupled analysis using ABAQUS and VISCOENDO but with no damage. The results are then the inputs of a post-processor damage calculation in order to perform the "a posteriori" time integration of the unified damage law.

The drawing of the specimen is given in Sect. 4.4.2. The material is a 2-1/4 CrMo steel. A first identification of the elasto-visco-plasticity and damage material parameters has been made at different temperatures (20°C, 100°C, . . . , 580°C). The temperature dependency of the parameters is then modelled through a fitting with nonlinear functions of the temperature given in Table 5.1. In the range 580°C–620°C, the viscosity parameters are assumed to remain constant (equal to those identified at 580°C). A thermomechanical validation is presented in Figs. 2.7 and 2.8.

Cases 2 to 6 are calculated first by the uncoupled method of a viscoplasticity ABAQUS computation of a few cycles followed by a damage calculation (see Sect. 2.1.3). On HP 700 working stations, each FE computation takes a few hours when about one minute is needed for the damage calculation at the most loaded point. Each case is also computed using the fully-coupled method VISCOENDO-ABAQUS but as all the cycles must be calculated with small time increments to ensure the convergence of the algorithm: each case takes about 1 week (!!) of computation on an HP 700.

The final results of the number of cycles-to-crack-initiation, both from experiments and from calculations, are summarized in Table 5.2 and in Fig. 5.10. The accuracy of the fully-coupled analysis is good for such complex loadings. The uncoupled analysis is of course less accurate (but so much cheaper!). It loses its feature of being conservative when the loadings become non-proportional.

Table 5.1. Material parameters for a 2-1/4 CrMo in the range 20°C–580°C

Poisson's ratio ν	$\nu = const$	$\nu = 0.3$
Young's modulus E	$E(T) = e_1 [1 - \exp(e_2 T)] + e_3$	$e_1 = 31800$ MPa $e_2 = 1.88\,10^{-3}$ °C^{-1} $e_3 = 199800$ MPa
Yield stress σ_y	$\sigma_y(T) = k_1 [1 - \exp(k_2 T)] + k_3$	$k_1 = 70$ MPa $k_2 = 1.58\,10^{-3}$ °C^{-1} $k_3 = 191.5$ MPa
Thermal expansion α	$\alpha(T) = \alpha_1 + \alpha_2 T$	$\alpha_1 = 1.1\,10^{-5}$ °C^{-1} $\alpha_2 = 3.66\,10^{-9}$ °C^{-2}
Isotropic hardening R	$R_\infty = r_1 [1 + \exp(r_2 T)] + r_3$	$r_1 = 0.78$ MPa $r_2 = 6\,10^{-3}$ °C^{-1} $r_3 = 4.5$ MPa $b = 2$
Kinematic hardening X	$\gamma(T) = \gamma_1 [1 + \exp(\gamma_2 T)] + \gamma_3$ $X_\infty(T) = x_1 - x_2 T$	$x_1 = 150$ MPa $x_2 = 0.216$ °C^{-1} $\gamma_1 = 0.392$ $\gamma_2 = 9.69\,10^{-3}$ °C^{-1} $\gamma_3 = 140.5$
Viscosity parameters	$K(T) = k_1 + k_2 T + k_3 \exp\left(\frac{T - k_4}{k_5}\right)$ $N(T) = n_1 \exp\left[-(n_2 T)^{n_3}\right] + n_4$	$k_1 = 190$ $k_2 = 0.25$ °C^{-1} $k_3 = 2.7$ $k_4 = 372.7$ $k_5 = 35.32$ $n_1 = 10.315$ $n_2 = 0.022$ °C^{-1} $n_3 = 4.89$ $n_4 = 2.41$
Damage D	$\epsilon_{pD}(T) = p_1 T + p_2$ $S(T) = s_1 [1 - \exp(s_2 T)] + s_3$	$s = 2$ $D_c = 0.2$ $p_1 = -0.000285$ $p_2 = 0.2657$ $s_1 = 0.108$ $s_2 = 0.0052$ $s_3 = 2.81$

266   5 Creep, Creep-Fatigue, and Dynamic Failures

To finish, note that the results are very sensitive to temperature. The same computation with a slight difference in the applied temperature gives a very different number of cycles to failure: for test 2, a temperature of 590°C instead of 580°C leads to a number of cycles-to-crack-initiation of 300 instead of 420.

Table 5.2. Calculated and experimental numbers of cycles at mesocrack initiation N_R

	Comparisons		
Test performed	Experiments	Fully-coupled analysis	Uncoupled analysis
Test 2 & 3 Creep-fatigue	$N_{R\,(580°C)} = 331$ $N_{R\,(620°C)} = 100$	$N_R = 420$ $N_R = 115$	$N_R = 266$
Test 4 Out of phase non-proportional loading	$N_R = 2356$	$N_R = 2135$	$N_R = 1130$
Test 5 Thermal fatigue	$N_R = 56$	$N_R = 48$	$N_R = 27$
Test 6 Sequential non-proportional loading	$N_R = 196$	$N_R = 230$	$N_R = 273$

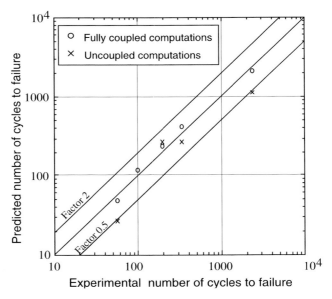

Fig. 5.10. Comparison between predicted numbers of cycles to mesocrack initiation and experimental results

5.4.4 Dynamic Analysis of Crash Problems[♂♂♂]
(F. Lauro, B. Bennani, S. Tison, P. Croix 2003)

Crash problems or dynamic failures on metallic or polymer structures generally involve a large range of plastic strain rates whose effects have to be taken into account through visco-plasticity and, furthermore, through visco-plasticity coupled with damage.

As an example, consider a notch extruded tube of rectangular cross section impacted in bending as represented in Fig. 5.11.

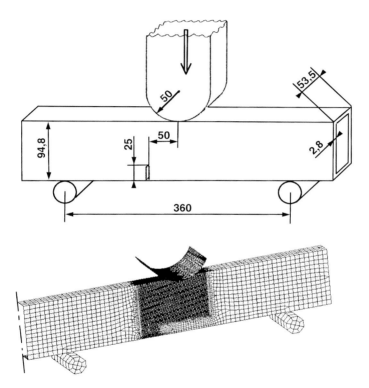

Fig. 5.11. Dimensions and finite element model of impacted tube (half by symmetry)

The constitutive equations are those of Table 1.3: elasto-(visco-)plasticity coupled with damage but associated to the Hill criterion due to the plastic anisotropy induced by the process of extrusion.

The material is an aluminum alloy 6014 T7 with the following values of the material parameter identified from tension-compression experiments described in Sects. 1.4.4 and 1.5.1:

- Elasticity: $E = 70000$ MPa, $\nu = 0.33$
- Plasticity:
  - Hill criterion $f = \left(\dfrac{3}{2}\dfrac{\sigma}{1-D} : \underline{\underline{h}} : \dfrac{\sigma}{1-D}\right)^{1/2} = \sigma_v$ with $\underline{\underline{h}}$ as the Hill fourth order tensor (see Sect. 3.4.8) with 6 independent material parameters,
  - In the longitudinal direction for $\dot{\epsilon} \approx 10^{-3}$ s^{-1}: yield stress $\sigma_y = 240.6$ MPa, ultimate stress $\sigma_u = 275$ MPa, rupture strain $\epsilon_R \approx 0.5$
  - The effect of the strain rate is assumed saturated in this crash application so that the viscous law is simply modelled by a $\sigma_v(p)$ law entered point by point in the FE input file:

σ_v (MPa)	240.6	264.5	273	278.8	340.3	410
p	0	0.034	0.05	0.065	0.5	1

- Damage: $\epsilon_{pD} = 0.05$, $s = 2$, $S = 1.22$ MPa, $D_c = 0.36$

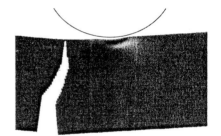

Fig. 5.12. Experimental and numerical simulation of the crack path

The calculation is performed with PAMCRASH 2000 code with 15614 shell elements and 7 integration points in the thickness. The propagation of the crack is simulated by element removal when the damage D reaches D_c at all its integration points. The full calculation up to rupture in 5 ms in the experiment takes 22 hr on an HP workstation.

The corresponding experiment is performed with an impactor reaching the tube at 3 m/s. The comparison between the numerical simulation and the experiment is shown in Fig. 5.12 for the failure path and in Fig. 5.13 for the load displacement curve. The agreement is satisfactory for the path and the maximum load. For the energy absorbed, the numerical simulation results are ≈ 20 % lower than the experiment.

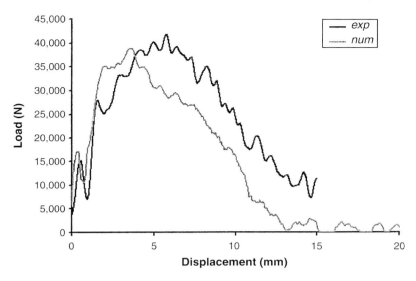

Fig. 5.13. Load-displacement curve

5.4.5 Ballistic Impact and Penetration of Projectiles
(T. Børvik, O.S. Hopperstad, T. Berstad and M. Langseth 2001)

The penetration of projectiles at high speed is a difficult problem as it involves adiabatic shear localization, thermal plastic instabilities, and high gradients of stresses, as shown in Fig. 5.14.

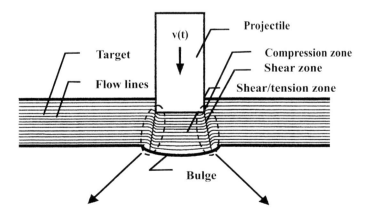

Fig. 5.14. Description of the mechanism of penetration

Nevertheless, it is possible to use elasto-visco-plasticity coupled with damage constitutive equations described in Sect. 1.5 and a finite element numerical procedure as long as the effects of temperature are represented by proper functions. For the latter one may use either the expression of the yield stress function of the strain rate and the temperature derived in Sect. 5.3.1,

$$\sigma_y(\dot{p}, T) \approx \sigma_y(\dot{\epsilon}_{p0}, T_0) \left(\frac{\dot{p}}{\dot{\epsilon}_{p0}}\right)^{1/N} \exp\left\{-\frac{H}{N}\left(\frac{1}{T_0} - \frac{1}{T}\right)\right\}, \quad (5.117)$$

or the Johnson–Cook model (1983) modified by M. Ortiz (1997),

$$\sigma_{eq} = \left(\sigma_{y0} + K_{JC}\, p^{1/M_{JC}}\right)\left(1 + \frac{\dot{p}}{\dot{\epsilon}_{p0}}\right)^{C_{JC}} \left[1 - \left(\frac{T - T_0}{T_m - T_0}\right)^{m_{JC}}\right], \quad (5.118)$$

where σ_{y0}, K_{JC}, M_{JC}, C_{JC}, $\dot{\epsilon}_{p0}$, and m_{JC} are material parameters, T_0 is the room temperature, and T_m is the melting temperature of the material.

In fact, the Johnson–Cook model is a particular case of the general framework of Sect. 1.5, where the yield function without kinematic hardening and damage reads

$$\sigma_{eq} = \sigma_y + R + \sigma_v \quad (5.119)$$

if the following functions for the yield stress σ_y, the isotropic hardening R, and the viscous stress σ_v are adopted together with:

$$T^\star = \frac{T - T_0}{T_m - T_0},$$

$$\sigma_y(T) = \sigma_{y0}\left(1 - T^{\star\, m_{JC}}\right),$$

$$R(p, T) = K_{JC}\, p^{1/M_{JC}} \left(1 - T^{\star\, m_{JC}}\right), \quad (5.120)$$

$$\sigma_v(\dot{p}, p, T) = \left[\left(1 + \frac{\dot{p}}{\dot{\epsilon}_{p0}}\right)^{C_{JC}} - 1\right] \left(\sigma_{y0} + K_{JC}\, p^{1/M_{JC}}\right)\left(1 - T^{\star\, m_{JC}}\right).$$

Introducing the isotropic damage variable D with $\dot{r} = \dot{p}(1 - D)$,

$$\sigma_{eq} = (1 - D)\left(\sigma_{y0} + K_{JC}\, r^{1/M_{JC}}\right)\left(1 + \frac{\dot{r}}{\dot{\epsilon}_{p0}}\right)^{C_{JC}} \left(1 - T^{\star\, m_{JC}}\right), \quad (5.121)$$

which is the constitutive relation used for numerical simulation in the following example, the damage evolution law derived from (3.15) and (3.16) considering ductile damage and proportional loading,

$$\dot{D} = \frac{D_c}{p_R - \epsilon_{pD}}\dot{p} \quad \text{if} \quad p > \epsilon_{pD}, \quad (5.122)$$

we can rewrite p_R as a modified version of the Johnson–Cook model:

$$p_{\rm R} = [a_1 + a_2 \exp(a_3 T_{\rm X})] \left[1 + \frac{\dot{r}}{\dot{\epsilon}_{\rm p0}}\right]^{a_4} [1 + a_5 T^\star] . \tag{5.123}$$

They are implemented for large strains in LS-DYNA Finite Element using a fully vectorized backward-Euler integration algorithm for 3D, shell, and 2D analysis. The crack growth is simulated by an element killing procedure which removes the element when the damage reaches its critical value $D_{\rm c}$. This makes the result dependent on mesh size.

All the material parameters may be identified from tension tests at different temperatures following the procedure described in Sect. 2.4. Their values characterizing elasto-visco-plasticity and damage for Weldox 460E steel are: $E_{(T_0)} = 200000$ MPa, $\nu = 0.33$, $\sigma_{\rm y0} = 490$ MPa, $K_{\rm JC} = 807$ MPa, $M_{\rm JC} = 1.37$, $\dot{\epsilon}_{\rm p0} = 5 \cdot 10^{-4}{\rm s}^{-1}$, $C_{\rm JC} = 0.0114$, $T_0 = 293°$K, $T_{\rm m} = 1800°$K, $m_{\rm JC} = 4$, $\epsilon_{\rm pD} = 0$, $s = 1$, and $D_{\rm c} = 0.3$.

The case we are taking into consideration is a blunt nose projectile with a diameter of 20 mm, a length of 80 mm and a mass of 0.197 kg, launched by a compressed gas gun, at velocities V just below and above the ballistic limit of non-penetration, against circular targets of thickness 12 mm, diameter 500 mm, and clamped in a rigid frame. Initial velocities $150 < V < 400$ m/s and residual velocities are measured and a digital high-speed camera is used for visualization and measurement of the penetration process, as shown in Fig. 5.15.

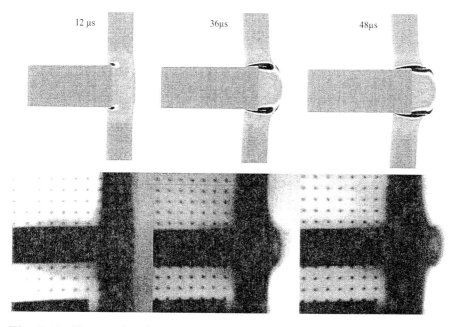

Fig. 5.15. Numerical and experimental kinetic perforation of a 12-mm thick Weldox 460E steel plate at initial velocity of 303.5 m/s

Numerical simulations are carried out to determine the ballistic limit curves and are compared with the experiments. The computations use 10280 4-node axisymmetric elements with one integration point. The mesh size in the impacted region giving the best result compared to experiments is 0.25 × 0.2 mm^2, giving 60 elements over the largest thickness of the plate. The projectile is modelled as an elasto-plastic material. The contact between the projectile and the target is made without friction by a penalty formulation. For the example of Fig. 5.15 the initial velocity of the projectile is 303.5 m/s, the residual velocity after perforation is 193 m/s, compared to 200 m/s measured in experiments.

A comparison between the numerical simulations and experiments is also given in Fig. 5.16 for the ballistic limit curve where a limit of 193 m/s is found against 185 m/s in experiments. Note that 220 m/s is found with a coarse meshing of 30 elements in the thickness instead of 60 for the fine mesh.

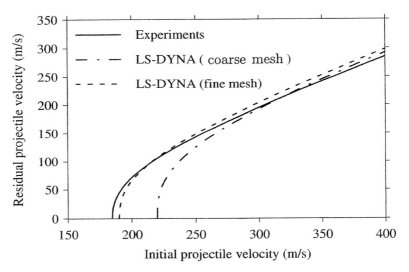

Fig. 5.16. Ballistic limit curve for a 12-mm thick target of Weldox 460E steel

5.4.6 Predeformed and Predamaged Initial Conditions

After the forming processes, accidents, or over loadings, the internal variables may have non-zero initial values for a following structure analysis: accumulated plastic strain p_0, corresponding values of strain hardening R_0 and X_0, stored energy w_s^0, damage D_0, and internal residual stresses σ_0. Introduced as initial values in the structure calculation and in the damage analysis, they modify the resulting time or number of cycles to mesocrack initiation.

- In pure 3D creep the integration of the unified damage law of Sect. 5.2.2 is modified as follows:

$$\dot{D} = \left[\frac{\sigma_{eq}^2 R_\nu}{2ES(1-D)^2}\right]^s \left[\frac{\sigma_{eq}}{K_{N_0}(1-D)}\right]^{N_0}. \qquad (5.124)$$

If $\sigma_{eq} = const$, an initial working time $t = 0$ such as $D_0 = D(t=0)$ makes $p_0 = p(t=0)$ larger than the damage threshold p_D, yielding

$$\int_{D_0}^{D} (1-D)^{2s+N_0} dD = \int_0^t \left[\frac{\sigma_{eq}^2 R_\nu}{2ES}\right]^s \left[\frac{\sigma_{eq}}{K_{N_0}}\right]^{N_0} dt \qquad (5.125)$$

$$D = 1 - \left[(1-D_0)^{2s+N_0+1} - (2s+N_0+1)\frac{\sigma_{eq}^{2s+N_0} R_\nu^s}{(2ES)^s (K_N^0)^{N_0}} t\right]^{\frac{1}{2s+N_0+1}}. \qquad (5.126)$$

The time-to-mesocrack-initiation t_{RC}^0 is reached for $D = D_c$ in

$$t_{RC}^0 = \frac{(1-D_0)^{2s+N_0+1} - (1-D_c)^{2s+N_0+1}}{2s+N_0+1} \frac{(2ES)^s (K_N^0)^{N_0}}{\sigma_{eq}^{2s+N_0} R_\nu^s}, \qquad (5.127)$$

which makes a decrease in the time to rupture in comparison to the time t_{RC} of mesocrack initiation without any predamaged initial condition:

$$t_{RC} = t_{DC} + \frac{1 - (1-D_c)^{2s+N_0+1}}{2s+N_0+1} \frac{(2ES)^s (K_N^0)^{N_0}}{\sigma_{eq}^{2s+N_0} R_\nu^s} \qquad (5.128)$$

as

$$\frac{t_{RC}^0}{t_{RC}} = \left(1 - \frac{t_{DC}}{t_{RC}}\right) \frac{(1-D_0)^{2s+N_0+1} - (1-D_c)^{2s+N_0+1}}{1 - (1-D_c)^{2s+N_0+1}} < 1. \qquad (5.129)$$

- In 3D creep-fatigue, with the same assumptions and notations as in Sect. 5.2.2, the difference is again in the time integration of the damage from D_0 to D_c,

$$N_R^0 = \frac{(1-D_0)^{2s+N_0+1} - (1-D_c)^{2s+N_0+1}}{(2s+N_0+1)\left[\frac{2}{2s+N_0+1}\frac{\sigma_{eq\,max}}{\dot{\sigma}_{eq}^f} + \Delta t\right]} \frac{(2ES)^s (K_N^0)^{N_0}}{\sigma_{eq}^{2s+N_0} R_\nu^s}, \qquad (5.130)$$

which gives, in the same manner as for creep, a reduced number of cycles to rupture:

$$\frac{N_R^0}{N_R} = \left(1 - \frac{N_D}{N_R}\right) \frac{(1-D_0)^{2s+N_0+1} - (1-D_c)^{2s+N_0+1}}{1 - (1-D_c)^{2s+N_0+1}} < 1. \qquad (5.131)$$

5.4.7 Hierarchic Approach up to Viscous Elastomers♂,♂♂♂

The accuracy of a prediction of the evolution of any system decreases as the number of causes increases! Then, predictions for creep, creep-fatigue, or dynamic failures involving temperature and time dependencies do not expect the same accuracy as for ductile or low cycle fatigue failures. The simplest models are nevertheless relevant in the comparison of different solutions in early design.

- The early Kachanov model may be used for uniaxial constant states of stress in order to draw a net or to interpolate some experimental isochronous curves,

$$\dot{D} = \left[\frac{\sigma}{A_\mathrm{D}(1-D)}\right]^{r_\mathrm{D}}. \tag{5.132}$$

- Likewise, the linear Taira rule of creep-fatigue interaction is quite easy to use for uniaxial periodic loadings with hold time,

$$\frac{t_\mathrm{R}}{t_\mathrm{RC}} + \frac{N_\mathrm{R}}{N_\mathrm{RF}} = 1. \tag{5.133}$$

- For more complicated loadings such as 3D, time-varying loadings and/or temperature-varying loadings, we advise using a numerical integration of the unified damage law,

$$\dot{\boldsymbol{D}} = \left(\frac{\overline{Y}}{S}\right)^s |\dot{\epsilon}^\mathrm{p}| \quad \text{if} \quad w_\mathrm{s} > w_\mathrm{D} \tag{5.134}$$

with the material parameters eventually functions of the temperature.

- In the case of proportional loading, closed-form solutions may be obtained and, if the state of stress is constant, simple relations relate the time or the number of cycles to rupture in 3D and 1D, provided the equality in stresses $\sigma = \sigma_\mathrm{eq}$ and the equality of the hold times Δt for creep-fatigue hold:

$$\begin{aligned} t_\mathrm{RC}(\sigma_{ij}) - t_\mathrm{DC}(\sigma_{ij}) &= [t_\mathrm{RC}(\sigma) - t_\mathrm{DC}(\sigma)]\, R_\nu^{-s} \\ N_\mathrm{R}(\sigma_{ij}) - N_\mathrm{D}(\sigma_{ij}) &= [N_\mathrm{R}(\sigma) - N_\mathrm{D}(\sigma)]\, R_\nu^{-s}. \end{aligned} \tag{5.135}$$

- For small scale yielding visco-plasticity, the complex numerical analysis may be avoided by applying the extended Neuber method using the Norton law. For large scale yielding, an approximate method which needs the full computation may also be used, but on a RVE only (see Sect. 5.2.4).
- In dynamic problems, if only the yield stress, function of the strain rate, and the temperature are needed, use:

$$\sigma_\mathrm{y}(\dot{p}, T) = \sigma_\mathrm{y}(\dot{\epsilon}_{\mathrm{p}0}, T_0) \left(\frac{\dot{p}}{\dot{\epsilon}_{\mathrm{p}0}}\right)^{1/N_0} \exp\left\{-\frac{H}{N_0}\left(\frac{1}{T_0} - \frac{1}{T}\right)\right\} \tag{5.136}$$

or the Johnson–Cook model.

- For complex problems, there is no way to escape from a numerical analysis but often the fully coupled analysis may be avoided by the post-processing of an elasto-visco-plastic calculation performed with $\boldsymbol{D} = 0$.

The previous hierarchic approach applies to metals or polymers which can be modelled by classical elasto-visco-plasticity as long as no damage occurs. For specific materials and loadings such as polymers loaded at high strain rates, the hardening, the viscosity laws, or even the whole model can be different from the laws described in this book. Most often, the models found in the literature for specific material behaviors do not consider damage. Sometimes they are built for low strain rates such as time-independent plasticity-like models. Model updating consists then in extending these models either to damage or to viscosity effects.

Recall that a simple way to make an existing model coupled with damage consists of using the effective stress concept: replace, for instance, the stress $\boldsymbol{\sigma}$ in the elasticity law and in the yield criterion of the undamaged material by the effective stress $\tilde{\boldsymbol{\sigma}}$ for the damaged material. Examples for the hardening laws are given in Sects. 3.4.8 and 4.4.5.

For plasticity-like models with a yield criterion defined in the stress space as $f = 0$, a simple way to introduce the viscosity is to replace the consistency condition $f = 0$ and $\dot{f} = 0$ by a viscosity law $\sigma_{\rm v} = \sigma_{\rm v}(\dot{p})$ such as $f = \sigma_{\rm v}$. This is how elasto-visco-plasticity can be build from the elasto-plasticity framework. As a last example, consider the model of hyperelasticity with internal sliding for elastomers of Sect. 4.4.3. With the notations of Sect. 4.4.3, the reversibility domain of undamaged elastomers is

$$f = \|\tilde{\boldsymbol{S}}^\pi - \boldsymbol{X}\| - \sigma_{\rm s} < 0 \tag{5.137}$$

and the internal sliding is gained from the condition $f = 0$ and $\dot{f} = 0$. A possible modelling for viscous elastomers submitted to creep or creep-fatigue is to consider a viscosity law $\sigma_{\rm v} = \sigma_{\rm v}(\dot{\pi})$ and to determine the internal sliding multiplier from $f = \sigma_{\rm v}$. For the Norton law this means

$$\sigma_{\rm v} = K_N \dot{\pi}^{1/N} \quad \longrightarrow \quad \dot{\pi} = \left\langle \frac{\|\tilde{\boldsymbol{S}}^\pi - \boldsymbol{X}\| - \sigma_{\rm s}}{K_N} \right\rangle^N. \tag{5.138}$$

Except for the consistency condition replaced by eq. (5.138), the whole set of constitutive equations remains unchanged, provided the stresses $\boldsymbol{S}$ and $\boldsymbol{S}^\pi$ are replaced by the effective stresses $\tilde{\boldsymbol{S}}$ and $\tilde{\boldsymbol{S}}^\pi$. The constitutive equations for quasi-incompressible materials are

$$\tilde{\boldsymbol{S}} = \frac{\boldsymbol{S}}{1-D} = \frac{\partial(w_1 + w_2)}{\partial \boldsymbol{E}},$$
$$\tilde{\boldsymbol{S}}^\pi = \frac{\boldsymbol{S}^\pi}{1-D} = \frac{\partial w_2}{\partial \boldsymbol{E}},$$

$$\begin{aligned}
\dot{\boldsymbol{E}}^{\pi} &= \dot{\pi}\frac{\tilde{\boldsymbol{S}}^{\pi} - \boldsymbol{X}}{\|\tilde{\boldsymbol{S}}^{\pi} - \boldsymbol{X}\|}, \\
\dot{\boldsymbol{X}} &= \left[C_x \dot{\boldsymbol{E}}^{\pi} - \gamma \boldsymbol{X}\dot{\pi}\right](1 - D), \\
Y &= w_1 + w_2, \\
\dot{D} &= \left(\frac{Y}{S}\right)^s \dot{\pi} \quad \text{if} \quad \pi > \pi_\text{D},
\end{aligned} \qquad (5.139)$$

with $f < 0 \longrightarrow$ hyperelasticity, $\pi = \int_0^t \|\dot{\boldsymbol{E}}^{\pi}\|\mathrm{d}t$ as the cumulative measure of the internal sliding, and $D = D_\text{c}$ as the mesocrack initiation condition.

6
High Cycle Fatigue

High cycle fatigue is probably the most difficult phenomenon to handle within solid mechanics and is, by consequence, the main cause of failures of mechanical components in service. The difficulty comes from the early stage of damage which initiates **defects** at a very small micro- or nanoscale under cyclic stresses **below the engineering yield stress**. Even during the evolution of the damage, there is no easy precursor to detect the danger of a failure! The purpose here is not to be pessimistic but to caution readers when dealing with the design of a component which can be subjected to cyclic stresses during 10^5, 10^6, 10^7 cycles or more. Nothing to see! But something to do!

From the physical point of view, the repeated variations of elastic stresses in metals induce micro-internal stresses above the local yield stress, with dissipation of energy via **microplastic strains** which arrest certain slips due to the increase of dislocations nodes. There is formation of permanent micro slip bands and decohesions, often at the surface of the material, to produce the mechanism of **intrusion-extrusion**. After this first stage located inside the grains, where the microcracks follow the planes of maximum shear stress, there is a second stage in which the microcracks cross the crystal boundaries to grow more or less perpendicular to the direction of the maximum principal stress up to coalescence to produce a mesocrack. For polymers or concrete, the physical mechanisms are different but are also characterized by debondings and microdecohesions which induce the same effects on the mesoscale.

The phenomenon of fatigue was recognized around 1830 by M. Albert in Germany about mine chains but it was A. Wöhler who really gave the starting point on research on fatigue by its "Wöhler curve" (1860). These researches were mainly stimulated at first by the safe design of bridges and railways: Fairbarn (1864) in England, Bauschinger (1886) in Germany, and Lechatelier (1909) in France until the first book on fatigue by Cough in 1926. After that came the need of more research in the aeronautics industry, then nuclear power plants and off shore structures, concrete structures, and then the automobile industry.

The specific items of this chapter are the **fatigue limit criteria**, the use of the **two-scale damage model** and **stochastic** approaches either for random loadings or for random distributions of initial microdefects in order to improve the knowledge of the remaining life of structures in service.

6.1 Engineering Considerations

High-cycle fatigue is considered when the cyclic loadings induce stresses close to but below the engineering yield stress so that the number of cycles to initiate a mesocrack is "high," that is larger than 10^5. The plastic strain is usually not measurable on a mesoscale but dissipation exists on a microscale to induce the phenomenon of damage.

The main cause of high cycle fatigue is vibrations of large amplitude at a zero mean stress or at small amplitude with a large mean stress. Then a major way to avoid high cycle fatigue failures is to try to avoid vibrations or at least to reduce their levels.

The second cause of high cycle fatigue is the **stress concentrations**. As the "meso" behavior of the materials is elastic, there is no possibility of plastic shakedown to reduce the stress level and the stress concentration coefficient is the one calculated by an elastic analysis ($K_T = 3$ for a hole in a large plate). Then the risk of high cycle fatigue failures is often decreased by a careful design of notches that is as smooth as possible and a careful machining that is as polished as possible to minimize roughness in the corresponding zones.

The corrosion is an external factor which increases the rate of fatigue damage by the phenomenon of passivation-depassivation at the tips of the microcracks. More generally, be careful each time a component is subjected to fatigue in an aggressive atmosphere; there is no suitable models to precisely quantify its effect.

The **fatigue limit**, that is the stress below which no damage fatigue occurs, is another snare because the "true" or **asymptotic fatigue limit** (σ_f^∞) for a given material historically decreases as the possibility to increase the number of cycles in experiments goes on: 10^8, 10^{10}, 10^{12} cycles with ultrasonic loadings! This is the reason to fix an **engineering fatigue limit** (σ_f) corresponding to the stress which yields a number of cycles to rupture of 10^6 or 10^7. Note that 10^7 cycles correspond to a test of almost 4 days at 30 Hertz. A minimum of 10 tests for a good mean value takes a month and a half!

Last, but not the least, is the scatter which is always large in high cycle fatigue. A factor of 10 on the number of cycles to rupture of "similar" tests is not abnormal! A probability analysis is advised each time it is possible.

6.2 Fast Calculation of Structural Failures

As the state of mesostrain is elastic, the calculation of the stress is easy in most cases and sometimes it may be found in handbooks of stress concentration coefficients from a nominal stress simply calculated on the basis of load equilibrium equations. Nevertheless, many specific effects of the phenomenon of fatigue must be taken into account in the damage evaluation which often needs to consider the effect of microdefects closure in compression.

6.2.1 Characteristic Effects in High-Cycle Fatigue

Stress intensity effects are quite complex because they include amplitude, mean value, and history. The effect of the stress amplitude $\Delta\sigma$ is represented by the **Wöhler curve** which gives, from experiments in tension-compression (or other load mode) at constant amplitude, $\Delta\sigma$ or $\sigma_{\max}$ as a function of the number of cycles to rupture N_R (Fig. 6.1).

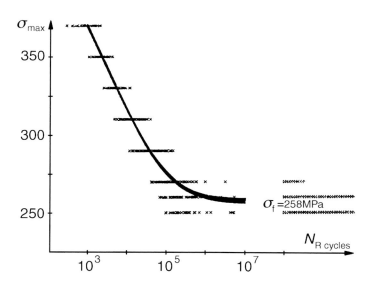

Fig. 6.1. Fatigue curve of a XC 10 steel drawn from a very large number of tests (Doc. CETIM 1992)

In the range of small stresses, the curve has a quasi-asymptotic shape which means that a small variation of stress corresponds to a large variation in the number of cycles to rupture and explains the large scatter.

In fact, this curve depends on the mean stress $\bar{\sigma} = (\sigma_{\max} + \sigma_{\min})/2$. The range of stress corresponding to a fixed value of the number of cycles to rupture decreases as the mean stress increases. This effect is represented by the **Goodman diagram** shown in Fig. 6.2. This diagram also shows the big difference of the fatigue strength of materials in tension and compression. While important in tension-compression, such effect of mean stress practically does not exist in torsion or shear.

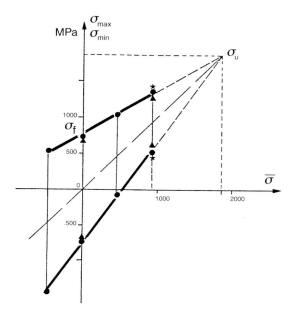

Fig. 6.2. Goodman diagram of 35 NCD 16 steel for $N_R \approx 10^6$ cycles (Doc. CETIM 1992)

When the loading is not periodic, there is an effect of the load history. The accumulation of damage due to different sequences of loading depends on the order of appearance of the sequences, as shown in Fig. 6.3 for a two-level loading: this is a sequence of n_1 cycles of range $\Delta\sigma_1$ corresponding on the Wöhler curve to a number of cycles to rupture N_{R1} and a sequence of n_2 cycles of range $\Delta\sigma_2$ corresponding to N_{R2} in such a way that $n_1 + n_2 = N_R$ is the number of cycles to rupture of the two-level test.

This dependence of the order of the stress amplitudes applied is called the **nonlinear accumulation** of fatigue damages. Nevertheless for many sequences of different amplitudes or for random loading, this effect is much less pronounced and the linear **Palmgreen–Miner rule** may apply at least as a rough approximation so that

$$\boxed{\sum \frac{n_i}{N_{Ri}} = 1\,.} \qquad (6.1)$$

This is the straight line $[(1,0),(0,1)]$ in Fig. 6.3.

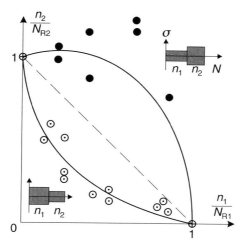

Fig. 6.3. Accumulation diagram of Maraging steel for two levels of stress range $\Delta\sigma = 1655$ MPa and 965 MPa (J.L. Chaboche 1985)

The fatigue behavior under multiaxial states of stress is even more complicated. Unfortunately not many experimental results exist. **Non-proportional loadings** seem to increase the fatigue strength in a way that is somewhat similar to cross hardening in plasticity. The **temperature** also plays a role but its softening effect is easier to model.

The state of the material itself also has an effect on the fatigue. The **defects** are sources of microcrack initiations and the **surfaces** are usually weaker than the plain material: within a uniform state of stress the first fatigue crack always initiates on the surface. This is due to easy plastic slips that glide close to the surface in the process of intrusion-extrusion. Initial conditions such as a **pre-hardening** or a **pre-damage**, due to metal forming for example, usually modify the fatigue life (see Sect. 6.4.5).

A **scale effect** is generally observed in such a way that the number of cycles to rupture of a thin specimen is higher than for a thicker specimen loaded by the same stress. This is due to the lower probability of having a large defect in the thin specimen than in the thicker one.

It seems also that the **gradient** of the stresses plays a role. For example, the number of cycles needed to initiate a mesocrack in pure bending is higher than in pure tension for the same level of maximum stress (see Sect. 6.4.6).

6.2.2 Fatigue Limit Criteria♂♂

In quick design, a safe criterion is to manage to have a state of stress below the fatigue limit where no damage may occur. This is easy to say! But it is not so easy to practice for at least two reasons.

6 High Cycle Fatigue

The notion of fatigue limit is not rigorous as nobody has waited an infinite number of cycles! As already mentioned earlier, only a conventional fatigue limit in tension (σ_f) can be defined objectively, at least from a statistics point of view, as the maximum stress which corresponds to 10^6 or 10^7 cycles. But experiments on metals show that a rupture may occur for number of cycles as high than 10^{10} or 10^{12} cycles (C. Bathias 2000). Then, another limit is defined as the asymptotic fatigue limit σ_f^∞ corresponding to an "estimated" asymptote of the Wöhler curve.

For multiaxial states of stress there is no definitively admitted scalar function of the stress components, a fatigue norm, to compare with the fatigue limit in tension. There is nothing like the von-Mises criterion in plasticity. An equivalent would be the damage equivalent stress $\sigma_{eq}^\star$ based on the total elastic energy density (see Sects. 1.2.2 and 7.2.1) but it does not take into account the effect of mean stress. Nevertheless, a fast estimation of fatigue safety is

$$\boxed{\sigma_{\max}^\star < \sigma_f,} \qquad (6.2)$$

with

$$\sigma^\star = \sigma_{eq} R_\nu^{1/2},$$

$$\sigma_{eq} = \sqrt{\frac{3}{2}\sigma_{ij}^D \sigma_{ij}^D}, \quad \sigma_H = \frac{1}{3}\sigma_{kk}, \qquad (6.3)$$

$$R_\nu = \frac{2}{3}(1+\nu) + 3(1-2\nu)\left(\frac{\sigma_H}{\sigma_{eq}}\right)^2.$$

6.2.2.1 Sines Criterion

Based on phenomenological considerations, it is a function of the range of octahedral shear A_{II} defined in proportional loading ($\boldsymbol{\sigma} = \sigma_\Sigma \boldsymbol{\Sigma}$ with $\Sigma_{eq} = 1$) by

$$A_{II} = \frac{1}{2}\sqrt{\frac{3}{2}(\sigma_{ij\,\max}^D - \sigma_{ij\,\min}^D)(\sigma_{ij\,\max}^D - \sigma_{ij\,\min}^D)} = \frac{\Delta\sigma_\Sigma}{2} \qquad (6.4)$$

and of the mean value of the hydrostatic stress $\bar{\sigma}_H = \frac{1}{2}(\sigma_{H\,\max} + \sigma_{H\,\min})$ (Crossland criterion uses the maximum value $\sigma_{H\,\max}$ of σ_H). Then the safety criterion is written as

$$\boxed{\frac{A_{II}}{1 - 3b_S\bar{\sigma}_H} < \sigma_f,} \qquad (6.5)$$

where b_S is a material-dependent parameter on the order of 10^{-3} to 10^{-2}.

6.2.2.2 Dang Van Criterion

Based on micromechanics, it is related in metals to the first plastic slip band which initiates in the weakest oriented crystal on which the elastic shakedown is assumed to happen. If $\sigma_{\rm f}$ is the fatigue limit in tension-compression and $\tau_{\rm f}$ the fatigue limit in shear or torsion, the criterion is written as a function of the maximum value of the shear stress over a cycle in time and in direction and of the hydrostatic stress $\sigma_{\rm H}$. Using σ_I as the principal stresses, we have

$$\max \left| \frac{\max_{I,J} (\sigma_I(t) - \sigma_J(t))}{2 \left[\tau_{\rm f} - \dfrac{\tau_{\rm f} - \dfrac{\sigma_{\rm f}}{2}}{\dfrac{\sigma_{\rm f}}{3}} \sigma_{\rm H}(t) \right]} \right| < 1 . \qquad (6.6)$$

It is represented in the graph of shear τ and hydrostatic stresses $\sigma_{\rm H}$ by the inner domain limited by two straight lines (Fig. 6.4).

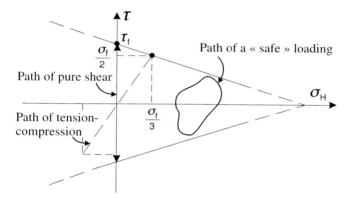

Fig. 6.4. Dang Van diagram criterion

In order to check if a loading, proportional or not, is below the conventional fatigue limit, verify that its path is fully in the inner domain.

6.2.3 Two-Scale Damage Model in Proportional Loading◊◊

For some precise but still quick evaluation of the safety for proportional and periodic loadings, it is possible to check by the two-scale damage model of Sect. 1.5.5 (see also Sect. 2.3) if the number of cycles to rupture will or will not reach a given limit of 10^5, 10^6, 10^7 or more. Recall that this model has the advantage to take into account the real history of the loading and can be used for multilevel fatigue loading as in Sects. 4.2.1 and 4.4.2 for low cycle fatigue or random fatigue.

On the mesoscopic scale, a high cycle fatigue proportional loading is characterised by

$$\boldsymbol{\sigma} = \sigma_\Sigma(t)\,\boldsymbol{\Sigma}, \qquad \boldsymbol{\epsilon}^\mathrm{p} = \mathbf{0}, \quad \text{and} \quad \sigma_\Sigma(t) \text{ between } \sigma_{\min} \text{ and } \sigma_{\max}, \qquad (6.7)$$

with the normalization $\Sigma_\mathrm{eq} = \sqrt{\frac{3}{2}\Sigma^\mathrm{D}_{ij}\Sigma^\mathrm{D}_{ij}} = 1$ (σ_Σ is the signed von Mises stress). For the model, high cycle fatigue corresponds to a stress range lower than twice the yield stress ($\Delta\sigma = \sigma_{\max} - \sigma_{\min} < 2\sigma_\mathrm{y}$) but, due to kinematic hardening, to a stress range larger than twice the asymptotic fatigue limit ($\Delta\sigma > 2\sigma_\mathrm{f}^\infty$).

On the microscopic scale, considering the two-scale damage model leads to a proportional loading only in terms of (micro) deviatoric stresses and strains with then:

$$\boldsymbol{\sigma}^{\mu\mathrm{D}} = \sigma^\mu_\Sigma(t)\boldsymbol{\Sigma}^\mathrm{D}, \qquad \boldsymbol{\epsilon}^{\mu\mathrm{p}} = \frac{3}{2}\epsilon^\mu_{\mathrm{p}\Sigma}(t)\,\boldsymbol{\Sigma}^\mathrm{D}, \quad \text{and} \quad \boldsymbol{X}^\mu = X^\mu_\Sigma(t)\,\boldsymbol{\Sigma}^\mathrm{D}. \quad (6.8)$$

Due to the Eshelby–Kröner localization law

$$\tilde{\boldsymbol{\sigma}}^\mu = \frac{\boldsymbol{\sigma}^\mu}{1-D} = \boldsymbol{\sigma} - 2G(1-\beta)\boldsymbol{\epsilon}^{\mu\mathrm{p}}, \qquad (6.9)$$

the hydrostatic stress at microscale becomes

$$\sigma^\mu_\mathrm{H} = (1-D)\tilde{\sigma}^\mu_\mathrm{H} = (1-D)\sigma_\mathrm{H} \qquad (6.10)$$

and is a function of the hydrostatic stress on the mesoscale ($\sigma_\mathrm{H} = \sigma_{kk}/3 = \sigma_\Sigma \Sigma_{kk}/3$) and the microdamage ($D^\mu = D$). The stress triaxiality at microscale is then

$$T^\mu_\mathrm{X} = \frac{\sigma^\mu_\mathrm{H}}{\sigma^\mu_\mathrm{eq}} = \frac{(1-D)\sigma_\mathrm{H}}{\sigma^\mu_\mathrm{eq}} = \frac{\sigma_\mathrm{H}}{\tilde{\sigma}^\mu_\mathrm{eq}}. \qquad (6.11)$$

It is a function of the von Mises stress on the microscale. Then, in order to perform the time integration of the unified damage law at microscale,

$$\dot{D} = \left(\frac{Y^\mu}{S}\right)^s \dot{p}^\mu, \qquad (6.12)$$

with Y^μ as the strain energy release rate (here without the microdefects closure effect) such that

$$Y^\mu = \frac{\sigma^{\mu 2}_\mathrm{eq} R^\mu_\nu}{2E(1-D)^2} \quad \text{and} \quad R^\mu_\nu = \frac{2}{3}(1+\nu) + 3(1-2\nu)T^{\mu 2}_\mathrm{X}, \qquad (6.13)$$

one needs to determine the effective stress $\tilde{\sigma}^\mu_\Sigma(t)$ and the plastic strain $\epsilon_{\mathrm{p}\Sigma}(t)$ on the microscale (see Fig. 6.5).

First, the localization law (6.9) considered within the yield criterion $f^\mu = (\tilde{\boldsymbol{\sigma}}^\mu - \boldsymbol{X}^\mu)_\mathrm{eq} - \sigma_\mathrm{f}^\infty = |\sigma_\Sigma - X^\mu_\Sigma| - \sigma_\mathrm{f}^\infty$ with linear kinematic hardening

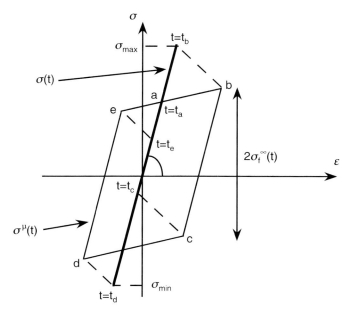

Fig. 6.5. Stress-strain cycles at mesoscale and at microscale

$\dot{X}_\Sigma = C_y(1-D)\dot{\epsilon}_{p\Sigma}$ leads to $\dot{\sigma}^\mu_\Sigma = C_y(1-D)\dot{\epsilon}_{p\Sigma}$ and a plastic strain rate $\dot{\epsilon}^\mu_{p\Sigma}$ that is linearly dependent on the meso stress rate (with generally $C_y \ll G$):

$$\dot{\epsilon}^\mu_{p\Sigma} = \frac{\dot{\sigma}_\Sigma}{\mathcal{G}} \quad \text{and} \quad \mathcal{G} = 3G(1-\beta) + C_y(1-D) \approx 3G(1-\beta). \tag{6.14}$$

The plastic strain increment over one cycle is then:

$$\begin{cases} \dfrac{\delta p^\mu}{\delta N} = \dfrac{2(\Delta\sigma - 2\sigma_f^\infty)}{\mathcal{G}} & \text{if} \quad \Delta\sigma > 2\sigma_f^\infty, \\ \dfrac{\delta p^\mu}{\delta N} = 0 & \text{if} \quad \Delta\sigma < 2\sigma_f^\infty, \end{cases} \tag{6.15}$$

and vanishes (as already mentioned) for any stress range smaller than $2\sigma_f^\infty$. In that case the model predicts no failure, this is the endurance domain.

Second, the maximum and minimum signed von Mises stresses do not vary much as

$$\begin{cases} \tilde{\sigma}^\mu_{\Sigma\,\text{max}} \approx \sigma_f^\infty + \dfrac{C_y(1-D)}{\mathcal{G}}(\sigma_{\text{max}} - \sigma_f^\infty) \approx \sigma_f^\infty \\ \tilde{\sigma}^\mu_{\Sigma\,\text{min}} \approx -\sigma_f^\infty + \dfrac{C_y(1-D)}{\mathcal{G}}(\sigma_{\text{min}} + \sigma_f^\infty) \approx -\sigma_f^\infty. \end{cases} \tag{6.16}$$

Considering then $\tilde{\sigma}^\mu_{\text{eq}} \approx \sigma_f^\infty$, the damage increment over one cycle is

$$\frac{\delta D}{\delta N} = \Delta D^+ + \Delta D^-, \tag{6.17}$$

with

$$\Delta D^+ = \int_{t_e(\sigma_e^\mu)}^{t_b(\sigma_b^\mu)} \dot{D} dt \approx \frac{\sigma_f^{\infty\,2s}}{(2ES)^s \mathcal{G}} \int_{\sigma_\Sigma(t_e)}^{\sigma_\Sigma(t_b)} \left[R_{\nu\star}\left(\xi = \Sigma_{kk}\frac{\sigma}{\sigma_f^\infty}\right)\right]^s d\sigma,$$

$$\Delta D^- = \int_{t_d(\sigma_d^\mu)}^{t_c(\sigma_c^\mu)} \dot{D} dt \approx \frac{\sigma_f^{\infty\,2s}}{(2ES)^s \mathcal{G}} \int_{\sigma_\Sigma(t_d)}^{\sigma_\Sigma(t_c)} \left[R_{\nu\star}\left(\xi = \Sigma_{kk}\frac{\sigma}{\sigma_f^\infty}\right)\right]^s d\sigma,$$
(6.18)

and

$$R_{\nu\star}(\xi) = \frac{2}{3}(1+\nu) + \frac{1}{3}(1-2\nu)\xi^2.$$
(6.19)

Define next the minimum and maximum normalized stresses as

$$\xi_{\min} = \Sigma_{kk}\frac{\sigma_{\min}}{\sigma_f^\infty} \quad \text{and} \quad \xi_{\max} = \Sigma_{kk}\frac{\sigma_{\max}}{\sigma_f^\infty}$$
(6.20)

and introduce the dimensionless function $\varphi_\star$,

$$\boxed{\varphi_\star(\xi_{\min}, \xi_{\max}) = \int_{\xi_{\min}+2}^{\xi_{\max}} R_{\nu\star}^s(\xi) d\xi + \int_{\xi_{\min}}^{\xi_{\max}-2} R_{\nu\star}^s(\xi) d\xi,}$$
(6.21)

which can be approximated by the following form that is conservative for the number of cycles to rupture, better for small s,

$$\boxed{\varphi_\star(\xi_{\min}, \xi_{\max}) \approx \langle \xi_{\max} - \xi_{\min} - 2 \rangle \left[R_{\nu\star}^s(\xi_{\min}) + R_{\nu\star}^s(\xi_{\max})\right].}$$
(6.22)

This allows us to write the damage increment per cycle as

$$\frac{\delta D}{\delta N} = \left[\frac{\sigma_f^{\infty\,2}(1+\nu)}{3ES}\right]^s \frac{2(\Delta\sigma - 2\sigma_f^\infty)}{\mathcal{G}} \quad \text{if} \quad \Sigma_{kk} = 0 \text{ (e.g. shear)}$$

$$\frac{\delta D}{\delta N} = \frac{\sigma_f^{\infty\,2s+1} \varphi_\star(\xi_{\min}, \xi_{\max})}{\Sigma_{kk}(2ES)^s \mathcal{G}} \quad \text{if} \quad \Sigma_{kk} \neq 0.$$
(6.23)

Damage will initiate after a number of cycles N_D when the damage threshold p_D is reached on the microscale. According to (6.14), the plastic strain increment for half a cycle is (for linear kinematic hardening and assuming $D = const$ over a cycle),

$$\Delta\epsilon_{p\Sigma}^\mu = \frac{1}{2}\frac{\delta p^\mu}{\delta N} = \frac{\Delta\sigma - 2\sigma_f^\infty}{\mathcal{G}} = \frac{\Delta\sigma_\Sigma^\mu - 2\sigma_f^\infty}{C_y(1-D)},$$
(6.24)

with $D = 0$ as long as:

$$p^\mu \leq p_D = \epsilon_{pD}\left(\frac{\sigma_u - \sigma_f^\infty}{\frac{\Delta\sigma_\Sigma^\mu}{2} - \sigma_f^\infty}\right)^m.$$
(6.25)

See Sect. 1.4.1 for the expression of the damage threshold in fatigue.

There is then no microdamage as long as $N < N_\mathrm{D}$. Considering a periodic loading, we have

$$N_\mathrm{D} = \frac{p_\mathrm{D}}{2|\Delta\epsilon_{p\Sigma}^\mu|} \quad \text{and} \quad p_\mathrm{D} = \epsilon_{\mathrm{pD}} \left(\frac{\mathcal{G}}{C_y}\right)^m \left[\frac{\sigma_u - \sigma_f^\infty}{\frac{\Delta\sigma}{2} - \sigma_f^\infty}\right]^m \tag{6.26}$$

or

$$N_\mathrm{D} = \frac{1}{4}\epsilon_{\mathrm{pD}} \frac{\mathcal{G}^{m+1}}{C_y^m} \frac{(\sigma_u - \sigma_f^\infty)^m}{\left(\frac{\Delta\sigma}{2} - \sigma_f^\infty\right)^{m+1}}. \tag{6.27}$$

The formulae for the number of cycles at crack initiation are finally (often with $\mathcal{G} \approx 3G(1-\beta)$):

$$\begin{aligned}
N_\mathrm{R} &= N_\mathrm{D} + \left[\frac{3ES}{\sigma_f^{\infty\,2}(1+\nu)}\right]^s \frac{\mathcal{G}D_c}{2(\Delta\sigma - 2\sigma_f^\infty)} && \text{if} \quad \Sigma_{kk} = 0 \\
N_\mathrm{R} &= N_\mathrm{D} + \frac{\Sigma_{kk}(2ES)^s \mathcal{G} D_c}{\sigma_f^{\infty\,2s+1} \varphi_\star\left(\frac{\sigma_{\min}}{\sigma_f^\infty}\Sigma_{kk}, \frac{\sigma_{\max}}{\sigma_f^\infty}\Sigma_{kk}\right)} && \text{if} \quad \Sigma_{kk} \neq 0.
\end{aligned} \tag{6.28}$$

They show that there is a mean stress effect in tension-compression (of course better described if the microdefects closure parameter h is introduced, see next part), no mean stress effect in shear as experimentally observed as for $\Sigma_{kk} = 0$, and the number of cycles to rupture depends only on the stress range. The formulae for the approximation (6.22) are detailed in Sects. 6.2.4 and 6.3.1. They show more explicitly that an increase of the stress triaxiality lowers the fatigue asymptote and reduces then (compared to tension-compression) the conventional fatigue limit of multiaxial loadings with $\Sigma_{kk} > 1$.

The accumulation of damages due to two successive loadings with different stress ranges $\Delta\sigma_1$ and $\Delta\sigma_2$ is obtained, as in Sect. 4.2.1. A bilinear damage accumulation is obtained for a non-zero damage threshold and the Miner rule is recovered if $\epsilon_{\mathrm{pD}} = 0$: if n_i is the number of cycles spent at the stress range $\Delta\sigma_i$, $N_\mathrm{R} = n_1 + n_2$ the total number of cycles to rupture corresponding to a two-level fatigue case, then

$$\begin{aligned}
\frac{n_1}{N_{\mathrm{R}1}} + \frac{n_2}{N_{\mathrm{R}2}} \frac{1 - \frac{N_{\mathrm{D}1}}{N_{\mathrm{R}1}}}{1 - \frac{N_{\mathrm{D}2}}{N_{\mathrm{R}2}}} &= 1 && \text{if} \quad n_1 \leq N_{\mathrm{D}1} \\
\frac{n_1}{N_{\mathrm{R}1}} \frac{N_{\mathrm{D}2}}{N_{\mathrm{D}1}} \frac{N_{\mathrm{R}1}}{N_{\mathrm{R}2}} + \frac{n_2}{N_{\mathrm{R}2}} &= 1 && \text{if} \quad n_1 > N_{\mathrm{D}1}.
\end{aligned} \tag{6.29}$$

Finally, for a symmetric fatigue loading with zero mean stress, i.e., with $\sigma_{\text{eq min}} = \sigma_{\text{eq max}} = \sigma_{\max}$, one can compare the number of cycles to rupture in 3D to the number of cycles in tension-compression (case $\Sigma_{kk} \neq 0$),

$$\boxed{\frac{N_R(\sigma_{ij}) - N_D(\sigma_{ij})}{N_R(\sigma_{\max}) - N_D(\sigma_{\max})} = \Sigma_{kk} \frac{\varphi_\star\left(-\frac{\sigma_{\max}}{\sigma_f^\infty}, \frac{\sigma_{\max}}{\sigma_f^\infty}\right)}{\varphi_\star\left(-\Sigma_{kk}\frac{\sigma_{\max}}{\sigma_f^\infty}, \Sigma_{kk}\frac{\sigma_{\max}}{\sigma_f^\infty}\right)}} \quad (6.30)$$

For 3D deviatoric states of stress, use the fatigue reference in shear with $\sigma_{\text{eq min}} = \sigma_{\text{eq max}} = \sqrt{3}\tau_{\max}$ (case $\Sigma_{kk} = 0$) to obtain

$$\frac{N_R(\sigma_{ij}^D) - N_D(\sigma_{ij}^D)}{N_R(\sqrt{3}\tau_{\max}) - N_D(\sqrt{3}\tau_{\max})} = 1. \quad (6.31)$$

6.2.3.1 Case with Microdefects Closure Effect

If the mean stress effect is of first importance, it is of course better to use the two-scale damage model with microcracks closure effect: introducing the parameter $h < 1$ leads to a damage growth larger in tension than in compression as

$$Y^\mu = \frac{1+\nu}{2E}\left[\frac{\langle\sigma_{ij}^\mu\rangle_+\langle\sigma_{ij}^\mu\rangle_+}{(1-D)^2} + h\frac{\langle\sigma_{ij}^\mu\rangle_-\langle\sigma_{ij}^\mu\rangle_-}{(1-hD)^2}\right] - \frac{\nu}{2E}\left[\frac{\langle\sigma_{kk}^\mu\rangle^2}{(1-D)^2} + h\frac{\langle-\sigma_{kk}^\mu\rangle^2}{(1-hD)^2}\right]. \quad (6.32)$$

For simplicity, consider here the case of a **tension-compression loading** on a mesoscale ((6.8) with $\sigma_\Sigma(t) = \sigma(t)$ and $\Sigma = \text{diag}[1,0,0]$) with $\tilde\sigma_{\text{eq}}^\mu \approx \sigma_f^\infty$ and $\tilde\sigma_H^\mu = \sigma_H = \sigma$. Neglecting the damage within the expression for the strain energy release rate gives

$$Y^\mu \approx \frac{\sigma_f^{\infty\,2}}{2E} R_{\nu h\star}, \quad (6.33)$$

with ((2.35)–(2.37) of Sect. 2.1.2)

$$R_{\nu h\star} = \frac{1+\nu}{9}$$
$$\times\left[\left\langle 2+\frac{\sigma}{\sigma_f^\infty}\right\rangle^2 + 2\left\langle-1+\frac{\sigma}{\sigma_f^\infty}\right\rangle^2 + h\left\langle-2-\frac{\sigma}{\sigma_f^\infty}\right\rangle^2 + 2h\left\langle 1-\frac{\sigma}{\sigma_f^\infty}\right\rangle^2\right]$$
$$-\nu\left\langle\frac{\sigma}{\sigma_f^\infty}\right\rangle^2 - \nu h\left\langle-\frac{\sigma}{\sigma_f^\infty}\right\rangle^2. \quad (6.34)$$

It also leads to the same expression as for the damage increment and the number of cycles to rupture as previously but with $R_{\nu\star}$ replaced by $R_{\nu h\star}$, i.e., with $\varphi_\star$ replaced by

6.2 Fast Calculation of Structural Failures 289

$$\varphi_{h\star}(\xi_{\min}, \xi_{\max}) = \int_{\xi_{\min}+2}^{\xi_{\max}} R^s_{\nu h\star}(\xi)\mathrm{d}\xi + \int_{\xi_{\min}}^{\xi_{\max}-2} R^s_{\nu h\star}(\xi)\mathrm{d}\xi, \qquad (6.35)$$

where $R^s_{\nu h\star}(\xi) = R^s_{\nu h\star}\left(\dfrac{\sigma}{\sigma_f^\infty} = \xi\right)$ or

$$\varphi_{h\star}(\xi_{\min}, \xi_{\max}) \approx \langle \xi_{\max} - \xi_{\min} - 2 \rangle [R^s_{\nu h\star}(\xi_{\min}) + R^s_{\nu h\star}(\xi_{\max})]. \qquad (6.36)$$

For tension compression ($\Sigma_{kk} = 1$, N_D unchanged),

$$\boxed{N_R = N_D + \dfrac{(2ES)^s \mathcal{G} D_c}{\sigma_f^{\infty\, 2s+1} \varphi_{h\star}\left(\dfrac{\sigma_{\min}}{\sigma_f^\infty}, \dfrac{\sigma_{\max}}{\sigma_f^\infty}\right)}.} \qquad (6.37)$$

6.2.4 Sensitivity Analysis♂♂

For ductile damage, low cycle fatigue, and creep failures (see Sects. 3.2.3, 4.2.3, and 5.2.3) the relative influence of the loading and the material parameters is calculated as an application of the general method of Sect. 2.4.6. Here this is the influence on the **number of cycles to rupture** in periodic proportional loading obtained in Sect. 6.2.3: Eq. (6.28) with the approximate formula (6.22) for $\varphi_\star(\xi_{\min}, \xi_{\max})$ gives

$$N_D = \dfrac{1}{4}\dfrac{\epsilon_{pD}}{C_y^m}\left(\dfrac{3E(1-\beta)}{2(1+\nu)}\right)^{m+1}\dfrac{(\sigma_u - \sigma_f^\infty)^m}{\left(\dfrac{\Delta\sigma}{2} - \sigma_f^\infty\right)^{m+1}} \qquad (6.38)$$

$$N_R = N_D + \dfrac{(2ES)^s 3E(1-\beta)D_c}{2(1+\nu)\sigma_f^{\infty\,2s}\left[\sigma_{\max} - \sigma_{\min} - \dfrac{2\sigma_f^\infty}{\Sigma_{kk}}\right][R^s_{\nu\,\min} + R^s_{\nu\,\max}]},$$

where the proportional loading is written as

$$\boldsymbol{\sigma} = \sigma\,\boldsymbol{\Sigma} \quad \text{with} \quad \Sigma_{eq} = 1 \quad \text{and} \quad T_X = \dfrac{1}{3}\Sigma_{kk} \qquad (6.39)$$

and where

$$\begin{aligned}
\Delta\sigma &= \sigma_{\max} - \sigma_{\min}, \\
\mathcal{G} &\approx 3G(1-\beta), \\
R_{\nu\,\min} &= R_{\nu\star}(\xi_{\min}), \\
R_{\nu\,\max} &= R_{\nu\star}(\xi_{\max}),
\end{aligned} \qquad (6.40)$$

with $\xi_{\min} = \Sigma_{kk}\dfrac{\sigma_{\min}}{\sigma_f^\infty}$, $\xi_{\max} = \Sigma_{kk}\dfrac{\sigma_{\max}}{\sigma_f^\infty}$.

Taking the logarithmic derivative gives $\dfrac{\mathrm{d}N_D}{N_D}$ and $\dfrac{\mathrm{d}(N_R - N_D)}{N_R - N_D}$.

Taking $\delta \mathcal{A}_k = |\mathrm{d}\,\mathcal{A}_k|$ for each material parameter $\mathcal{A}_k$, the sensitivity on the number of cycles to rupture is $\dfrac{\delta N_\mathrm{R}}{N_\mathrm{R}} = \sum_k S_{\mathcal{A}k}^{N_\mathrm{R}} \dfrac{\delta \mathcal{A}_k}{\mathcal{A}_k}$, where the coefficients $S_{\mathcal{A}k}^{N_\mathrm{R}}$ are the sensitivity coefficients of the parameters $\mathcal{A}_k$ on the number of cycles to rupture N_R:

$$S_{T_\mathrm{X}}^{N_\mathrm{R}} = S_{\Sigma_{kk}}^{N_\mathrm{R}} = \frac{N_\mathrm{R} - N_\mathrm{D}}{N_\mathrm{R}} \left[\frac{\dfrac{2\sigma_\mathrm{f}^\infty}{\Sigma_{kk}}}{\Delta\sigma - \dfrac{2\sigma_\mathrm{f}^\infty}{\Sigma_{kk}}} + \frac{2s}{3}(1-2\nu)\frac{\xi_\mathrm{min}^2 R_{\nu\,\mathrm{min}}^{s-1} + \xi_\mathrm{max}^2 R_{\nu\,\mathrm{max}}^{s-1}}{R_{\nu\,\mathrm{min}}^s + R_{\nu\,\mathrm{max}}^s} \right],$$

$$S_{\sigma_\mathrm{min}}^{N_\mathrm{R}} = \left| -\frac{N_\mathrm{D}}{N_\mathrm{R}} \frac{(m+1)\sigma_\mathrm{min}}{\Delta\sigma - 2\sigma_\mathrm{f}^\infty} + \frac{N_\mathrm{R} - N_\mathrm{D}}{N_\mathrm{R}} \left[\frac{-\sigma_\mathrm{min}}{\Delta\sigma - \dfrac{2\sigma_\mathrm{f}^\infty}{\Sigma_{kk}}} + \frac{2s}{3}\frac{(1-2\nu)\xi_\mathrm{min}^2 R_{\nu\,\mathrm{min}}^{s-1}}{R_{\nu\,\mathrm{min}}^s + R_{\nu\,\mathrm{max}}^s} \right] \right|,$$

$$S_{\sigma_\mathrm{max}}^{N_\mathrm{R}} = \left| \frac{N_\mathrm{D}}{N_\mathrm{R}} \frac{(m+1)\sigma_\mathrm{max}}{\Delta\sigma - 2\sigma_\mathrm{f}^\infty} + \frac{N_\mathrm{R} - N_\mathrm{D}}{N_\mathrm{R}} \left[\frac{\sigma_\mathrm{max}}{\Delta\sigma - \dfrac{2\sigma_\mathrm{f}^\infty}{\Sigma_{kk}}} + \frac{2s}{3}\frac{(1-2\nu)\xi_\mathrm{max}^2 R_{\nu\,\mathrm{max}}^{s-1}}{R_{\nu\,\mathrm{min}}^s + R_{\nu\,\mathrm{max}}^s} \right] \right|,$$

$$S_E^{N_\mathrm{R}} = \frac{N_\mathrm{D}}{N_\mathrm{R}}(m+1) + \frac{N_\mathrm{R} - N_\mathrm{D}}{N_\mathrm{R}}(s+1),$$

$$S_\nu^{N_\mathrm{R}} = \left| \left(1 + \frac{N_\mathrm{D}}{N_\mathrm{R}}m\right)\frac{\nu}{1+\nu} + \frac{N_\mathrm{R} - N_\mathrm{D}}{N_\mathrm{R}} \frac{2s\nu}{3} \frac{(1-\xi_\mathrm{min}^2)R_{\nu\,\mathrm{min}}^{s-1} + (1-\xi_\mathrm{max}^2)R_{\nu\,\mathrm{max}}^{s-1}}{R_{\nu\,\mathrm{min}}^s + R_{\nu\,\mathrm{max}}^s} \right|,$$

$$S_\beta^{N_\mathrm{R}} = \left(1 + \frac{N_\mathrm{D}}{N_\mathrm{R}}m\right)\frac{\beta}{1-\beta},$$

$$S_{C_\mathrm{y}}^{N_\mathrm{R}} = \frac{N_\mathrm{D}}{N_\mathrm{R}} m,$$

$$S_{\epsilon_\mathrm{pD}^{t_\mathrm{R}}}^{N_\mathrm{R}} = \frac{N_\mathrm{D}}{N_\mathrm{R}},$$

$$S_{\sigma_\mathrm{f}^\infty}^{N_\mathrm{R}} = \frac{N_\mathrm{D}}{N_\mathrm{R}} \left[\frac{(m+1)\sigma_\mathrm{f}^\infty}{\dfrac{\Delta\sigma}{2} - \sigma_\mathrm{f}^\infty} - \frac{m\sigma_\mathrm{f}^\infty}{\sigma_\mathrm{u} - \sigma_\mathrm{f}^\infty} \right] + 2\frac{N_\mathrm{R} - N_\mathrm{D}}{N_\mathrm{R}}$$
$$\times \left[\frac{(1-2\nu)s}{3} \frac{\xi_\mathrm{min}^2 R_{\nu\,\mathrm{min}}^{s-1} + \xi_\mathrm{max}^2 R_{\nu\,\mathrm{max}}^{s-1}}{R_{\nu\,\mathrm{min}}^s + R_{\nu\,\mathrm{max}}^s} - s + \frac{\sigma_\mathrm{f}^\infty}{\Sigma_{kk}\Delta\sigma - 2\sigma_\mathrm{f}^\infty} \right],$$

$$S_{\sigma_\mathrm{u}}^{N_\mathrm{R}} = \frac{N_\mathrm{D}}{N_\mathrm{R}} \frac{m\sigma_\mathrm{u}}{\sigma_\mathrm{u} - \sigma_\mathrm{f}^\infty},$$

$$S_m^{N_\mathrm{R}} = \frac{N_\mathrm{D}}{N_\mathrm{R}} m \ln\left(\frac{\mathcal{G}}{C_\mathrm{y}} \frac{\sigma_\mathrm{u} - \sigma_\mathrm{f}^\infty}{\dfrac{\Delta\sigma}{2} - \sigma_\mathrm{f}^\infty} \right),$$

$$S_S^{N_\mathrm{R}} = \frac{N_\mathrm{R} - N_\mathrm{D}}{N_\mathrm{R}} s,$$

$$S_s^{N_\mathrm{R}} = \frac{N_\mathrm{R} - N_\mathrm{D}}{N_\mathrm{R}} s \left| \ln \frac{\sigma_\mathrm{f}^{\infty\, 2}}{2ES} + \frac{R_{\nu\,\min}^s \ln R_{\nu\,\min} + R_{\nu\,\max}^s \ln R_{\nu\,\max}}{R_{\nu\,\min}^s + R_{\nu\,\max}^s} \right|,$$

$$S_{D_\mathrm{c}}^{N_\mathrm{R}} = \frac{N_\mathrm{R} - N_\mathrm{D}}{N_\mathrm{R}}.$$

As in the other chapters, the values of those coefficients are represented by the height of the boxes in Fig. 6.6 for the following set of parameters chosen as an average of many materials and identical for the mesoscopic parameters to those used for ductile and low cycle fatigue failures: $\sigma_{\max} = 250$ MPa, $\sigma_{\min} = -250$ MPa, $T_\mathrm{X} = 3\Sigma_{kk} = 1$, $R_{\nu\,\max} = R_{\nu\,\min} = 2.75$, $E = 200000$ MPa, $\nu = 0.3$, $\sigma_\mathrm{f}^\infty = 200$ MPa, $\sigma_\mathrm{u}/\sigma_\mathrm{f}^\infty = 2$, $C_\mathrm{y} = E/100$, $\beta = \frac{2}{15}(4 - 5\nu)/(1 - \nu)$, $\mathcal{G} \approx 3G(1 - \beta) = 121000$ MPa, $N_\mathrm{D}/N_\mathrm{R} = 1/2$, $m = 2$, $S = 2$ MPa, and $s = 5$.

For ductile, low cycle fatigue, and creep failures, the quality of the result, i.e., the number of cycles to rupture in this case, is mainly influenced by the accuracy of the stress, here twice 5.8 mitlplied by its relative error. Using the two-scale damage model gives more emphasis to the exponents m and s and increases the uncertainties by a larger number of parameters involved. Only ϵ_pD and D_c are of less importance as in the previous chapters.

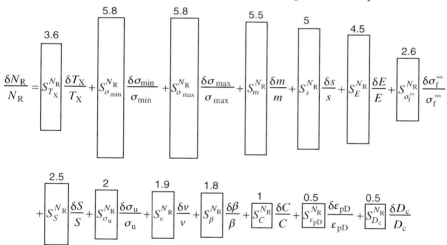

Fig. 6.6. Relative importance of each parameter in high cycle fatigue

If the sensitivity in terms of stress range $\Delta\sigma = \sigma_{\max} - \sigma_{\min}$ and mean stress $\bar{\sigma} = (\sigma_{\max} + \sigma_{\min})/2$ are preferred to the sensitivity in terms of $\sigma_{\min}$ and $\sigma_{\max}$, calculate, then

$$S_{\bar{\sigma}}^{N_\mathrm{R}} = \left| \frac{\bar{\sigma}}{\sigma_{\min}} S_{\sigma_{\min}}^{N_\mathrm{R}} + \frac{\bar{\sigma}}{\sigma_{\max}} S_{\sigma_{\max}}^{N_\mathrm{R}} \right|$$

$$S_{\Delta\sigma}^{N_\mathrm{R}} = \frac{1}{2} \left| \frac{\Delta\sigma}{\sigma_{\max}} S_{\sigma_{\max}}^{N_\mathrm{R}} - \frac{\Delta\sigma}{\sigma_{\min}} S_{\sigma_{\min}}^{N_\mathrm{R}} \right|.$$

(6.41)

For the previous set of material parameters, $S_{\overline{\sigma}}^{N_R} = 0$ as $\overline{\sigma} = 0$ and $S_{\Delta\sigma}^{N_R} = 11.6$ here.

6.2.5 Safety Margin and Crack Growth⚙⚙

The proper parameter for the evaluation of the risk of fatigue failure is the **number of cycles**, as already mentioned in Sect. 4.2.5 for low cycle fatigue.

- If the number of cycles of loading service N_{service} is known, a safe design is defined from the calculated number of cycles to rupture N_R for the in service history of loading, and a safety factor Saf given by rules or by the state of the art of the domain of application Saf $= 100, 10, 5, \ldots$:

$$\boxed{N_{\text{service}} < \frac{N_R}{\text{Saf}}.} \qquad (6.42)$$

- Due to a large scatter in high cycle fatigue, a better definition of the safety margin uses the probability concept. It may include the uncertainties of the loading and the material properties but it needs information on their stochastic character. Instead of the previous equation, consider as safety design criterion:

$$\boxed{\text{Prob}\,(N_R > N_{\text{service}}) < \frac{1}{\text{Saf}_{\text{Pr}}},} \qquad (6.43)$$

where $\frac{1}{\text{Saf}_{\text{Pr}}}$ must be very small ($\text{Saf}_{\text{Pr}} = 10^2, 10^3, \ldots 10^5$ or more). This means a stochastic analysis is difficult to perform as the accuracy is always poor for the low values of the probability density (nevertheless see Sect. 6.4.4).

Finally, if a mesocrack initiates, it is important to check if it is dangerous regarding a possible fast propagation by instability or fatigue growth.

- The criterion for a fracture by instability is when the structure strain energy release rate G reaches the toughness of the material G_c. G is calculated from the fracture mechanics concepts as a function of the far field loading $\sigma_\infty > 0$ and the length of the mesocrack initiated δ_0 stated in Sect. 1.6.3.

$$\delta_0 = \frac{G_c}{\frac{\sigma_u^2}{2E} D_c + \sigma_u \epsilon_{\text{pR}}}. \qquad (6.44)$$

Classically, $G \approx \kappa \frac{\sigma_\infty^2 \pi \delta_0}{E}$ where κ is a shape factor, so one has to check $G < G_c$ or

$$\kappa \frac{\sigma_\infty^2 \pi}{E} \frac{G_c}{\frac{\sigma_u^2}{2E} D_c + \sigma_u \epsilon_{\text{pR}}} < G_c. \qquad (6.45)$$

The toughness G_c disappears and

$$\sigma_\infty < \sigma_u \sqrt{\frac{1}{\pi\kappa}\left(\frac{D_c}{2} + \frac{E\epsilon_{pR}}{\sigma_u}\right)}. \tag{6.46}$$

Due to the numerical value of the material parameters, this criterion is satisfied most of the time, at least for metals.

- But the crack may continue to grow under the same fatigue loading. The simplest model to evaluate the fatigue crack growth rate is the **generalized Paris law** written for a crack of area A in a 3D medium:

$$\frac{\delta A}{\delta N} = \frac{G_{\max}^{\eta_P/2} - G_{\min}^{\eta_P/2}}{C_P^{\eta_P/2}}, \tag{6.47}$$

where $G_{\max} \approx \kappa\sigma_{\infty\max}^2 \frac{\pi A^{1/2}}{E}$, $G_{\min} \approx \kappa\sigma_{\infty\min}^2 \frac{\pi A^{1/2}}{E}$, and where η_P and C_P are material-dependent parameters determined by experiments and are different from the parameters of the Paris law written for the crack length in 2D problems.

If the loading is periodic, $\sigma_{\max} = const$, $\sigma_{\min} = const$, and in order to obtain the evolution of the crack surface A as a function of the number of cycles N one has to solve ($\eta_P \neq 4$)

$$\frac{\delta A}{\delta N} = \kappa^{\eta_P/2} \frac{\sigma_{\infty\max}^{\eta_P} - \sigma_{\infty\min}^{\eta_P}}{C_P^{\eta_P/2}} \left(\frac{\pi}{E}\right)^{\eta_P/2} A^{\eta_P/4}, \tag{6.48}$$

with the initial condition $N = 0 \to A = A_0 \approx \delta_0^2$,

$$\int_{A_0}^{A} A^{-\frac{\eta_P}{4}} \delta A = \kappa^{\eta_P/2} \frac{\sigma_{\infty\max}^{\eta_P} - \sigma_{\infty\min}^{\eta_P}}{C_P^{\eta_P/2}} \left(\frac{\pi}{E}\right)^{\eta_P/2} N, \tag{6.49}$$

$$\boxed{A = \left[\delta_0^{\frac{4-\eta_P}{2}} + \left(\frac{4-\eta_P}{4}\right) \kappa^{\eta_P/2} \frac{\sigma_{\infty\max}^{\eta_P} - \sigma_{\infty\min}^{\eta_P}}{C_P^{\eta_P/2}} \left(\frac{\pi}{E}\right)^{\eta_P/2} N\right]^{\frac{4}{4-\eta_P}}.} \tag{6.50}$$

One may also calculate the number of cycles $N^\star$ to reach the complete failure by instability:

$$G_{\max} = G_c \quad \text{or} \quad \kappa\frac{\sigma_{\infty\max}^2 \pi A^{\star 1/2}}{E} = G_c. \tag{6.51}$$

That is

$$A^\star = \left(\frac{EG_c}{\kappa\pi\sigma_{\infty\max}^2}\right)^2, \tag{6.52}$$

$$N^\star = \left(\frac{EG_c}{\kappa\pi\sigma_{\infty\max}^2}\right)^{\frac{4-\eta_P}{2}} \frac{1}{\delta_0^{\frac{4-\eta_P}{2}} + \left(\frac{4-\eta_P}{4}\right) \kappa^{\eta_P/2} \frac{\sigma_{\infty\max}^{\eta_P} - \sigma_{\infty\min}^{\eta_P}}{C_P^{\eta_P/2}} \left(\frac{\pi}{E}\right)^{\eta_P/2}}. \tag{6.53}$$

6.3 Basic Engineering Examples

High-cycle fatigue occurs for mesostresses below or close to the yield stress so that they can be calculated in elasticity. At least something simple! For many simple geometries, the stress concentration coefficient at the critical points K_T may be found in handbooks (see bibliography). In the following, K_T is supposed to be known.

6.3.1 Plates or Members with Holes or Notches

The weakness comes from a sharp variation of the geometry where the maximum stress is at the surface border, uniaxial, and related to the normal stress σ_n by

$$\sigma = K_T \sigma_n \tag{6.54}$$

or for cyclic loading

$$\Delta\sigma = K_T \Delta\sigma_n . \tag{6.55}$$

- For periodic loading at zero mean stress, one may use the solution given by the two-scale damage model in proportional loading with $\Sigma_{kk} = 1$ (eq. (6.28) with the approximate formula (6.22)),

$$N_R = N_D + \frac{(2ES)^s \mathcal{G} D_c}{2\sigma_f^{\infty\,2s} (K_T \Delta\sigma - 2\sigma_f^\infty) R_{\nu\,\text{max}}^s}$$

$$N_D = \frac{1}{4}\epsilon_{pD} \frac{\mathcal{G}^{m+1}}{C_y^m} \frac{(\sigma_u - \sigma_f^\infty)^m}{\left(K_T \frac{\Delta\sigma}{2} - \sigma_f^\infty\right)^{m+1}} , \tag{6.56}$$

where $\mathcal{G} \approx 3G(1-\beta)$ and

$$R_{\nu\,\text{max}} = R_{\nu\star}(\xi_{\text{max}}) = \frac{2}{3}(1+\nu) + \frac{1}{3}(1-2\nu)\left[\frac{K_T \sigma_{\text{max}}}{\sigma_f^\infty}\right]^2 . \tag{6.57}$$

This is nothing more than directly using the experimental Wöhler curve at zero mean stress for the stress range $K_T \Delta\sigma$.
- But if the mean stress $\overline{\sigma}$ is not zero it helps to derive the number of cycles to rupture $N_R(\overline{\sigma})$ from the Wöhler curve known at $\overline{\sigma} = 0$:

$$N_R(\Delta\sigma, \overline{\sigma}) = N_D(\Delta\sigma) + \frac{(2ES)^s \mathcal{G} D_c}{\sigma_f^{\infty\,2s} (K_T \Delta\sigma - 2\sigma_f^\infty)(R_{\nu\,\text{min}}^s + R_{\nu\,\text{max}}^s)} , \tag{6.58}$$

with $\dfrac{(2ES)^s \mathcal{G} D_c}{\sigma_f^{\infty\,2s}(K_T \Delta\sigma - 2\sigma_f^\infty)} = 2\left[N_R(\Delta\sigma, \overline{\sigma}=0) - N_D(\Delta\sigma)\right] R_{\nu\,\text{max}}^s .$

Note that the number of cycles to damage initiation N_D does not depend on the mean stress. Then,

$$\boxed{\frac{N_R(\Delta\sigma, \overline{\sigma}) - N_D(\Delta\sigma)}{N_R(\Delta\sigma, \overline{\sigma}=0) - N_D(\Delta\sigma)} = \frac{2R_{\nu\,\max}^s}{R_{\nu\,\min}^s + R_{\nu\,\max}^s}.} \qquad (6.59)$$

The ratio of the triaxiality functions represents the effect of mean stress already described by the Goodman diagram. For a better description of the mean stress effect, replace $R_{\nu\star}$ by $R_{\nu\star h}$. Figure 6.7 gives $R_{\nu\star}$ and $R_{\nu\star h}$ versus $K_T\sigma/\sigma_f^\infty$ (recall that Eq. (6.34) and therefore the curve for $R_{\nu\star h}$ apply only to the tension-compression case). To help a bit more, take $N_D = N_R/2$ as a rough approximation.
- If the cyclic loading is piecewise periodic with a number of cycles $n_i(\Delta\sigma_i, \overline{\sigma}_i)$ for the i-th sequence of loading of stress range $\Delta\sigma_i$ and of mean stress $\overline{\sigma}_i$, $\sum n_i = N_R$, use a cumulation equation similar to eq. (6.29).

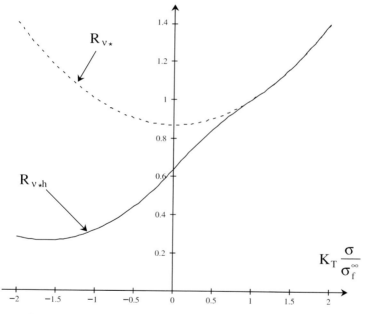

Fig. 6.7. Effect of mean stress through the triaxiality function $R_{\nu\star}$ or $R_{\nu\star h}$ for $\nu = 0.3$ and $h = 0.2$

6.3.2 Pressurized Shallow Cylinders

This is the same problem as in Sect. 4.3.2, and Fig. 4.7, except the damage equation (here damage acts on a microscale). The state of stress, function of the radius of the cylinder R_{cyl}, its thickness t_{cyl}, and the internal relative pressure $P(t)$ varying between $P_{\min}$ and $P_{\max}$ is

$$\boldsymbol{\sigma} = \begin{bmatrix} 0 & 0 & 0 \\ 0 & \dfrac{PR_{\text{cyl}}}{t_{\text{cyl}}} & 0 \\ 0 & 0 & \dfrac{PR_{\text{cyl}}}{2t_{\text{cyl}}} \end{bmatrix} \tag{6.60}$$

or $\boldsymbol{\sigma} = \dfrac{\sqrt{3}}{2}\dfrac{PR_{\text{cyl}}}{t_{\text{cyl}}}\boldsymbol{\Sigma}$ with $\boldsymbol{\Sigma} = \begin{bmatrix} 0 & 0 & 0 \\ 0 & \frac{2}{\sqrt{3}} & 0 \\ 0 & 0 & \frac{1}{\sqrt{3}} \end{bmatrix}$ in order to have $\Sigma_{\text{eq}} = 1$.

Then $\Sigma_{kk} = \frac{3}{\sqrt{3}} = \sqrt{3}$.

Now it is the same problem as in the previous section (Sect. 6.3.1) as well. For periodic loading, $\sigma_{\min} = \frac{\sqrt{3}}{2} P_{\min} \dfrac{R_{\text{cyl}}}{t_{\text{cyl}}}$, $\sigma_{\max} = \frac{\sqrt{3}}{2} P_{\max} \dfrac{R_{\text{cyl}}}{t_{\text{cyl}}}$, $\Delta\sigma = \frac{\sqrt{3}}{2}(P_{\max} - P_{\min})\dfrac{R_{\text{cyl}}}{t_{\text{cyl}}}$,

$$N_{\text{R}} = N_{\text{D}} + \dfrac{(2ES)^s \mathcal{G} D_{\text{c}}}{\sigma_{\text{f}}^{\infty\,2s}\left(\Delta\sigma - \dfrac{2\sigma_{\text{f}}^{\infty}}{\Sigma_{kk}}\right)\left[R_{\nu\star}^s\!\left(\Sigma_{kk}\dfrac{\sigma_{\min}}{\sigma_{\text{f}}^{\infty}}\right) + R_{\nu\star}^s\!\left(\Sigma_{kk}\dfrac{\sigma_{\max}}{\sigma_{\text{f}}^{\infty}}\right)\right]},$$

$$N_{\text{D}} = \dfrac{1}{4}\epsilon_{\text{pD}}\dfrac{\mathcal{G}^{m+1}}{C_{\text{y}}^m}\dfrac{(\sigma_{\text{u}} - \sigma_{\text{f}}^{\infty})^m}{\left(\dfrac{\Delta\sigma}{2} - \sigma_{\text{f}}^{\infty}\right)^{m+1}}.$$

(6.61)

The rupture condition can also be expressed as a function of the number of cycles to rupture read on the Wöhler curve at zero mean stress using Fig. 6.7:

$$\boxed{\dfrac{N_{\text{R}}(\Delta\sigma,\bar{\sigma}) - N_{\text{D}}(\Delta\sigma)}{N_{\text{R}}(\Delta\sigma,\bar{\sigma}=0) - N_{\text{D}}(\Delta\sigma)} = \dfrac{2R_{\nu\star}^s\!\left(\dfrac{\sqrt{3}\Delta\sigma}{2\sigma_{\text{f}}^{\infty}}\right)}{R_{\nu\star}^s\!\left(\sqrt{3}\dfrac{\sigma_{\min}}{\sigma_{\text{f}}^{\infty}}\right) + R_{\nu\star}^s\!\left(\sqrt{3}\dfrac{\sigma_{\max}}{\sigma_{\text{f}}^{\infty}}\right)}.} \tag{6.62}$$

For two-level fatigue loading, apply eq. (6.29). For cyclic loading, stepwise periodic, perform as in Sect. 4.4.2. If damage initiates during the first level, then

$$D = (n_1 - N_{\text{D}})\dfrac{\delta D}{\delta N}^{(1)} + n_2 \dfrac{\delta D}{\delta N}^{(2)} + \ldots + n_k \dfrac{\delta D}{\delta N}^{(k)} = D_{\text{c}} \;\rightarrow\; N_{\text{R}} = \sum_{i=1}^{k} n_i, \tag{6.63}$$

with the damage increment per cycle on a microscale given by (6.23).

6.3.3 Bending of Beams

This is the same problem as in Sect. 4.3.3 (Fig. 4.9) except that the beam remains elastic, within domain of application of the two-scale damage model.

The simple Bernoulli hypothesis applied to pure circular bending of a part of beam of height h, of inertia moment I gives the maximum elastic stress as a function of the momentum applied M ($I = bh^3/12$ for a rectangular cross section of width b),

$$\sigma_{\max} = \frac{M_{\max} h}{I}. \tag{6.64}$$

Then for a periodic movement of the beam described by $M = M_{\max} \sin \omega t$, the problem of fatigue reduces to a simple tension compression case between $\sigma_{\max}$ and $\sigma_{\min} = -\sigma_{\max}$ at the two surfaces of the beam where the stress is maximum. Equations (6.56) directly apply with $K_T = 1$.

For a two-level piece wise periodic movement, the equations in (6.29) also apply. For multilevel high cycle fatigue loading, use eq. (6.63).

6.3.4 Random Loadings

Sometimes the problem to solve is of a random nature known only by stochastic properties. This is the case of the wheel suspensions of a car driven on a rough road. This is the case of planes flying in turbulent clouds. This is also the case of any mechanical component for which the in service loading is not precisely known.

The only way to find statistics on the number of cycles to rupture by the two-scale damage model in which the loading is a random variable is to use the numerical method of Monte Carlo, as explained in Sect. 6.4.4. Nevertheless some indications may be obtained in a closed-form if a very simple damage model is used.

Consider here the very crude model restricted to loadings at zero mean value,

$$\frac{\delta D}{\delta N} = \left\langle \frac{\sigma_M^\star - \sigma_f^\infty}{\sigma_u - \sigma_f^\infty} \right\rangle^c \quad \text{and} \quad D = D_c \to N = N_R, \tag{6.65}$$

where $\frac{\delta D}{\delta N}$ is the damage increment per cycle of maximum damage equivalent stress $\sigma_M^\star = (\sigma_{eq} R_\nu^{1/2})_{\max}$ and of range $\Delta \sigma^\star = 2\sigma_M^\star$, σ_f^∞ is the asymptotic fatigue limit, σ_u is the ultimate stress, and c is also a material parameter. The number of cycles to rupture N_R for a periodic loading $\sigma_M^\star = const$ is

$$\int_0^{D_c} \delta D = \left\langle \frac{\sigma_M^\star - \sigma_f^\infty}{\sigma_u - \sigma_f^\infty} \right\rangle^c \int_0^{N_R} \delta N \tag{6.66}$$

$$N_R = \left\langle \frac{\sigma_M^\star - \sigma_f^\infty}{\sigma_u - \sigma_f^\infty} \right\rangle^{-c} D_c \quad \text{and} \quad \delta D = \frac{\delta N}{N_R} D_c. \tag{6.67}$$

Consider now a random proportional loading $\boldsymbol{\sigma} = \sigma_\Sigma(t)\boldsymbol{\Sigma}$ (with $\Sigma_{\text{eq}} = 1$) where the signed damage equivalent stress $\sigma^\star = \sigma_\Sigma(t)R_\nu^{1/2}$ is a Gaussian process given by its probability density $P(\sigma^\star)$ which fits many applications with large numbers of random loading parameters,

$$P(\sigma^\star) = \frac{1}{\bar{\bar{\sigma}}^\star\sqrt{2\pi}} \exp-\frac{\sigma^{\star\,2}}{2\bar{\bar{\sigma}}^{\star\,2}}, \qquad (6.68)$$

where $\bar{\bar{\sigma}}^\star$ is the standard deviation associated with $\sigma^\star$ and $\sigma^{\star\,2} = \sigma_\Sigma^2 R_\nu$. The stress triaxiality function is constant equal to $R_\nu = \frac{2}{3}(1+\nu)+3(1-2\nu)\left(\frac{\Sigma_{kk}}{3}\right)^2$.

For a narrow band process, the probability density of the maximum values of the stress $\sigma^\star$ is given by

$$P(\sigma_M^\star) = -\frac{1}{P(0)}\frac{\partial P(\sigma^\star)}{\partial \sigma^\star}(\sigma^\star = \sigma_M^\star), \qquad (6.69)$$

which is a Rayleigh's law:

$$P(\sigma_M^\star) = \frac{\sigma_M^\star}{\bar{\bar{\sigma}}^{\star\,2}}\exp-\frac{\sigma_M^{\star\,2}}{2\bar{\bar{\sigma}}^{\star\,2}}. \qquad (6.70)$$

It will give rise to rupture only if the standard deviation $\bar{\bar{\sigma}}^\star$ is large to ensure enough stresses $\sigma_M^\star$ above the asymptotic fatigue limit.

Within a set of N cycles the number of cycles $\mathrm{d}N$ for which $\sigma_M^\star \leqslant \sigma_M^\star \leqslant \sigma_M^\star + \mathrm{d}\sigma_M^\star$ is $\mathrm{d}N = NP(\sigma_M^\star)\mathrm{d}\sigma_M^\star$ or

$$\mathrm{d}N = N\frac{\sigma_M^\star}{\bar{\bar{\sigma}}^{\star\,2}}\exp-\frac{\sigma_M^{\star\,2}}{2\bar{\bar{\sigma}}^{\star\,2}}\mathrm{d}\sigma_M^\star, \qquad (6.71)$$

but the mean damage $\overline{D}$ at the cycle N is

$$\overline{D} = \int_0^\infty D_{\mathrm{c}}\frac{\mathrm{d}N}{N_{\mathrm{R}}} = \int_0^\infty \left\langle\frac{\sigma_M^\star - \sigma_{\mathrm{f}}^\infty}{\sigma_{\mathrm{u}} - \sigma_{\mathrm{f}}^\infty}\right\rangle^c \mathrm{d}N \qquad (6.72)$$

or

$$\overline{D} = \frac{N}{(\sigma_{\mathrm{u}} - \sigma_{\mathrm{f}}^\infty)^c}\int_{\sigma_{\mathrm{f}}^\infty}^\infty (\sigma_M^\star - \sigma_{\mathrm{f}}^\infty)^c \cdot \frac{\sigma_M^\star}{\bar{\bar{\sigma}}^{\star\,2}}\exp-\frac{\sigma_M^{\star\,2}}{2\bar{\bar{\sigma}}^{\star\,2}}\mathrm{d}\sigma_M^\star, \qquad (6.73)$$

or with $x = \dfrac{\bar{\bar{\sigma}}^\star}{\sigma_M^\star}$ and $J = \int_0^{\frac{\bar{\bar{\sigma}}^\star}{\sigma_{\mathrm{f}}^\infty}}\left(\dfrac{\bar{\bar{\sigma}}^\star}{\sigma_{\mathrm{f}}^\infty}\dfrac{1}{x} - 1\right)^c\dfrac{1}{x^3}\exp-\dfrac{1}{2x^2}\mathrm{d}x$:

$$\overline{D} = \frac{N\sigma_{\mathrm{f}}^{\infty\,c}}{(\sigma_{\mathrm{u}} - \sigma_{\mathrm{f}}^\infty)^c}J. \qquad (6.74)$$

The integral $J = J\left(c, \dfrac{\bar{\bar{\sigma}}^\star}{\sigma_{\mathrm{f}}^\infty}\right)$ can be numerically integrated by mathematical software as a function of the Wöhler exponent c and the ratio $\bar{\bar{\sigma}}^\star/\sigma_{\mathrm{f}}^\infty$ of the standard deviation $\bar{\bar{\sigma}}^\star$ of the loading process to the fatigue limit $\sigma_{\mathrm{f}}^\infty$.

Finally, the number of cycles $\overline{N}_{\mathrm{R}}$ which corresponds to the rupture for the mean value of the damage $\overline{D} = D_{\mathrm{c}}$ is

$$\overline{N}_{\mathrm{R}} = \left(\frac{\sigma_{\mathrm{u}} - \sigma_{\mathrm{f}}^{\infty}}{\sigma_{\mathrm{f}}^{\infty}}\right)^{c} \frac{D_{\mathrm{c}}}{J}. \qquad (6.75)$$

It is interesting to calculate the value of the maximum stress $\sigma^{\star}_{\mathrm{M\,equiv}}$ for a constant amplitude process which would give the same number of cycles to rupture,

$$\left\langle \frac{\sigma^{\star}_{\mathrm{M\,equiv}} - \sigma_{\mathrm{f}}^{\infty}}{\sigma_{\mathrm{u}} - \sigma_{\mathrm{f}}^{\infty}} \right\rangle^{-c} D_{\mathrm{c}} = \left(\frac{\sigma_{\mathrm{u}} - \sigma_{\mathrm{f}}^{\infty}}{\sigma_{\mathrm{f}}^{\infty}}\right)^{c} \frac{D_{\mathrm{c}}}{J}, \qquad (6.76)$$

$$\sigma^{\star}_{\mathrm{M\,equiv}} = \sigma_{\mathrm{f}}^{\infty} \left\{ 1 + \left[J\left(\frac{\bar{\bar{\sigma}}^{\star}}{\sigma_{\mathrm{f}}^{\infty}}\right) \right]^{1/c} \right\}. \qquad (6.77)$$

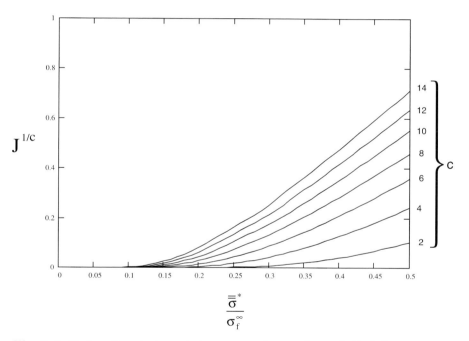

Fig. 6.8. Factor of equivalence between a random and a periodic fatigue process

The graph of Fig. 6.8 gives the value of $J^{1/c}$ as a function of the ratio $\bar{\bar{\sigma}}^{\star}/\sigma_{\mathrm{f}}^{\infty}$ for different c. The curves are obtained for a zero mean stress with a maximum stress range directly related to the value of the standard deviation $\bar{\bar{\sigma}}^{\star}$. When it is too low ($\bar{\bar{\sigma}}^{\star}/\sigma_{\mathrm{f}}^{\infty} <\approx 0.1$ to 0.3), there is no rupture. It is interesting to note that the scatter correction factor $J^{1/c}$ varies only from 0 to less than 1 in a large range of both the ratio $\bar{\bar{\sigma}}^{\star}/\sigma_{\mathrm{f}}^{\infty}$ and the Wöhler curve nonlinearity parameter c.

Then, knowing the standard deviation $\bar{\bar{\sigma}}^\star$ of a Gaussian process for the signed damage equivalent stress $\sigma^\star = \sigma_\Sigma R_\nu^{1/2}$, it is easy to calculate the maximum stress $\sigma^\star_{M\,equiv}$ of an equivalent periodic process (the exponent c always fits a Wöhler curve in the large number of cycles to rupture). Report this equivalent stress directly on the Wöhler curve in order to obtain the number of cycles to rupture corresponding to the random process. Nice, isn't it?

Of course more accurate results may be obtained with a better fatigue model but the price to pay is numerical calculations with the Monte Carlo method.

6.4 Numerical Failure Analysis

As pointed out many times, high cycle fatigue occurs in the elastic range. The nice thing is that the computations needed for design purposes are performed in elasticity. For most materials, elasticity linearity allows for the superposition of solutions: if the loading is proportional (see Sect. 2.1.2), i.e., if all applied loads and displacements are proportional to the same scalar function of time $\alpha(t)$, perform only one computation for the reference loading α_{ref}. Store the stress and strain fields obtained ($\boldsymbol{\sigma}_{\mathrm{ref}}(M)$ and $\boldsymbol{\epsilon}_{\mathrm{ref}}(M)$) and have in mind that there is no need for additional computation to get the stress and strain fields at any time t as

$$\boldsymbol{\sigma}(M,t) = \frac{\alpha(t)}{\alpha_{\mathrm{ref}}}\boldsymbol{\sigma}_{\mathrm{ref}}(M) \quad \text{and} \quad \boldsymbol{\epsilon}(M,t) = \frac{\alpha(t)}{\alpha_{\mathrm{ref}}}\boldsymbol{\epsilon}_{\mathrm{ref}}(M). \quad (6.78)$$

If a few applied loads or displacements vary independently, compute the reference elastic solutions for each load or displacement. The stress and strain fields history are obtained by superposition, yielding

$$\boldsymbol{\sigma}(M,t) = \sum_i \frac{\alpha_i(t)}{\alpha_{\mathrm{ref}\,i}}\boldsymbol{\sigma}_{\mathrm{ref}\,i}(M) \quad \text{and} \quad \boldsymbol{\epsilon}(M,t) = \sum_i \frac{\alpha_i(t)}{\alpha_{\mathrm{ref}\,i}}\boldsymbol{\epsilon}_{\mathrm{ref}\,i}(M). \quad (6.79)$$

These elastic fields are then the inputs of a fatigue structure analysis, either to check the safety by use of fatigue limit criteria or to estimate the number of cycles to failure by a damage analysis.

The bad thing is that the straightforward use of the Wöhler curve to read the number of cycles to rupture (or at least to mesocrack initiation) is not always an easy task for real structures. The use of a damage post-processor (such as DAMAGE 2000 described in Sect. 2.3.3) is often necessary. The calculations are made at the most loaded point only or at the structure's Gauss points to draw damage maps.

6.4.1 Effects of Loading History♂♂
(J.P. Sermage 1999)

Most of the characteristic effects of high cycle fatigue described in Sect. 6.2.1 may be predicted by the **unified damage law within the two-scale dam-**

age model of Sect. 6.2.3. Some numerical results obtained with the post-processor DAMAGE 2000 are shown in the following for the values of the material parameters characteristic of a steel at room temperature: $E = 200000$ MPa, $\nu = 0.3$, $\sigma_f^\infty = 200$ MPa, $C_y = 2000$ MPa, $\sigma_u = 600$ MPa, $\epsilon_{pD} = 0.05$, $m = 1$, $S = 0.3$ MPa, $s = 2$, $h = 0.2$, $D_c = 1$.

6.4.1.1 Influence of the Mean Stress

The consideration of the microdefects closure parameter h in the model allows us to obtain the influence of the mean stress in compression. This difference of behavior between tension and compression leads to larger numbers of cycles to rupture in the same range as that for the decreasing mean stress $\bar{\sigma}$ (Fig. 6.9).

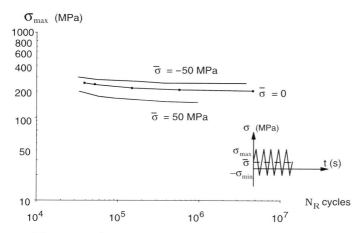

Fig. 6.9. Effect of mean stresses on the Wöhler curve

The same results are also plotted for tension-compression and shear loadings at a fixed number of cycles to rupture $N_R = 10^5$ (Fig. 6.10). As already pointed out, a remarkable result is that the shear mean stress $\bar{\tau}$ does not have any effect in shear in the range considered, as observed experimentally.

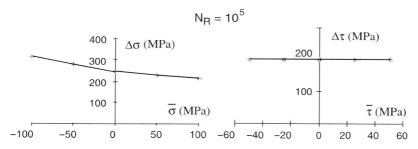

Fig. 6.10. Computed effect of mean stress in tension-compression and shear

6.4.1.2 Nonlinear Damage Accumulation

Another important feature of fatigue is the nonlinearity of the accumulation of damages due to loadings of different stress amplitudes. Consider here a two-level loading at zero mean stress as represented in Fig. 6.11: n_1 cycles at $\sigma_{\max} = \sigma_1$ followed by n_2 cycles at $\sigma_{\max} = \sigma_2$ such as $N_R = n_1 + n_2$. N_{R1} and N_{R2} are the numbers of cycles to rupture corresponding to constant amplitude loadings $\Delta\sigma_1$ and $\Delta\sigma_2$. The results are compared with the linear Palmgreen–Miner rule $n_1/N_{R1} + n_2/N_{R2} = 1$ in the n_2/N_{R2} vs n_1/N_{R1} plot in Fig. 6.11. The nonlinear damage accumulation is here obtained as a bilinear accumulation rule.

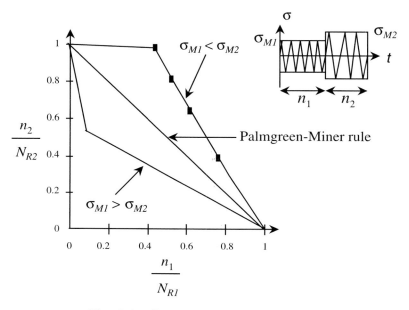

Fig. 6.11. Damage accumulation diagram

6.4.1.3 Biaxial High-Cycle Fatigue

Consider next a plate under a biaxial fatigue loading. Plane stress conditions and proportional loading are assumed and the in-plane stresses σ_1 and σ_2 vary proportionally between $(-\sigma_{1\max}$ and $+\sigma_{1\max})$ and $(-\sigma_{2\max}$ and $+\sigma_{2\max})$. Figure 6.12 shows the stress contours corresponding to numbers of cycles to rupture $N_R = 10^4, 10^5, 10^6$ cycles for in-phase cyclic loadings. The contour corresponding to the damage equivalent stress criterion $\sigma^\star = \sigma_{eq} R_\nu^{1/2} = \sigma_f = 210$ MPa is also drawn and fits well with the two-scale damage model for $N_R = 10^6$.

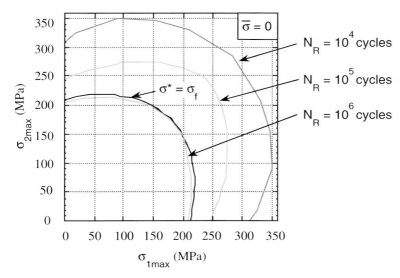

Fig. 6.12. Biaxial high cycle fatigue

6.4.1.4 Effect of Non-Proportional Loading

The effect of a non-proportional loading is considered here for a number of cycles to rupture, $N_R = 10^5$ cycles. First we consider an alternate loading of stress range $\Delta\sigma_1 = 2\sigma_{1\,\text{max}}$ until about $N_R/2$ cycles in a first direction (named 1) and then a second alternate loading of stress range $\Delta\sigma_2 = 2\sigma_{2\,\text{max}}$ in the orthogonal direction (named 2) corresponding to N_R.

Figure 6.13 shows that to obtain the same number of cycles ($N_R = 10^5$) in non-proportional loading (square points) as in proportional loading (contour reproduced from Fig. 6.12), a higher state of stress is needed. This is an effect similar to that in plasticity where non-proportional loading induces an over strain-hardening or cross hardening.

6.4.2 Non-Proportional Loading of a Thinned Structure♂♂♂
(M. Sauzay and A. Carmet 2000)

As an example, consider a tube of total length $L = 250$ mm, external radius $R_e = 28.7$ mm, internal radius $R_i = 27.5$ mm, and thickness $t = 1.2$ mm. The middle part of the tube has been thinned with a grindstone on one side, leading to a gradually thinner zone of stress concentration, where the fatigue cracks will initiate (see Fig. 6.14). The thickness of the thinner part varies from 1.2 to 0.6 mm.

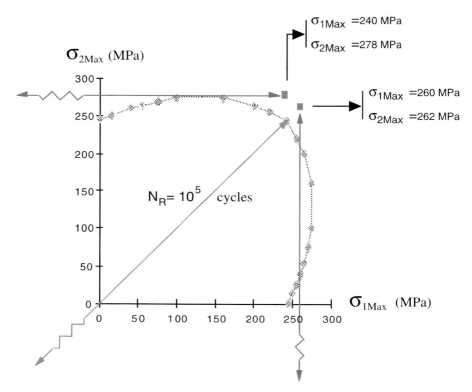

Fig. 6.13. Effect of non-proportional loading

6.4.2.1 Elastic FE Analysis

The structure has been numerically analyzed and tested in tension, torsion, and tension-torsion fatigue loadings. Elastic computations in tension and in torsion give the meso stress fields σ_{ref1} and σ_{ref2} to be used as inputs for DAMAGE 2000 post-processing analysis, with the tension-torsion treated as a linear combination of σ_{ref1} and σ_{ref2}.

The meshing comprises 5120 3 and 4 node Kirchhoff–Love shell elements (full and reduce integrations here give the same results). For the structure loaded in torsion, the local state of stress in the thinned zone is shear with an overstress coefficient $\sigma_{z\theta}/\sigma_{z\theta\,\text{n}} = 1.76$ (where the nominal stress $\sigma_{z\theta\,\text{n}}$ is the shear stress in the original tube of constant thickness t). For the structure loaded in tension, the state of stress is uniaxial in the center of the thinned zone with an over stress coefficient $\sigma_{zz}/\sigma_{zz\,\text{n}} = 1.66$ (again, "nominal" is related to the tube of constant thickness). The axial stress is even larger at the beginning of the transition zone with $\sigma_{zz}/\sigma_{zz\,\text{n}} = 2.03$ but also with a hoop stress.

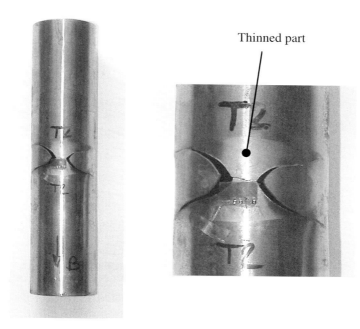

Fig. 6.14. Picture (M. Grange) of the broken thinned shell

6.4.2.2 Proportional Fatigue Loadings

The numbers of cycles to mesocrack initiation, N_R, are calculated by use of DAMAGE 2000 post-processor. The material is a ductile steel with material parameters: $E = 200000$ MPa, $\nu = 0.3$, $\sigma_y = 380$ MPa, $\sigma_u = 474$ MPa, $C_y = 50000$ MPa, $\sigma_f^\infty = 180$ MPa, $\epsilon_{pD} = 0.025$, $m = 3$, $S = 2.6$ MPa, $s = 2$, $h = 0.2$, $D_c = 0.3$.

The experiments have been performed at a 5 Hz frequency. The numerical simulations compare well with the experimental results in Fig. 6.16 where the two straight lines represent a relative error $N_R/N_R^{\exp}$ of 4 and 1/4. Be happy with a factor of 4 in fatigue!

6.4.2.3 Non-Proportional Fatigue Loadings

Four tension-torsion high cycle fatigue tests have also been performed at LEDEPP-ARCELOR at a frequency of 3 Hz. The non-proportional loading paths for the axial force F and the torsion torque C are shown in Fig. 6.15. The maximum applied force is $F_{\max} = 14000$ N and the maximum (or minimum) applied torque $C_{\max} = 420$ Nm (or $C_{\min} = -C_{\max}$). The results are given in Table 6.1.

Table 6.1. Tension-torsion of the thinned shell

Test	$\dfrac{F_{\min}}{F_{\max}}$	Type	Location of crack initiation	$N_R^{\exp}$	N_R^{comp}
a	-1	in phase	transition zone	$1.13 \cdot 10^5$	$4 \cdot 10^5$
b	0.1	in phase	thinner part	$4.86 \cdot 10^5$	$4.4 \cdot 10^5$
c	0.1	$90°$ out of phase	thinner part	$3.72 \cdot 10^5$	$6.9 \cdot 10^5$
d	-1	$90°$ out of phase	thinner part	$2.3 \cdot 10^5$	$11.6 \cdot 10^5$

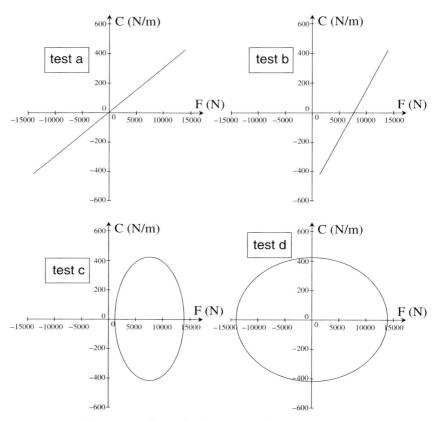

Fig. 6.15. Cyclic loading paths of tension-torsion

The model gives a correct estimation of the number of cycles to meso crack initiation with a factor of at most 5 between computed and experimental numbers of cycles, N_R^{comp} and $N_R^{\exp}$, even for out-of-phase fatigue loadings. The corresponding 4 points are also reported in Fig. 6.16. This shows the ability of the model to predict high cycle fatigue failures in complex loadings.

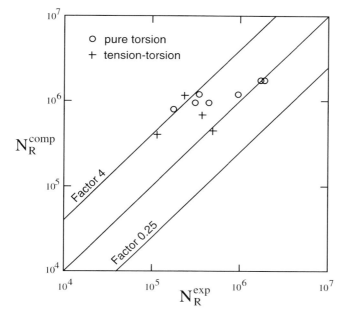

Fig. 6.16. Comparison between computations and experiments

6.4.3 Random Distribution of Initial Defects

The scatter in high cycle fatigue tests or in fatigue failures is always very large. A factor of 10 (1000%!) on the numbers of cycles to rupture around 10^6 for "identical" tests may happen. Since W. Weibull's studies (1939), independent of the experimental conditions, we know that the main reason is the initiation of the fatigue phenomenon on microdefects of different sizes. The number of cycles to rupture is as high as the probability of existence of a "big" defect is small. This is why small components have a larger lifetime than bigger ones.

In the two-scale damage model, the only two material parameters introduced on a microscale are the kinematic hardening coefficient (C_y) which is taken identical to its value on a mesoscale and the yield stress which is the asymptotic fatigue limit (σ_f^∞) on a mesoscale. The fatigue limit σ_f^∞ represents the local weakness due to hidden defects in the material, therefore it is the natural candidate for the random parameter responsible for the high cycle fatigue scatter.

Instead of guessing or measuring (but how?) the characteristics of microinternal defects to identify a random distribution of initial damage, it is much easier to use or perform a series of fatigue tests loaded identically in tension-compression to obtain numbers of cycles to rupture (N_{Ri}) around 10^6 cycles. The distribution of N_R for the stress considered as the engineering fatigue limit σ_f is thus recorded. To go further we have to find the distribution of σ_f^∞ which would give the N_{Ri} recorded. This is a deterministic calculation if σ_f^∞ is expressed as a function of N_R.

Taking the result of the simplified two-scale damage model of Sect. 6.2.3, it is not possible to invert the formula (6.28) (even with the simplifying assumption of (6.22)) to derive σ_f^∞ in a closed-form as

$$N_D = \frac{1}{4}\epsilon_{pD}\frac{\mathcal{G}^{m+1}}{C_y^m}\frac{(\sigma_u - \sigma_f^\infty)^m}{(\sigma_{max} - \sigma_f^\infty)^{m+1}}$$

$$N_R = N_D + \frac{(2ES)^s \mathcal{G} D_c}{2\sigma_f^{\infty\, 2s}\left[\sigma_{max} - \dfrac{\sigma_f^\infty}{\Sigma_{kk}}\right][R_{\nu\,min}^s + R_{\nu\,max}^s]}, \quad (6.80)$$

where the numbers of cycles are calculated for the engineering fatigue limit so that $\sigma_{max} = \sigma_f$ and $R_{\nu\,min} = R_{\nu\,max} = R_{\nu\star}\left(\Sigma_{kk}\dfrac{\sigma_{max}}{\sigma_f^\infty}\right)$.

A numerical analysis similar to the Monte Carlo method is needed while assuming a fixed N_{Ri} characterizing the scatter of the engineering fatigue limit σ_f and corresponding to a number of specimens i_{max} ($i_{max} = 20$ in Fig. 6.17 for the example of the ductile steel of of Sect. 6.4.2).

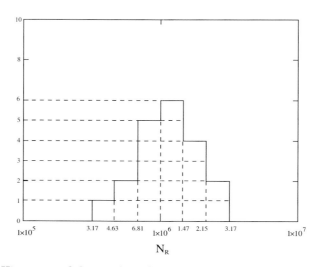

Fig. 6.17. Histogram of the number of cycles to rupture corresponding to $\sigma_{max} = \sigma_f = 230$ MPa at 10^6 cycles

First, the material parameters are determined by the procedure of Sects. 1.4.4 and 2.4 applied on the experimental Wöhler curve drawn from the mean value of the number of cycles for each stress range. Then $\sigma_{\rm f}^\infty$ is calculated for the 20 measured $N_{{\rm R}i}$ from (6.80) using mathematical software. The $N_{{\rm R}i}$ distribution is represented by the histogram of Fig. 6.18. The abscissas are discrete values $\sigma_{{\rm f}\,p}^\infty$ of $\sigma_{\rm f}^\infty$ and each ordinate is the number of values $\sigma_{\rm f}^\infty$ in between each $\sigma_{{\rm f}\,p}^\infty$ and $\sigma_{{\rm f}\,p+1}^\infty$. As the number of tests is most often low (≤ 20) statistically, a difficulty arises in the choice of the number of $\sigma_{{\rm f}\,p}^\infty$ that determine the shape of the probability curve. A **number of intervals** $p_{\max} \approx 2 + \sqrt{i_{\max}}$ is considered, a good compromise between exceedingly low (flat distribution) and exceedingly large numbers of values (also a flat distribution).

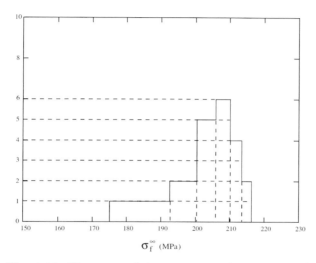

Fig. 6.18. Histogram of the asymptotic fatigue limit $\sigma_{\rm f}^\infty$

Once fitted, this **distribution** is considered a **characteristic of the material** and may be introduced as input together with the values of the material parameters in any fatigue calculation by the Monte Carlo method.

6.4.4 Stochastic Resolution by Monte Carlo Method♂♂

The Monte Carlo method is a numerical method to determine the stochastical response of any nonlinear model to random inputs by solving a large number of deterministic realizations of the inputs and by performing a statistical analysis on the results. Let's consider the histogram of the fatigue limit $\sigma_{\rm f}^\infty$ in Fig. 6.18. It is an approximation of the statistical distribution of $\sigma_{\rm f}^\infty$ and the first task is to regularize this discrete histogram by a known probability law or by curve fitting. For example, the histogram of Fig. 6.19 is a regularization with 25 intervals of the histogram of Fig. 6.18 and the graph $P_{\sigma_f}(\sigma_f^\infty)$ in Fig. 6.19 is the corresponding probability density fitted by a truncated β–function:

$$P_{\sigma_f}(\sigma_f^\infty) = \frac{1}{220}\beta\left(\frac{\sigma_f^\infty}{220}, 30, 2.5\right) \quad \text{if} \quad \sigma_f^\infty \geq 160 \text{ MPa} \quad (6.81)$$

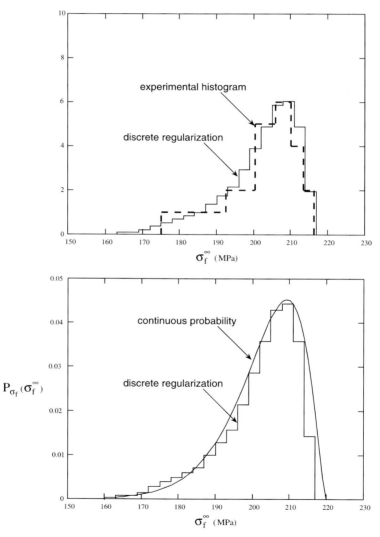

Fig. 6.19. Regularized histograms of σ_f^∞ and corresponding probability density

and $P_{\sigma_f} = 0$ elsewhere, with

$$\beta(x, a, b) = \frac{\Gamma(a+b)}{\Gamma(a) \cdot \Gamma(b)} x^{a-1}(1-x)^{b-1} \quad \text{and} \quad \Gamma(z) = \int_0^\infty t^{z-1} \exp(-t) \, dt. \tag{6.82}$$

Then, the general procedure for a structural analysis is as follows:

1. An elastic structural analysis gives the state of the 3D stress history at the critical point(s) (Gauss point(s) for a finite element analysis).
2. The damage parameters are supposed to be known, including the statistics of the asymptotic fatigue limit σ_f^∞ such as its histogram or its probability density (Fig. 6.19).
3. Consider a number $p_{\max}$ of discrete possible values $\sigma_{f\,p}^\infty$ of σ_f^∞. In the following example, it is 25 but a larger number (up to 100) is better.
4. Calculate the $p_{\max}$ values of the number of cycles to rupture $N_{R\,p}$ corresponding to the stress history with a post-processor such as DAMAGE 2000 for the two-scale damage model. They represent the failure statistics.
5. Draw the histogram of the results with the same ordinate n_p for $N_{R\,p}$ and $\sigma_{f\,p}^\infty$.
6. Calculate for the considered structure the mean number of cycles to rupture,

$$\overline{N}_R = \frac{\sum n_p N_{R\,p}}{\sum n_p}, \tag{6.83}$$

its standard deviation,

$$\overline{\overline{N}}_R = \sqrt{\frac{\sum n_i (N_{R\,i} - \overline{N}_R)^2}{\sum n_p}}, \tag{6.84}$$

its histogram, or its probability law.

This procedure has been applied to many values of unidimensional cyclic stresses corresponding to the Wöhler curve of ductile steel whose damage parameters are those used in Sect. 6.4.3. The numbers of cycles to rupture are calculated by means of the simplified two-scale damage model of Sect. 6.2.3 (also Eq. (6.80) of Sect. 6.4.3). Figure 6.20 shows the results:

- For the mean value $\overline{N}_R$ of the number of cycles to rupture as a function of the maximum stress $\sigma_{\max}$. The line fits the experimental Wöhler curve drawn almost exactly, with only one test for each stress.
- For the histogram of $\overline{N}_R$ corresponding to the three stresses.
- For the standard deviation $\overline{\overline{N}}_R$ where it is interesting to see how it increases with the number of cycles to rupture as it is always found by experiments.

This shows that it is not so complicated to take into account the scatter in high cycle fatigue failures predictions!

312   6 High Cycle Fatigue

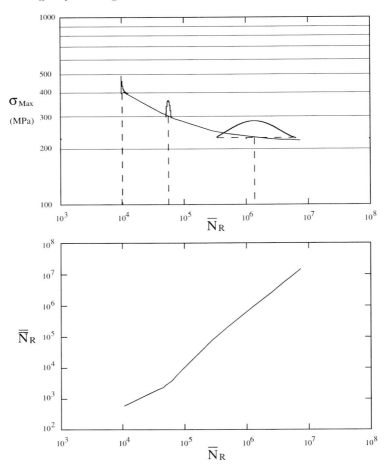

Fig. 6.20. Mean value and standard deviation of the number of cycles to rupture

6.4.5 Predeformed and Predamaged Initial Conditions♂♂

Other types of damage high cycle fatigue failures are influenced by the initial state of the materials (A. Galtier 1998). It is mainly an initial plastic strain p_0, an initial damage D_0, and an initial residual stress σ^{res} induced by the thermomechanical history of casting, metal forming, welding, and also damages by accident.

Let's assume that the initial or residual quantities p_0, D_0, and σ_{ij}^{res} exist but that they have been obtained for states of stresses larger than the in-service states of stresses that lead to high cycle fatigue (otherwise it would be low cycle fatigue!). The reference structure analysis is still elastic and those values are introduced in the two-scale damage model as initial values of the differential scheme. The conditions $D(t=0) = D_0$, $p(t=0) = p_0$, $\epsilon^{\text{p}}(t=$

$0) = \epsilon_0^p$, and $\boldsymbol{X}(t=0) = \boldsymbol{X}_0$ satisfy the equilibrium equation $div\ \boldsymbol{\sigma}^{\text{res}} = 0$ (or $\sigma_{ij,j}^{\text{res}} = 0$) for a static problem. For a plastic state, local tension leads to

$$\epsilon_0^p = p_0 \begin{bmatrix} 1 & 0 & 0 \\ 0 & -\frac{1}{2} & 0 \\ 0 & 0 & -\frac{1}{2} \end{bmatrix} \quad \text{and} \quad \boldsymbol{X}_0 = C_y p_0 \begin{bmatrix} 1 & 0 & 0 \\ 0 & -\frac{1}{2} & 0 \\ 0 & 0 & -\frac{1}{2} \end{bmatrix}. \tag{6.85}$$

Figures 6.21 and 6.22 describe the effect of an initial plastic strain p_0 that is smaller than the damage threshold p_D and an initial damage D_0 associated with the corresponding plastic strain, respectively, on steel at room temperature. The Wöhler curves are calculated by the damage post-processor DAMAGE 2000 with the following values of the material parameters: $E = 200000$ MPa, $\nu = 0.3$, $\sigma_f^\infty = 200$ MPa, $C_y = 2000$ MPa, $\sigma_u = 600$ MPa, $\epsilon_{pD} = 0.05$, $m = 1$, $S = 0.3$ MPa, $s = 2$, $h = 0.2$, and $D_c = 1$. This influence may be of a factor 10 on N_R.

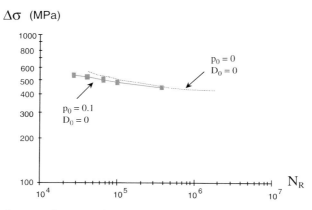

Fig. 6.21. Influence of an initial plastic strain on the Wöhler curve (J.-P. Sermage 1998)

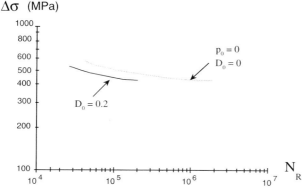

Fig. 6.22. Influence of an initial damage $D_0 = D_c/5 = 0.2$ on the Wöhler curve (J.-P. Sermage 1998)

314   6 High Cycle Fatigue

If their values are known, residual stresses σ_{ij}^{res} can be added to the loading stresses.

6.4.6 Hierarchic Approaches up to Surface and Gradient Effects^{♂,♂♂♂}

For other types of failure, we advise beginning any prediction of high cycle fatigue rupture using the simplest model, especially if only a few material parameters are known.

- In early design it is often sufficient to check if the state of stress is below the fatigue limit conditions. In 1D, the only effect to take into account is the effect of mean stress and the simplest model is the Goodman diagram shown in Fig. 6.2. An elementary geometric calculation gives the following as non-failure conditions:

$$\sigma_{\max} < \sigma_{\text{f}} + \bar{\sigma}\left(1 - \frac{\sigma_{\text{f}}}{\sigma_{\text{u}}}\right)$$
$$\text{or} \quad \sigma_{\min} > -\sigma_{\text{f}} + \bar{\sigma}\left(1 + \frac{\sigma_{\text{f}}}{\sigma_{\text{u}}}\right). \tag{6.86}$$

- For 3D state of stress, use one of the fatigue limit criteria of Sect. 6.2.2: the damage equivalent stress $\sigma^\star$ or, better, the Sines or Dang Van criterion for non-proportional loading.
- Again, in early design, the Palmgreen–Miner rule is not so bad in modelling sequences of different levels of stress range. Together with the Wöhler curve, it allows for the determination of the number of cycles to rupture corresponding to a given history of unidimensional stress.
- For 3D proportional loadings use the two-scale damage model of Sect. 6.2.3 integrated either numerically or in a closed form for periodic loading.
- For random loading there is, at least for a first approach, a nice simple formula in Sect. 6.3.4, eq. (6.77). Otherwise the Monte Carlo method applies together with the two-scale damage model used as a post-processor.
- For complex loadings such as 3D, non-cyclic and/or non-proportional loadings, the Wöhler curve can be used only to identify the two-scale damage model. The number of cycles N_{R} or the time t_{R} to mesocrack initiation is calculated by post-processing the elastic computation and by performing the time integration of the two-scale damage model constitutive equations (see Sect. 2.3). The most loaded point is where the damage D is maximum. The number of cycles to rupture N_{R} corresponds to the reach of the critical damage D_{c} at this point, as described in Sect. 6.4.2.
- It is also possible to consider a random distribution of initial defects as explained in Sect. 6.4.3, which allows for the possibility to express the results in terms of probability modelling of the large scatter observed in high cycle fatigue. Then, as in Sect. 6.4.4, the Monte Carlo method may be applied to any kind of loading.

- For more accurate predictions, one has to consider the **fatigue weakness on any surface** as plastic slips occur more easily on the surface grains that are well oriented with respect to the loading direction. **Intrusion-extrusion** mechanism takes place; it leads to microplastic strains that are larger on free surfaces than inside and then to lower numbers of cycles to rupture.

This surface effect may be taken into account in the two-scale damage model through the **localization law**. The microscale stress used in the initial two-scale damage model (without meso plastic strain),

$$\sigma^{\mu}_{\text{Eshelby}} = \sigma - 3G(1-\beta)\epsilon^{\mu p}, \tag{6.87}$$

is calculated by means of the Eshelby–Kröner localization law for a spherical inclusion fully embedded in an RVE. It has to be changed **at the free surface**.

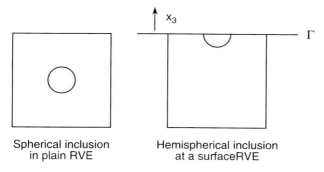

Fig. 6.23. Inclusions for the two-scale damage model

On the basis of the work of A. Cox (1989) and A. Deperrois & K. Dang Van (1990) a surface localization law can be built and used to describe the scale transition of the two-scale damage model (M. Sauzay 2000).

Following these authors, consider a hemispherical inclusion in an RVE located along a free edge (Fig. 6.23) and for the sake of simplicity take $D = 0$, $\tilde{\sigma}^{\mu} = \sigma^{\mu}$, and $\epsilon^{p} = 0$. If $\vec{x}_3$ is the normal to the free surface (denoted Γ), $\mathbf{x}$ the coordinates of a point in the RVE, the microscale stresses taking into account the existence of the free edge ($\sigma \vec{x}_3 = 0$ on Γ) are not uniform anymore, i.e.,

$$\sigma^{\mu}(\mathbf{x}) = \sigma^{\mu}_{\text{Eshelby}} - \sum_{r=1}^{3} \int_{\Gamma} \sigma^{\mu}_{r3\,\text{Eshelby}}(\mathbf{x}')C^{(r)}(\mathbf{x}',\mathbf{x})\mathrm{d}\Gamma', \tag{6.88}$$

with $\boldsymbol{C}^{(1)}(\mathbf{x}',\mathbf{x})$ and $\boldsymbol{C}^{(2)}(\mathbf{x}',\mathbf{x})$ as the Cerruti tensors and $\boldsymbol{C}^{(3)}(\mathbf{x}',\mathbf{x}) = \boldsymbol{B}(\mathbf{x}',\mathbf{x})$ as the Boussinesq tensor. Using the expression (6.87) for $\sigma^{\mu}_{\text{Eshelby}}$ allows us to obtain the formal localization law on a free surface as

$$\sigma^{\mu}(\mathbf{x}) = \sigma + \underline{\boldsymbol{M}}(\mathbf{x}) : \epsilon^{\mu p}, \tag{6.89}$$

with $\boldsymbol{M}$ as a fourth order localization tensor that is non-symmetric (the direction 3 plays a particular role) and dependent on the position of the considered point inside the inclusion. Using indexical notation, we have

$$M_{ijkl} = \left(E_{ijpq} - \int_\Gamma C^{(r)}_{ij}(\mathbf{x}', \mathbf{x}) E_{r3pq} \mathrm{d}\Gamma \right) (\underline{\boldsymbol{S}}^{\mathrm{Eshelby}} - \underline{\boldsymbol{I}})_{pqkl}, \qquad (6.90)$$

with $\underline{\boldsymbol{S}}^{\mathrm{Eshelby}}$ as the Eshelby tensor. A dimensionless tensor $\underline{\boldsymbol{m}}$ is also defined, such as

$$M_{ijkl} = -2G(1-\beta) m_{ijkl} \qquad (6.91)$$

and M. Sauzay (2000) gives, for $\nu = 0.3$, the average value of the inclusion of coefficients m_{ijkl}:

$$\begin{aligned}
& m_{1111} = m_{2222} = 1.24\,, \\
& m_{1122} = m_{2211} = -m_{3311} = -m_{3322} = 0.24\,, \\
& m_{1212} = m_{1221} = m_{2112} = m_{2121} = 0.5\,, \\
& m_{1313} = m_{1331} = m_{3113} = m_{3131} = 0.26\,, \\
& m_{2323} = m_{2332} = m_{3223} = m_{3232} = 0.26\,.
\end{aligned} \qquad (6.92)$$

The other coefficients are equal to zero. For $0.2 \leq \nu \leq 0.4$, the value for m_{ijkl} does not depend much on the Poisson ratio. The localization law to be used at surfaces is finally

$$\boxed{\sigma^\mu_{ij\,\mathrm{surf}} = \sigma_{ij} - 2G(1-\beta) m_{ijkl} \epsilon^{\mu\mathrm{p}}_{kl}\,,} \qquad (6.93)$$

also written as

$$\begin{aligned}
\sigma^\mu_{11\,\mathrm{surf}} &= \sigma_{11} - 2G(1-\beta)\left(m_{1111}\epsilon^{\mathrm{p}}_{11} + m_{1122}\epsilon^{\mathrm{p}}_{22}\right), \\
\sigma^\mu_{22\,\mathrm{surf}} &= \sigma_{22} - 2G(1-\beta)\left(m_{1122}\epsilon^{\mathrm{p}}_{22} + m_{1111}\epsilon^{\mathrm{p}}_{22}\right), \\
\sigma^\mu_{12\,\mathrm{surf}} &= \sigma_{12} - 4G(1-\beta) m_{1212}\epsilon^{\mathrm{p}}_{12}, \\
\sigma^\mu_{13\,\mathrm{surf}} &= -4G(1-\beta) m_{1313}\epsilon^{\mathrm{p}}_{13}, \\
\sigma^\mu_{23\,\mathrm{surf}} &= -4G(1-\beta) m_{2323}\epsilon^{\mathrm{p}}_{23}, \\
\sigma^\mu_{33\,\mathrm{surf}} &= -2G(1-\beta) m_{3311} \left(\epsilon^{\mathrm{p}}_{11} + \epsilon^{\mathrm{p}}_{22}\right).
\end{aligned} \qquad (6.94)$$

For FE computation of structures loaded in high cycle fatigue, one has then different localization laws for surface elements and inside elements. For a homogeneous mesoscopic stress field it leads to an increase of the microscale plastic strain on free surfaces and, by consequence, a significant decrease of the number of cycles to mesocrack initiation (of about 30%) in surface than in plain material. From a mechanical point of view this explains why the mesocrack initiation on free surfaces is always observed in uniform stress field specimens.

- Another characteristic effect of high cycle fatigue is the sensitivity to the gradient of the stress field. This **gradient effect** is illustrated here in cyclic bending where the most loaded points are on the upper and lower edges of a beam: the number of cycles to mesocrack initiation is always larger in cyclic bending than in the equivalent homogeneous tension compression case (I. Findley 1956). This is nothing peculiar as the chance to find a large defect in the upper or lower parts of the beam cross section is lower than the chance to find the same defect in the whole cross section. Such a gradient effect is observed in bending, but not in torsion (I.V. Papadopoulos and V.P. Panoskaltsis 1996), and can be taken into account within the two-scale damage model by considering the nonlocal localization law,

$$\tilde{\boldsymbol{\sigma}}^\mu = \boldsymbol{\sigma} - c_0 ||\nabla \sigma_H|| \mathbf{1} + c_1 \nabla^2 \boldsymbol{\sigma} - 2G(1-\beta)\boldsymbol{\epsilon}^{\mu p}, \qquad (6.95)$$

instead of the Eshelby–Kröner law, where $\sigma_H = \sigma_{kk}/3$ is the mesoscale hydrostatic stress, and $||\nabla \sigma_H||$ is the norm of its gradient (vanishing in shear). Two new material parameters are introduced:

- c_0, homogeneous to a length, is a phenomenological parameter identified from experimental results (such as high cycle fatigue bending tests).
- the parameter c_1, homogeneous to a square length, is naturally introduced from a homogenization procedure applied to non-uniform stress fields. It comes from the second order term of Taylor development of the stress tensor on the mesoscale. c_1 is found to be equal to $\delta_0^2/40$ (E. Aifantis 1995) where δ_0 is the size of the RVE (see Sect. 1.6.3). The effect of c_1 on the crack initiation conditions is not clear (it does not act for linear fields) and $c_1 = 0$ is probably a good choice to start with.

Consider as **reference** the tension-compression loading between $-\sigma_{max}$ and $+\sigma_{max}$. The meso stress field is homogeneous with vanishing gradient terms in the localization law (6.95). The number of cycles to rupture is then given by the initial two-scale damage model (numerical computation or analytical solutions of Sect. 6.2.3). For simplicity, consider the approximate expression (6.38) for zero mean stress with no damage threshold $\epsilon_{pD} = 0$:

$$N_R = \frac{(2ES)^s \mathcal{G} D_c}{4\sigma_f^{\infty\, 2s} \left[\sigma_{max} - \dfrac{\sigma_f^\infty}{\Sigma_{kk}}\right] R_{\nu\,max}^s}, \qquad (6.96)$$

where $\Sigma_{kk} = 1$ and $R_{\nu\,max} = \frac{2}{3}(1+\nu) + \frac{1}{3}(1-2\nu)\dfrac{\sigma_{max}^2}{\sigma_f^{\infty\,2}}$.

Consider then a circular bending test in the elastic range for which the stress field is locally uniaxial, with the same maximum and minimum applied stresses as for the tensile test:

$$\sigma = \frac{M}{I} y \quad \text{and} \quad \sigma_{max} = -\sigma_{min} = \frac{M}{I}\frac{d}{2}, \qquad (6.97)$$

with d either the diameter or the height of the beam, such as $y = d/2$ and $y = -d/2$, correspond to the upper and lower sides of the beam, and with I as its bending inertia momentum. As the stress field is linear in y, the stress gradient is

$$\nabla\sigma = \begin{bmatrix} 0 \\ \dfrac{2\sigma_{\max}}{d} \\ 0 \end{bmatrix} \quad \text{where} \quad ||\nabla\sigma|| = \dfrac{2\sigma_{\max}}{d}. \tag{6.98}$$

Apply the localization law with the gradient effect (6.95) at the most loaded point $y = \pm d/2$ (the second order terms vanish):

$$\tilde{\boldsymbol{\sigma}}^{\mu} = \boldsymbol{\sigma}^{\nabla} - 2G(1-\beta)\boldsymbol{\epsilon}^{\mu p} \quad \text{and} \quad \boldsymbol{\sigma}^{\nabla} = \boldsymbol{\sigma} - c_0||\nabla\sigma_{\mathrm{H}}||\mathbf{1}, \tag{6.99}$$

which is identical to the Eshelby–Kröner localization law but where the gradient effect is taken into account through the new loading term $\boldsymbol{\sigma}^{\nabla}$. Writing $\boldsymbol{\sigma}^{\nabla} = \sigma_{\Sigma}^{\nabla}\boldsymbol{\Sigma}$ (proportional loading) gives

$$\sigma_{\Sigma}^{\nabla} = \sigma_{\max}, \quad \boldsymbol{\Sigma} = \begin{bmatrix} 1 - \dfrac{2c_0}{d} & 0 & 0 \\ 0 & -\dfrac{2c_0}{d} & 0 \\ 0 & 0 & -\dfrac{2c_0}{d} \end{bmatrix}, \quad \Sigma_{\mathrm{eq}} = 1, \tag{6.100}$$

and allows us to use (6.96) directly, but with $\Sigma_{kk} = 1 - \dfrac{6c_0}{d}$ instead of $\Sigma_{kk} = 1$ to obtain the number of cycles to rupture in bending N_{R}^{∇}. Developing the expression for $R_{\nu\max}$ we end up with

$$\dfrac{N_{\mathrm{R}}^{\nabla}}{N_{\mathrm{R}}} = \dfrac{\dfrac{\sigma_{\max}}{\sigma_{\mathrm{f}}^{\infty}} - 1}{\dfrac{\sigma_{\max}}{\sigma_{\mathrm{f}}^{\infty}} - \dfrac{1}{1 - \dfrac{6c_0}{d}}} \left[\dfrac{\dfrac{2}{3}(1+\nu) + \dfrac{1}{3}(1-2\nu)\dfrac{\sigma_{\max}^2}{\sigma_{\mathrm{f}}^{\infty\,2}}}{\dfrac{2}{3}(1+\nu) + \dfrac{1}{3}(1-2\nu)\left(1 - \dfrac{6c_0}{d}\right)\dfrac{\sigma_{\max}^2}{\sigma_{\mathrm{f}}^{\infty\,2}}} \right]^s. \tag{6.101}$$

Then for the same maximum applied stress $\sigma_{\max}$ in tension-compression, we have

$$N_{\mathrm{R}}^{\nabla} > N_{\mathrm{R}} \tag{6.102}$$

in bending, i.e., a number of cycles to rupture increased in bending compared to the homogeneous tension-compression case. If the fatigue limits are defined for a given number of cycles for which no failure is observed, this corresponds to a fatigue limit in bending that is larger than the fatigue limit in tension-compression, as observed experimentally.

The ratio $N_{\mathrm{R}}^{\nabla}/N_{\mathrm{R}}$ is drawn in Fig. (6.24) as a function of the ratio $\sigma_{\max}/\sigma_{\mathrm{f}}^{\infty}$, with s as the damage exponent and c_0/d as the material and

6.4 Numerical Failure Analysis

geometry parameters. Knowing c_0 and d, $\sigma_{\max}$, σ_f^∞, and s, the Fig. 6.24 allows us to determine the factor with which to multiply the number of cycles to rupture classically calculated to obtain a better accuracy in bending. Inversely, from experiments made on specimens of height or diameter d, the knowledge of the ratios N_R^∇/N_R, $\sigma_{\max}/\sigma_f^\infty$, and the damage exponent s allows us to determine the material parameter c_0.

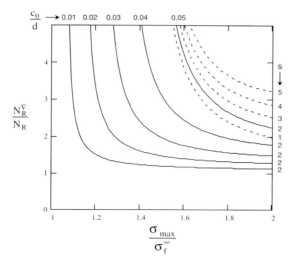

Fig. 6.24. Gradient effect in bending function of $\sigma_{\max}/\sigma_f^\infty$

For more general cases of stress gradients, use the localization law (6.95) within the two-scale damage model.

7
Failure of Brittle and Quasi-Brittle Materials

A material is considered **brittle** when it brakes **without any irreversible strains** and **without any dissipation** prior to cracking (e.g., glass and some ceramics). It is considered **quasi-brittle** when a **dissipation** prior to cracking exists with no or negligible permanent strains (e.g., concrete and some ceramics).

The main mechanism is atomic decohesions to induce a fast propagation of a crack by instability. Nevertheless, the quasi-brittle materials may have some reversible slidings of nano- or microcracks considered as initial defects to induce a loss of energy by friction. Both mechanisms are dangerous because there is no precursor to wake up the attention of observers.

The first scientific study of brittle failure was carried out by A.A. Griffith around 1920, introducing the concept of **energy of decohesion** used in fracture mechanics by cracking through the strain energy release rate variable G. But the study contained nothing about damage prior to crack initiation! In 1939 W. Weibull introduced the idea of **statistical distribution** of initial defects, giving rise to a model of probability of failure written in terms of stress and of the volume considered. Since then, many models of quasi-brittle materials are based on initial **microdefects** (F. McClintock 1973, S.B. Batdorf 1974, A.G. Evans 1978, A. Pineau 1983) or are damage models that are deterministic (Z.P. Bažant 1984, J. Mazars 1985, G.Z. Voyiadjis 2002) or stochastic (F. Hild 1990).

The specific topics of this chapter are the overall rupture criteria for brittle materials, **debonding**, **probabilistic approaches** for brittle and quasi-brittle materials, **delamination** of composites, and **dynamic failures** of concrete and ceramic structures. Long term, time-dependent behavior related to viscous effects of creep and relaxation is addressed in Chap. 5.

7.1 Engineering Considerations

The materials subjected to brittle or quasi-brittle failures are ceramics, concretes, cements, glasses, rocks, brittle matrix composites, quenched steels...

Their strain to rupture is small ($\epsilon_R < 2 \cdot 10^{-2}$) and their toughness is at most on the order of a few MPa$\sqrt{m}$. The design of structures made of brittle materials is difficult because there is no place for plastic shakedown in case of overloading, but reinforcements help.

For rough estimations meso stress criteria may be applied, but for precise estimations the statistical distribution of internal defects must be taken into consideration. Unfortunately, they cannot be precisely evaluated by non-destructive methods. A possible alternative is to deduce a probabilistic information from the scatter of test results by an inverse method (see Sect. 6.4.3) for high cycle fatigue (F. Hild 1994). To illustrate this point, let us determine the probability density of the relative size of defects in a brittle material on which many rupture tests have been performed.

The following simplified assumptions are made:

- 10 to 20 rupture tests are available in simple tension on the same geometry.
- For each specimen, the area density of the initial defects in the plane normal to the stress where the failure will occur is D_0. Then the rupture stress σ_R is simply given (see Sect. 1.4.1) by the effective stress concept. For brittle failures, σ_R is related to an initial damage,

$$\sigma_R = \sigma_u(1 - D_0) \quad \text{or} \quad D_0 = 1 - \frac{\sigma_R}{\sigma_u}, \tag{7.1}$$

where σ_u is the rupture stress of the material without any defect. On a practical level, it is the maximum value of σ_R measured, assuming that in the set of specimens, at least one has no defects (or only some very small ones).

The damage D_0 is now a random variable (as is σ_R) for which the probability density resulting from the tests is $P(\sigma_R)$. The probability for σ_R to have values bounded by σ_a and σ_b is

$$P(\sigma_a < \sigma_R < \sigma_b) = \int_{\sigma_a}^{\sigma_b} P(\sigma_R) d\sigma_R. \tag{7.2}$$

It is also the probability for the decreasing function $D_0(\sigma_R) = 1 - \sigma_u/\sigma_R$ to have values bounded by $D_0(\sigma_b) = D_{0b}$ and $D_0(\sigma_a) = D_{0a}$. Considering the inverse function $\sigma_R = \sigma_u(1 - D_0)$, we then have

$$P(D_{0b} < D_0(\sigma_R) < D_{0a}) = \int_{\sigma_u(1-D_{0b})}^{\sigma_u(1-D_{0a})} P(\sigma_R) d\sigma_R, \tag{7.3}$$

or by the change of variable $\sigma_R = \sigma_u(1 - D_0)$,

$$P(D_{0b} < D_0(\sigma_R) < D_{0a}) = -\int_{D_{0a}}^{D_{0b}} P(\sigma_u(1 - D_0)) \cdot (-\sigma_u dD_0), \tag{7.4}$$

which shows that the probability density of initial damage D_0 is

$$P(D_0) = \sigma_u P\left(\sigma_u(1 - D_0)\right). \tag{7.5}$$

For example, if σ_R obeys a Gaussian distribution,

$$P(\sigma_R) = \frac{1}{\bar{\bar{\sigma}}_R \sqrt{2\pi}} \exp -\frac{(\sigma_R - \bar{\sigma}_R)^2}{2\bar{\bar{\sigma}}_R^2}, \tag{7.6}$$

where $\bar{\sigma}_R$ is the mean value of the rupture stress and $\bar{\bar{\sigma}}_R$ its standard deviation, the probability density of D_0 is given by

$$P(D_0) = \frac{\sigma_u}{\bar{\bar{\sigma}}_R \sqrt{2\pi}} \exp -\frac{(\sigma_u(1 - D_0) - \bar{\sigma}_R)^2}{2\bar{\bar{\sigma}}_R^2} \tag{7.7}$$

or

$$P(D_0) = \frac{1}{\frac{\bar{\bar{\sigma}}_R}{\sigma_u} \sqrt{2\pi}} \exp -\frac{\left(D_0 - 1 + \frac{\bar{\sigma}_R}{\sigma_u}\right)^2}{2\frac{\bar{\bar{\sigma}}_R^2}{\sigma_u}}, \tag{7.8}$$

also a Gaussian distribution for D_0. The mean value of D_0 is:

$$\overline{D}_0 = 1 - \frac{\bar{\sigma}_R}{\sigma_u} \tag{7.9}$$

and its standard deviation is

$$\overline{\overline{D}}_0 = \frac{\bar{\bar{\sigma}}_R}{\sigma_u}. \tag{7.10}$$

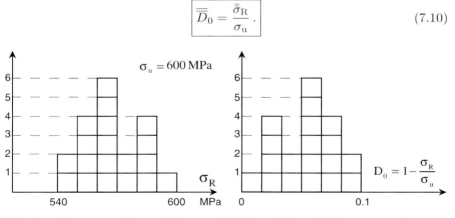

Fig. 7.1. Histogram of initial damage D_0 deduced from the histogram of rupture stress

There is an easy way to obtain a statistical input for stochastic analysis of failure. Instead of using a continuous probability law (such as a Gaussian law), a discrete numerical treatment is possible, as shown in Fig. 7.1, where the histogram of initial damage is simply deduced from a set of stresses to rupture by $D_{0i} = 1 - \sigma_{Ri}/\sigma_u$. If $i_{\max}$ is the number of tests (usually low, here $i_{\max} = 20$), using $p_{\max} \approx 2 + \sqrt{i_{\max}}$ (here $p_{\max} = 6$) as the number of intervals for the histograms is a good compromise between exceedingly low (flat distribution) and exceedingly large numbers of values (also a flat distribution).

7.2 Fast Calculations of Structural Failures

Structural calculations for brittle failures, fortunately, do not involve any plastic strain. They may be performed in elasticity or simply taken in handbooks of stress concentration coefficients. The bad news, as already mentioned, is the random character of brittle failures due to initial defects considered (or not considered) as initial damage. Nevertheless, some meso stress criteria give orders of magnitude of the conditions of brittle failures and the Weibull model is a first useful approach of the statistical characteristic of brittle rupture.

7.2.1 Damage Equivalent Stress Criterion♂

The simplest criterion to be used for the rupture of brittle or quasi-brittle materials is the damage equivalent stress σ^* deduced from the thermodynamics framework in Sect. 1.2.2. It is based on the **total elastic energy** whose amount at rupture is supposed to be a characteristic value for each material regardless of the state of stress:

$$\sigma^* = \sigma_{eq} R_\nu^{1/2} \quad \text{with} \quad R_\nu = \frac{2}{3}(1+\nu) + 3(1-2\nu)\left(\frac{\sigma_H}{\sigma_{eq}}\right)^2, \qquad (7.11)$$

where R_ν is the triaxiality function, σ_H is the hydrostatic stress, $\sigma_H = \sigma_{kk}/3$, σ_{eq} is the von Mises stress, $\sigma_{eq} = \sqrt{\frac{3}{2}(\sigma_{ij} - \sigma_H \delta_{ij})(\sigma_{ij} - \sigma_H \delta_{ij})}$.

If σ_u is the ultimate stress in pure tension, the rupture criterion is written as

$$\boxed{\sigma^* = \sigma_u \qquad \text{in three dimensions}} \qquad (7.12)$$

and simply $\sigma = \sigma_u$ in the uniaxial case as $R_\nu = 1$. Here we assume the same behavior in tension and compression.

Figure 7.2 shows the large difference due to the triaxiality between the damage equivalent stress σ^* and the von Mises (plastic) equivalent stress σ_{eq}, often used by mistake as a rupture criterion.

An improvement consists of taking into consideration the quasi-unilateral condition of microdefects closure. Back to Sect. 1.2.4 and eq. (1.46), the complementary elastic energy is

$$w_{\rm e}^{\star}(\sigma_{ij}) = \rho\psi_{\rm e}^{\star} = \frac{1+\nu}{2E}\left[\frac{\langle\boldsymbol{\sigma}\rangle_{ij}^{+}\langle\boldsymbol{\sigma}\rangle_{ij}^{+}}{1-D} + \frac{\langle\boldsymbol{\sigma}\rangle_{ij}^{-}\langle\boldsymbol{\sigma}\rangle_{ij}^{-}}{1-hD}\right]$$
$$-\frac{\nu}{2E}\left[\frac{\langle\sigma_{kk}\rangle^{2}}{1-D} + \frac{\langle-\sigma_{kk}\rangle^{2}}{1-hD}\right], \quad (7.13)$$

where $\langle.\rangle$ are the Macaulay brackets.

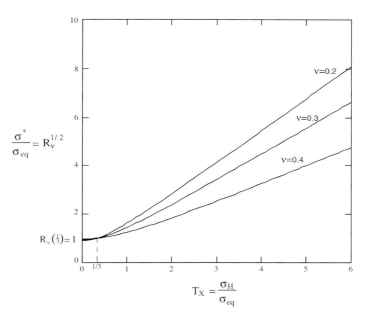

Fig. 7.2. Damage equivalent stress criterion function of the triaxiality ratio

For a tensile stress $\sigma^{\star}$,

$$w_{\rm e}^{\star}(\sigma^{\star}) = \frac{1}{2E(1-D)}\left[(1+\nu)\sigma^{\star 2} - \nu\sigma^{\star 2}\right] = \frac{\sigma^{\star 2}}{2E(1-D)}. \quad (7.14)$$

This tensile stress is equivalent to the 3D case if

$$w_{\rm e}^{\star}(\sigma^{\star}) = w_{\rm e}^{\star}(\sigma_{ij}). \quad (7.15)$$

In intrinsic notations, this is equivalent to

$$\boxed{\sigma^{\star} = \sqrt{(1+\nu)\langle\boldsymbol{\sigma}\rangle^{+}{:}\langle\boldsymbol{\sigma}\rangle^{+} - \nu\langle\operatorname{tr}\boldsymbol{\sigma}\rangle^{2} + \frac{1-D}{1-hD}\left[(1+\nu)\langle\boldsymbol{\sigma}\rangle^{-}{:}\langle\boldsymbol{\sigma}\rangle^{-} - \nu\langle-\operatorname{tr}\boldsymbol{\sigma}\rangle^{2}\right]}.}$$
$$(7.16)$$

Remember that h is on the order of 0.2 and the rupture criterion is still

$$\sigma^{\star} = \sigma_{\rm u}, \quad (7.17)$$

where D is here the relative surfacic fraction of microdefects at rupture D_c. The stress at rupture in tension is still $\sigma^* = \sigma_R^+ = \sigma_u$, but the stress at rupture in pure compression is now $|\sigma_R^-| = \sigma_u \left(\frac{1-hD_c}{1-D_c}\right)^{1/2}$, a small difference indeed. For materials like concrete, a much larger difference is obtained in the following definition of a damage equivalent strain ϵ^* and also in Sect. 7.4.1.

Sometimes a simpler criterion is used for the fracture of concrete where it is observed that the microcracks are often oriented in the direction normal to the positive (tension) principal strains. As a consequence, the **damage equivalent strain** ϵ^* is defined by (J. Mazars, 1984):

$$\epsilon^* = \sqrt{\langle\boldsymbol{\epsilon}\rangle^+ : \langle\boldsymbol{\epsilon}\rangle^+} = \sqrt{\langle\epsilon_1\rangle^2 + \langle\epsilon_2\rangle^2 + \langle\epsilon_3\rangle^2}, \qquad (7.18)$$

where ϵ_1, ϵ_2, and ϵ_3 are the principal strains and where $\langle\epsilon_i\rangle = \epsilon_i$ if $\epsilon_i > 0$, $\langle\epsilon_i\rangle = 0$ elsewhere.

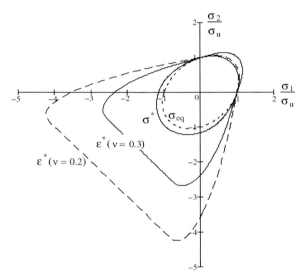

Fig. 7.3. Damage criteria in a plane stress case: σ_{eq}, $\sigma^*(\nu = 0.3, h = 0.2, D_c = 0.3)$, ϵ^*

In tension, $\epsilon^* = \epsilon^+$. In compression, $\epsilon^* = \sqrt{2}\nu|\epsilon^-|$ (ν is the Poisson ratio). The criterion normalized to the rupture stress in pure tension σ_u^+ or in pure compression σ_u^- is obtained as

$$\epsilon^* = \frac{\sigma_u^+}{E} = \sqrt{2}\,\nu\,\frac{|\sigma_u^-|}{E}, \qquad (7.19)$$

which gives $\sigma_u^+ = 0.28\,|\sigma_u^-|$ for $\nu = 0.2$. Figure 7.3 shows the difference between the last two criteria compared to the von Mises criterion in plane stress and proportional loading.

7.2.2 Interface Debonding Criterion

Consider a composite or two rigid elements A and B bound together by a film of glue I. It is represented by a 2D surface medium of "zero" thickness normal to the direction $\vec{x}_3$, as shown in Fig. 7.4.

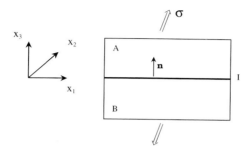

Fig. 7.4. Interface at mesoscale

The continuity of the strain and stress vectors through the interface ensure a state of plane strains $\epsilon_{13} = \epsilon_{23} = 0$ and a state of antiplane stresses $\sigma_{11} = \sigma_{22} = \sigma_{12} = 0$. The occurrence of two damage mechanisms, a normal debonding in mode I due to the normal stress σ_{33} and a shear debonding in mode II due to the shear stresses σ_{13} and σ_{23}, allows us to consider the thermodynamic potentials of Sects. 1.2 and 1.4 written as functions of the following interface equivalent stress,

$$\sigma_{eq}^I = \left\langle \sigma_{33}|\sigma_{33}| + \left(\frac{\sigma_R^I}{\tau_R^I}\right)^2 (\sigma_{13}^2 + \sigma_{23}^2) \right\rangle^{1/2}, \quad (7.20)$$

where $\langle . \rangle$ are the Macaulay brackets and $|.|$ is the absolute value.

It takes into account the unilateral character of interfaces which do not exhibit any damage in compression. The ratio of the ultimate tension stress σ_R^I to the ultimate shear stress τ_R^I represents the large difference in strength which may exist between tension and shear of interfaces. In supermarkets, you may buy glues with large σ_R^I or large τ_R^I depending on their use!

The criterion of debonding of interfaces written $\sigma_{eq}^I = \sigma_R^I \longrightarrow$ interface crack initiation is

$$\boxed{\left\langle \frac{\sigma_{33}|\sigma_{33}|}{\sigma_R^{I\,2}} + \frac{\sigma_{13}^2 + \sigma_{23}^2}{\tau_R^{I\,2}} \right\rangle = 1\,.} \quad (7.21)$$

It gives:

- Rupture in pure tension for $\sigma_{eq}^I = \sigma_{33} = \sigma_R^I$
- No rupture in pure compression as $\sigma_{eq}^I = 0$
- Rupture in pure shear for $\sigma_{eq}^I = \frac{\sigma_R^I}{\tau_R^I}|\sigma_{13}| = \sigma_R^I$ (or $\frac{\sigma_R^I}{\tau_R^I}|\sigma_{23}|$), i.e., $|\sigma_{13}| = \tau_R^I$

328   7 Failure of Brittle and Quasi-Brittle Materials

For mixed modes loading:
- A tension decreases the strength in shear
- A compression increases the strength in shear

This criterion requires the knowledge of two material parameters: the stresses to rupture in pure tension σ_R^I and in pure shear τ_R^I. It is represented in Fig. 7.5.

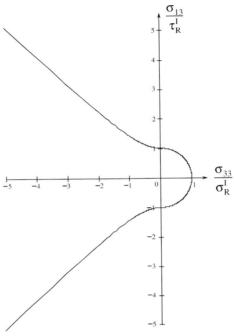

Fig. 7.5. Interface criterion in the tension and shear stress plane

7.2.3 The Weibull Model♂♀

This model takes into consideration the volume effect for which the probability to find a large defect is greater in a large volume than in a small one.

- The probability density of the defects considered as initial damages D_0 is supposed to be a **power function** where m_w is the Weibull modulus,

$$P(D_0) = m_w (1 - D_0)^{m_w - 1} . \qquad (7.22)$$

- The rupture criterion is related to the damage equivalent stress $\sigma^\star = \sigma_{eq} R_\nu^{1/2}$ (it was the maximum principal stress in original Weibull model),

$$\sigma^\star = \sigma_w (1 - D_0) \quad \text{or} \quad D_0 = 1 - \frac{\sigma^\star}{\sigma_w}, \qquad (7.23)$$

where σ_w is the ultimate stress to rupture of a hypothetic sample without any defect. It is also called Weibull stress.
- Then, the probability of rupture P_{F0} of any RVE loaded by a stress $\sigma^\star$ is the probability for which $\sigma^\star > \sigma_w(1-D_0)$ or $D_0 > 1-\sigma^\star/\sigma_w$. Considering the damage D_0 bound by 1,

$$P_{F0} = \int_{1-\frac{\sigma^\star}{\sigma_w}}^{1} P(D_0)\mathrm{d}D_0 = \left(\frac{\sigma^\star}{\sigma_w}\right)^{m_w}. \tag{7.24}$$

- The Weibull theory considers the **weakest link** hypothesis which states that the rupture of the whole structure of volume V is achieved as soon as the rupture of an elementary volume V_0 is initiated.
- Considering the probability of rupture P_{F0} of the RVE on which σ_w is defined as the probability of rupture of the volume V is expressed as (W. Weibull 1939)

$$P_F = 1 - \exp\left\{-\frac{V}{V_0}P_{F0}\right\} \tag{7.25}$$

or

$$P_F = 1 - \exp\left\{-\frac{V}{V_0}\left[\frac{\sigma^\star}{\sigma_w}\right]^{m_w}\right\}. \tag{7.26}$$

- For structures subjected to non-uniform stress fields another effect that arises is related to the probability that the largest defect exists precisely where the stress concentration lies. To take into account this effect, it is convenient to introduce the concept of an effective volume V_{eff} related to the elastic energy as it governs the phenomenon of rupture (F. Hild 1994). If $w_e = \sigma^{\star 2}(M)/2E$ is the elastic energy density field and $w_{e\,\text{max}} = \sigma^{\star 2}_{\text{max}}/2E$, the elastic energy density at the stress concentration point of the structure is

$$V_{\text{eff}} = \frac{\int_V \sigma^{\star 2}(M)\mathrm{d}V}{\sigma^{\star 2}_{\text{max}}} \leq V. \tag{7.27}$$

- Finally, the Weibull formula to use is

$$\boxed{P_F = 1 - \exp\left\{-\frac{V_{\text{eff}}}{V_0}\left[\frac{\sigma^\star}{\sigma_w}\right]^{m_w}\right\}.} \tag{7.28}$$

7.2.3.1 Use of the Weibull Model

To apply the model, one needs the value of the elasticity parameters as well as of the Weibull parameters:

- An elastic analysis of a structure of volume V gives the stress field $\sigma^\star(M)$ and the maximum stress at the stress concentration point. It gives also the energy density field $w_e(M)$ and then V_{eff}.

- Use Weibull formula (7.28) where V_0 is the volume of the uniform stress part of the specimens tested to identify Weibull parameters σ_w and m_w.
- The identification of Weibull parameters needs $i_{max} = 10$ to 20 experiments on specimens of cross section S_0, length L, volume $V_{eff} = V_0 = S_0 L$, and corresponds to a failure probability $P_F = P_{FL}$. To complete the identification, proceed as follows:
  1. Draw the cumulative histogram of the rupture stresses σ_R. Use $p_{max} \approx 2 + \sqrt{i_{max}}$ intervals and put in the ordinates the number of specimens $p = p(\sigma_R < \sigma_{Rp})$ broken under the maximum stress σ_{Rp} of the pth interval.
  2. Use $P_{FL} \approx p/(i_{max} + 1)$ (or any other) as the probability estimator to change the histogram into an experimental probability density $P_{FL}(\sigma_R)$.
  3. Identify $\sigma_w = \sigma_{wL}$ and m_w by curve fitting of the previous experimental $P_{FL}(\sigma_R)$ law,

$$P_{FL} = 1 - \exp\left\{-\left[\frac{\sigma_R}{\sigma_w}\right]^{m_w}\right\}. \tag{7.29}$$

Do it by performing a linear regression in the $\ln[-\ln(1-P_{FL})]$ vs $\ln \sigma_R$ diagram. The slope of the curve is then the Weibull modulus m_w and the ordinate at $\ln \sigma_R = 0$ is $-\ln \sigma_w^{m_w}$ which yields σ_w.

7.2.3.2 From Brittle to Quasi-Brittle Failure (F. Hild 1994)

A structure made of "brittle" material components often exhibits a quasi-brittle behavior (B. Coleman 1958, D. Krajcinovic 1981). To illustrate this point, consider here as a "structure" a fiber bundle made of carbon or ceramics fibers and modelled as a set of n fibers in parallel, loaded with the same applied displacement u. The fibers of length L, cross section S_0, and volume V behave elastically with a Young's modulus E. The failure probability of a fiber P_{FL} is given by the Weibull law,

$$P_{FL} = 1 - \exp\left\{-\left[\frac{\sigma_R}{\sigma_{wL}}\right]^{m_w}\right\}, \quad \text{where} \quad \sigma_{wL} = \sigma_w \left(\frac{V}{V_0}\right)^{1/m_w}. \tag{7.30}$$

The loading is displacement-controlled. For large n, the number of broken fibers is given by

$$n_b \approx P_{FL} n, \tag{7.31}$$

which corresponds to a number, $n - n_b \approx (1 - P_{FL})n$, of unbroken fibers.

The applied load is

$$F = \sum_i \sigma_i S_0, \tag{7.32}$$

with $\sigma_i = 0$ for the broken fibers and $\sigma_i = E\epsilon$ for the remaining ones, i.e.,

$$F = (1 - P_{FL})nS_0 E\epsilon \qquad (7.33)$$

or

$$\frac{F}{S} = E(1 - P_{FL})\epsilon, \qquad (7.34)$$

with $S = nS_0$ as the total cross section area. Setting $D = P_{FL}$ defines the damage variable of Continuum Damage Mechanics as the failure probability of fibers. The mean applied stress $\bar{\sigma} = F/S$ is then

$$\boxed{\bar{\sigma} = E(1 - D)\epsilon} \qquad (7.35)$$

and the bundle has quasi-brittle behavior.

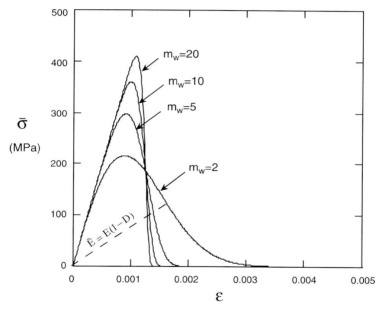

Fig. 7.6. Quasi-brittle stress-strain response of a ceramic fiber bundle ($E = 400000$ MPa, $\sigma_{wL} = 500$ MPa)

The damage evolution is obtained by this analysis. For monotonic loading,

$$D = D(\epsilon) = 1 - \exp\left\{-\left[\frac{E\epsilon}{\sigma_{wL}}\right]^{m_w}\right\} \qquad (7.36)$$

or if the strain energy release rate $Y = \frac{1}{2}E\epsilon^2$ is introduced,

$$\boxed{D = D(Y) = 1 - \exp\left\{-\left[\frac{2EY}{\sigma_{wL}^2}\right]^{\frac{m_w}{2}}\right\}}. \qquad (7.37)$$

The thermodynamics framework of such a damage model (called Marigo model) is given in Sect. 7.4.1.

The stress-strain curve obtained for a large number of ceramics fibers is given in Fig. 7.6 and the corresponding damage evolution in Fig. 7.7. If Weibull parameters are unknown, the stress-strain curve can be used to identify them using simple (nonlinear) curve fitting. Note also that the unloadings are elastic with no permanent strain: the Young's modulus of the damaged bundle is the effective modulus $\tilde{E} = E(1 - D)$.

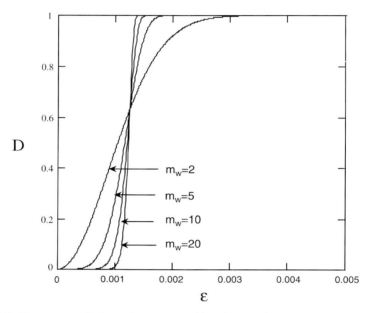

Fig. 7.7. Damage evolution of a ceramic fiber bundle ($E = 400000$ MPa, $\sigma_{\text{wL}} = 500$ MPa)

7.2.4 Two-Scale Damage Model for Quasi-Brittle Failures♂♂

It is possible to calculate quasi-brittle failure conditions by use of the two-scale damage model of Sect. 1.5.5. Consider here the case of a proportional monotonic loading with $\boldsymbol{\sigma} = \sigma_\Sigma(t)\boldsymbol{\Sigma}$, $\Sigma_{\text{eq}} = 1$, no plasticity on the mesoscale ($\boldsymbol{\epsilon}^{\text{p}} = 0$), and no damage threshold ($\epsilon_{\text{pD}} = 0$).

A scalar $\epsilon_{\text{p}\Sigma}^{\mu}$ is defined to quantify the microplasticity, as follows:

$$\boldsymbol{\epsilon}^{\mu\text{p}} = \frac{3}{2}\epsilon_{\text{p}\Sigma}^{\mu}\boldsymbol{\Sigma}^{\text{D}} \quad \text{and} \quad \dot{\sigma}_\Sigma^{\mu} = C_{\text{y}}\dot{\epsilon}_{\text{p}\Sigma}^{\mu} \approx \frac{C_{\text{y}}\dot{\sigma}_\Sigma}{\mathcal{G}}. \tag{7.38}$$

Here, $C_{\text{y}}/\mathcal{G} \ll 1$ and $\mathcal{G} \approx 3G(1-\beta)$. Once the yield stress on the microscale (equal to the asymptotic fatigue limit $\sigma_{\text{f}}^{\infty}$) is reached, we can assume $\dot{\sigma}^{\mu} \approx 0$

and the von Mises stress on the microscale remains quasi equal to σ_f^∞: $\sigma^\mu \approx \sigma_f^\infty$.

The time integration of the damage law (coupled with (7.38)) allows us to calculate the damage D at a given mesostress σ and then the quasi-brittle failure conditions when D reaches the critical damage D_c, such that

$$D \approx \frac{\sigma_f^{\infty\, 2s}}{\mathcal{G}(2ES)^s} \int_{\sigma_f^\infty}^{\sigma_\Sigma} R_{\nu\star}^s\left(\Sigma_{kk}\frac{\sigma_\Sigma}{\sigma_f^\infty}\right) d\sigma_\Sigma, \qquad (7.39)$$

with $R_{\nu\star}(\xi) = \frac{2}{3}(1+\nu) + \frac{1}{3}(1-2\nu)\xi^2$ as in Sect. 6.2.3.

The function $R_{\nu\star}(\xi)$ varies significantly over the integration interval and an approximate formula for the stress to failure is derived for small values of s:

$$\boxed{\sigma_{\Sigma R} \approx \sigma_f^\infty + \frac{3G(1-\beta)(2ES)^s D_c}{\sigma_f^{\infty\, 2s} R_{\nu\star}^s(\Sigma_{kk})}.} \qquad (7.40)$$

For compression-like loading ($\sigma < 0$), considering the parameter h within the damage law allows us to calculate the stress at failure in compression (σ_R^-) that is much larger than in tension (σ_R^+) and to show that

$$\frac{\sigma_R^+ - \sigma_f^\infty}{|\sigma_R^-| - \sigma_f^\infty} \approx h^s. \qquad (7.41)$$

Coupled with the previous knowledge of the asymptotic fatigue limit σ_f^∞ and the damage exponent s (from fatigue experiments), this result can be used to identify the microdefects closure parameter h. For concrete, for example, σ_f^∞ is neglected, $s = 1$, and $\sigma_R^-/\sigma_R^+ = 10$ gives $h \approx 0.1$.

7.2.5 Sensitivity Analysis

The sensitivity analysis of brittle failure is a special case as the rupture is not defined by a number like the plastic strain, time, or number of cycles in Sects. 3.2.3, 4.2.3, 5.2.3, and 6.2.4 of previous chapters but by the stress itself, through the damage equivalent stress $\sigma^\star$. For example,

$$\sigma^\star = \sigma_u \longrightarrow \text{rupture.} \qquad (7.42)$$

In order to perform the calculation in the spirit of the general method described in Sect. 2.4.6, let us consider the ratio $\sigma^\star/\sigma_u$ and derive how much it differs from 1. Furthermore, consider also the influence of initial defects in writing $\sigma_u(1-D_0)$ as the previous ultimate stress. Rupture corresponds then to $Rupt = 1$, with

$$Rupt = \frac{\sigma^\star}{\sigma_u(1-D_0)}, \qquad (7.43)$$

where $\sigma^\star = \sigma_{eq} R_\nu^{1/2}$, $R_\nu = \frac{2}{3}(1+\nu) + 3(1-2\nu)T_X^2$, and $T_X = \frac{\sigma_H}{\sigma_{eq}}$.

Taking the logarithm and the derivative with absolute values of the relative errors gives

$$\frac{\delta Rupt}{Rupt} = \frac{\delta\sigma_{\text{eq}}}{\sigma_{\text{eq}}} + \frac{(3T_X^2 - \frac{1}{3})\nu}{R_\nu}\frac{\delta\nu}{\nu} + \frac{3(1-2\nu)T_X^2}{R_\nu}\frac{\delta T_X}{T_X} + \frac{\delta\sigma_u}{\sigma_u} + \frac{D_0}{1-D_0}\frac{\delta D_0}{D_0},\tag{7.44}$$

which defines $S_{A_k}^{Rupt}$ as the coefficients of the sensitivity matrix of Sect. 2.4.6. Therefore,

$$\frac{\delta Rupt}{Rupt} = \sum_k S_{A_k}^{Rupt}\frac{\delta A_k}{A_k},\tag{7.45}$$

with

$$\begin{aligned}S_{\sigma_{\text{eq}}}^{Rupt} &= 1,\\ S_{T_X}^{Rupt} &= \frac{3(1-2\nu)T_X^2}{R_\nu},\\ S_{\sigma_u}^{Rupt} &= 1,\\ S_{\nu}^{Rupt} &= \frac{|3T_X^2 - \frac{1}{3}|\nu}{R_\nu},\\ S_{D_0}^{Rupt} &= \frac{D_0}{1-D_0}.\end{aligned}\tag{7.46}$$

As in the other chapters, the diagram in Fig. 7.8 shows the relative values of the influence of all the parameters by the height of the boxes for a set of parameters representing mean values for brittle materials: $T_X = 1 \Longrightarrow R_\nu = 2.07$, $\nu = 0.2$, and $D_0 = 0.05$.

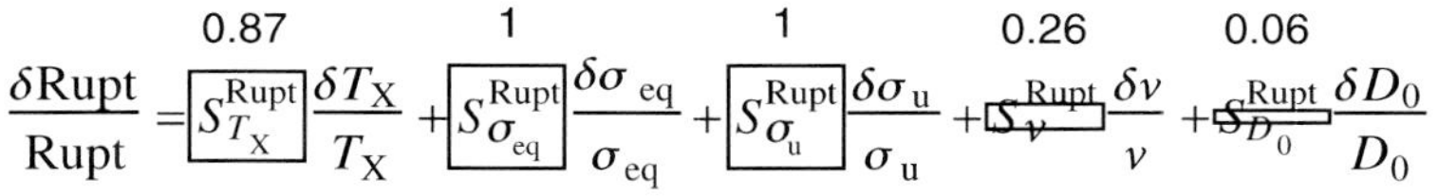

$$\frac{\delta Rupt}{Rupt} = \underbrace{\boxed{S_{T_X}^{Rupt}}}_{0.87}\frac{\delta T_X}{T_X} + \underbrace{\boxed{S_{\sigma_{\text{eq}}}^{Rupt}}}_{1}\frac{\delta\sigma_{\text{eq}}}{\sigma_{\text{eq}}} + \underbrace{\boxed{S_{\sigma_u}^{Rupt}}}_{1}\frac{\delta\sigma_u}{\sigma_u} + \underbrace{S_{\nu}^{Rupt}}_{0.26}\frac{\delta\nu}{\nu} + \underbrace{S_{D_0}^{Rupt}}_{0.06}\frac{\delta D_0}{D_0}$$

Fig. 7.8. Relative importance of each parameter in brittle failures

In comparison to other failures, the sensitivity is much lower. But the random value of the initial defects is not here taken into consideration: in practice, this is this parameter that governs the uncertainties.

7.2.6 Safety Margin and Crack Propagation

Most of the time, brittle failures occur by a fast propagation of a crack once the mesocrack is initiated. To avoid such an event, the state of stress anywhere in the structure must have a value below the rupture stress with a safety factor Saf. We advise checking to ensure that

$$\sigma_{\max}^\star < \frac{\sigma_u}{\text{Saf}}.\tag{7.47}$$

Due to the statistical effect of initial defects, a probabilistic safety margin is preferred each time it is possible:

$$\text{Prob}(\sigma^\star_{\max} \geq \sigma_u) < \frac{1}{\text{Saf}_{\text{Pr}}}, \qquad (7.48)$$

where σ_u is a random variable and $\sigma^\star_{\max}$ may also be a random variable due to the uncertainties of the loadings.

The statistically admissible risk defined by $1/\text{Saf}_{\text{Pr}}$ depends on many factors: knowledge of the material, knowledge of the loading, the quality of the mechanical analysis, the consequences of an eventual accident, and also subjective human factors. Therefore, $\text{Saf}_{\text{Pr}} = 10^2, 10^3, \ldots, 10^5$ or more.

The safest design cannot have $\text{Saf}_{\text{Pr}} = \infty$. It is, however, a design which minimizes the consequences of a failure. In this spirit, the "fail-safe" design consists of two or three parallel systems for one major resisting function.

Once a mesocrack is initiated, it is difficult to avoid the complete failure. Its size calculated in Sect. 1.6.3 does not contain the plastic term (1.252) as the material is considered fully elastic, i.e.,

$$\delta_0 \approx \frac{G_c}{\frac{\sigma_u^2}{2E} D_c}. \qquad (7.49)$$

A fast crack propagation will be avoided if the strain energy release rate G for this mesocrack is smaller than the toughness G_c:

$$G < G_c \quad \text{with} \quad G \approx \kappa \frac{\sigma_\infty^2 \pi \delta_0}{E}, \qquad (7.50)$$

where κ is a shape factor and σ_∞ is the far field stress,

$$\kappa \frac{\sigma_\infty^2 \pi \delta_0}{E} < G_c. \qquad (7.51)$$

The result does not depend on the toughness anymore but it depends on the critical damage, as in

$$\frac{\sigma_\infty}{\sigma_u} < \sqrt{\frac{D_c}{2\pi\kappa}}, \qquad (7.52)$$

with $0.2 < D_c < 0.5$ and $1 \leq \kappa < 10$ so that

$$0.06 < \frac{\sigma_\infty}{\sigma_u} < 0.28. \qquad (7.53)$$

These are indeed small values due to the large crack length δ_0 initiated by damage in brittle materials. They are often exceeded in practice, which is safe as long as no crack exists but not otherwise. Anyway, this explains why it is so difficult to design crack arrests in structures made of brittle materials!

7.3 Basic Engineering Examples

The same cases of stress concentrations in cylinders or beams as in the previous chapters are treated here, assuming the structures are made of brittle materials. It is somewhat simpler as plasticity is ignored but any defect may play an important role. The design of notches or geometry variations and their machining or forming must be done with much care. Furthermore, any scratch or surface damage in service must be controlled and examined from the point of view of residual strength.

7.3.1 Plates or Members with Holes and Notches

The uniaxial local stress at the critical point of stress concentration is supposed to be determined via a structural calculation or from the elastic stress concentration coefficient K_T and the nominal stress σ_n (see Fig. 3.4 and handbooks),
$$\sigma = K_T \sigma_n . \tag{7.54}$$
The stress condition for a mesocrack initiation is simply
$$\sigma_R = K_T \sigma_{nR} = \sigma_u . \tag{7.55}$$
In case of a 3D state of stress, the damage equivalent stress with or without the microdefects closure effect associated with an initial damage may be used:
$$\sigma^\star = \sigma_u . \tag{7.56}$$

For design purposes, it is interesting to express the admissible stress $\sigma^\star_{\max}$ or the ratio $\sigma^\star_{\max}/\sigma_u$ corresponding to a given probability of failure derived from the Weibull model of Sect. 7.2.3:

$$\boxed{\frac{\sigma^\star_{\max}}{\sigma_u} = \left[-\frac{V_0}{V_{\text{eff}}} \ln(1 - P_F)\right]^{\frac{1}{m_w}}} . \tag{7.57}$$

Figure 7.9 shows how it varies with the probability P_F, the ratio V_0/V_{eff}, and m_w as parameters.

This graph helps us choose the safety factor Saf of Sect. 7.2.6. For example, in order to ensure a failure probability below 10^{-5} with $m_w = 7$ and $V_0/V_{\text{eff}} = 10^{-1}$, the safety factor Saf $= \sigma_u/\sigma^\star_{\max}$ must be larger than 7.2. It shows also that the variation of Saf is on the order of 2 when the Weibull modulus m_w varies from 7 to 15 or when the volume fraction V_0/V_{eff} varies from 10^{-1} to 10^{-3}.

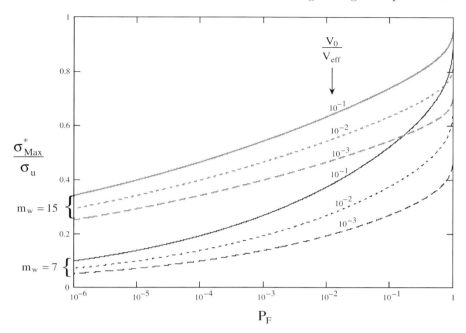

Fig. 7.9. Admissible damage equivalent stress for a given failure probability

7.3.2 Pressurized Shallow Cylinders ⚲

The structural analysis of Sect. 3.3.2 and Fig. 3.6 is again recalled for a long circular cylinder of radius R_{cyl} and thickness t_{cyl} that is submitted to an increasing relative pressure P, yielding

$$\sigma_{\text{eq}} = \frac{\sqrt{3}}{2} \frac{P R_{\text{cyl}}}{t_{\text{cyl}}},$$

$$\sigma_{\text{H}} = \frac{P R_{\text{cyl}}}{2 t_{\text{cyl}}},$$

$$T_X = \frac{\sigma_{\text{H}}}{\sigma_{\text{eq}}} = \frac{1}{\sqrt{3}},$$

$$R_\nu = \frac{5 - 4\nu}{3}.$$

(7.58)

The condition of rupture is still $\sigma^\star = \sigma_{\text{u}}$ except if some defect D_0 must be taken into consideration. In that case,

$$\sigma^\star = \sigma_{\text{eq}} \sqrt{R_\nu} = \sigma_{\text{u}}(1 - D_0),$$

(7.59)

and the pressure to bursting P_R is

$$P_R = \frac{2}{\sqrt{5-4\nu}} \frac{t_{cyl}}{R_{cyl}} \sigma_u (1-D_0). \tag{7.60}$$

Using $\nu = 0.3$, this gives a pressure that is $\approx 12\%$ lower (even with $D_0 = 0$) than the von Mises stress criterion $\sigma_{eq} = \sigma_u$ sometimes used!

7.3.3 Fracture of Beams in Bending♂

Sorry, this question is as simple as the first undergraduate course about strength of materials!

Assume a portion of a rectangular beam of width b and height h loaded in pure bending by a bending moment M (Fig. 3.7). According to the Bernoulli hypothesis, the maximum stress at the upper and lower part of the height is

$$\sigma_{max} = \frac{M}{I} \frac{h}{2}, \tag{7.61}$$

where $I = bh^3/12$ is the inertia moment.

Writing the condition of crack initiation with an eventual defect D_0,

$$\sigma_{max} = \sigma_u (1 - D_0) \tag{7.62}$$

gives the value of the bending moment at crack initiation M_R,

$$M_R = \frac{2I}{h} \sigma_u (1 - D_0), \tag{7.63}$$

or for the rectangular cross section,

$$M_R = \frac{bh^2}{6} \sigma_u (1 - D_0). \tag{7.64}$$

For considering fixed value of the moment and $D_0 = 0$,

$$\sigma_{max} = \frac{6 M_R}{bh^2}. \tag{7.65}$$

Once the crack initiated reaches a length of δ_0, a (very) rough estimation of the maximum stress is

$$\sigma'_{max} = \frac{6 M_R}{b(h - \delta_0)^2}. \tag{7.66}$$

If $\sigma'_{max} > \sigma_u$ (except if D_0 is large and localized only on the edge), the crack length increases, leading to a total fracture by instability.

Conclusion: If you try to initiate a small crack in a brittle component stress controlled by bending, you will obtain two components! But it may work if you try it by strain or controlled displacement.

7.4 Numerical Failure Analysis

The high level of damage encountered in components or parts of structures made of quasi-brittle materials is usually made acceptable by the presence of reinforcements: laminate composites are multi-layered, concrete is reinforced by bars or cables, and ceramic matrix composites (CMC) are fiber-reinforced.

Modern design solutions need to live with this damage and the corresponding computations are necessary for evaluations. By chance, reinforcements often prevent the strain localization phenomenon and one can expect nonlinear, but stable, damage computations.

As quasi-brittle materials behave quasi elastically, specific damage models may be formulated.

7.4.1 Quasi-Brittle Damage Models[6,66]

Quasi-brittle materials are usually modelled by elasticity coupled with damage (no plasticity!) and with a damage law of the form $D = D(Y)$ or $D = D(\epsilon)$, as justified in Sect. 7.2.3 from a Weibull analysis of a fiber bundle. Such damage models mainly apply to ceramics and composites.

The quasi-brittle material of main interest in civil engineering is concrete which exhibits very different behaviors in tension and in compression. An engineering design assumption is allowing its tensile strength to be equal to zero! This is of course not fully satisfactory and such an assumption is difficult to introduce in numerical computations, with the classical question: what is tension and what is compression for 3D states of stress? Fortunately, the use of the mathematical tools of Sect. 1.2.4 allows us to build suitable damage models for quasi-brittle materials and with a reduced number of material parameters.

7.4.1.1 Marigo Model of Elasticity Coupled with Damage (1981)

The Marigo model is written in the thermodynamics framework with the total strain ϵ associated with the stress σ and the damage D associated with the opposite of the strain energy release rate Y as state variables.

The Helmholtz free energy has the simple expression

$$\rho\psi = \frac{1}{2}(1 - D)\epsilon : \underline{\boldsymbol{E}} : \epsilon, \qquad (7.67)$$

with $\underline{\boldsymbol{E}}$ as the elasticity tensor.

The damage criterion (such as $f < 0 \Rightarrow$ elasticity) is written as

$$f = Y - \kappa(D), \qquad (7.68)$$

where $\kappa(D)$ is a function of the damage and $\kappa(0) = Y_D$ is the damage threshold in terms of elastic energy density. The term $S_M = \frac{d\kappa}{dD}$ is the consolidation modulus and linear consolidation corresponds to $\kappa(D) = SD + Y_D$, with $S_M = S = const$.

The law of elasticity coupled with damage derives from the Helmholtz potential as

$$\boxed{\boldsymbol{\sigma} = \underline{\boldsymbol{E}}(1-D) : \boldsymbol{\epsilon} \quad \text{or} \quad \boldsymbol{\epsilon} = \frac{1+\nu}{E(1-D)}\boldsymbol{\sigma} - \frac{\nu}{E(1-D)}\text{tr}\,\boldsymbol{\sigma}\,\mathbf{1}.} \quad (7.69)$$

The strain energy release rate is

$$Y = \frac{1}{2}\boldsymbol{\epsilon} : \underline{\boldsymbol{E}} : \boldsymbol{\epsilon} = \frac{\sigma^{\star\,2}}{2E(1-D)^2}, \quad (7.70)$$

where $\sigma^\star = \sigma_{eq} R_\nu^{1/2}$ is the damage equivalent stress.

The damage evolution law derives from the dissipative potential, taken here as $F_D = f$ (associated model), through the normality law

$$\dot{D} = \dot{\mu}\frac{\partial F_D}{\partial Y} = \dot{\mu}, \quad (7.71)$$

with $\dot{\mu}$ as the damage multiplier calculated by means of the consistency condition $f = 0$ and $\dot{f} = 0$. The damage evolution law is then

$$\dot{D} = \frac{\dot{Y}}{S_M}. \quad (7.72)$$

For monotonic loading, $D = 0$ as long as $Y \leq Y_D$. The integration of the previous law gives, of course,

$$D = \kappa^{-1}(Y) \quad \text{if} \quad Y > Y_D, \quad (7.73)$$

where $\kappa^{-1}(Y)$ is the damage function. For complex loadings such as non-monotonic and/or 3D loadings, considering the damage surface $F_D = f = 0$ gives a damage simply related to the maximum value of the strain energy release rate $Y_{\max}$ over the loading,

$$Y_{\max} = \sup_{\tau \in [0,t]} Y(\tau), \quad (7.74)$$

and

$$D = \kappa^{-1}(Y_{\max}) = D(Y_{\max}) \quad \text{such as} \quad Y < Y_D \Longrightarrow D = 0. \quad (7.75)$$

A simple but general damage evolution law is:

$$\boxed{D = \kappa^{-1}(Y_{\max}) = \left\langle \frac{Y_{\max} - Y_D}{S} \right\rangle^s,} \quad (7.76)$$

with Y_D, S, and s as the damage parameters.

A damage law adapted to concrete either in tension or compression is (C. Laborderie 1991)

$$D = \kappa_{\text{Lab}}^{-1}(Y_{\max}) = 1 - \frac{1}{1 + \left\langle \dfrac{Y_{\max} - Y_{\text{D}}}{S} \right\rangle^s} \,. \tag{7.77}$$

A damage law adapted to composite materials is (P. Ladevèze and E. Le Dantec 1992)

$$D = \kappa_{\text{Lad}}^{-1}(Y_{\max}) = \left\langle \frac{\sqrt{Y_{\max}} - \sqrt{Y_{\text{D}}}}{S} \right\rangle \tag{7.78}$$

and the Weibull damage law of Sect. 7.2.3 may be written as

$$D = \kappa_{\text{w}}^{-1}(Y_{\max}) = 1 - \exp\left\{-\left[\frac{Y_{\max}}{S_{\text{w}}}\right]^{s_{\text{w}}}\right\}, \tag{7.79}$$

with S, s, or S_{w}, s_{w} as material parameters.

Note that these models yield symmetric behavior in tension and in compression.

7.4.1.2 Marigo Model with Microdefects Closure Effect

This section gives a simple extension of the Ladevèze–Lemaitre isotropic damage framework of Sect. 1.2.4 to quasi-brittle materials. The model uses a single damage variable D corresponding to a damage state that is either tension-like or compression-like (or a mix).

Take as state potential the following:

$$\rho\psi^\star = \frac{1+\nu}{2E}\left[\frac{\langle\boldsymbol{\sigma}\rangle^+ : \langle\boldsymbol{\sigma}\rangle^+}{1-D} + \frac{\langle\boldsymbol{\sigma}\rangle^- : \langle\boldsymbol{\sigma}\rangle^-}{1-hD}\right] - \frac{\nu}{2E}\left[\frac{\langle\text{tr }\boldsymbol{\sigma}\rangle^2}{1-D} + \frac{\langle-\text{tr }\boldsymbol{\sigma}\rangle^2}{1-hD}\right], \tag{7.80}$$

from which

$$\boxed{\boldsymbol{\epsilon} = \frac{1+\nu}{E}\left[\frac{\langle\boldsymbol{\sigma}\rangle^+}{1-D} + \frac{\langle\boldsymbol{\sigma}\rangle^-}{1-hD}\right] - \frac{\nu}{E}\left[\frac{\langle\text{tr }\boldsymbol{\sigma}\rangle}{1-D} - \frac{\langle-\text{tr }\boldsymbol{\sigma}\rangle}{1-hD}\right]\mathbf{1}} \tag{7.81}$$

and

$$\boxed{Y = \frac{1+\nu}{2E}\left[\frac{\langle\boldsymbol{\sigma}\rangle^+ : \langle\boldsymbol{\sigma}\rangle^+}{(1-D)^2} + h\frac{\langle\boldsymbol{\sigma}\rangle^- : \langle\boldsymbol{\sigma}\rangle^-}{(1-hD)^2}\right] - \frac{\nu}{2E}\left[\frac{\langle\text{tr }\boldsymbol{\sigma}\rangle^2}{(1-D)^2} + h\frac{\langle-\text{tr }\boldsymbol{\sigma}\rangle^2}{(1-hD)^2}\right],} \tag{7.82}$$

with the elasticity law and the strain energy release rate density as functions of the microdefects closure parameter h.

Keeping the damage criterion unchanged $f = Y - \kappa(D)$ still leads to

$$D = \kappa^{-1}(Y_{\max}) = D(Y_{\max})\,. \tag{7.83}$$

Due to the introduction of $h < 1$, the elasticity law is different in tension and in compression:

$$\sigma = \sigma^+ = E(1-D)\epsilon \quad \text{in tension}$$
$$\sigma = \sigma^- = E(1-hD)\epsilon \quad \text{in compression.} \quad (7.84)$$

The damage growth is also different for tension and compression. For the same loading intensity $|\epsilon|$,

$$Y = Y^+ = \frac{\sigma^{+\,2}}{(1-D)^2} = \frac{1}{2}E\epsilon^2 \longrightarrow D = \kappa^{-1}\left(\frac{1}{2}E\epsilon_{\max}^2\right) \quad \text{(tension)}$$

$$Y = Y^- = h\frac{\sigma^{-\,2}}{(1-hD)^2} = \frac{h}{2}E\epsilon^2 \longrightarrow D = \kappa^{-1}\left(\frac{h}{2}E\epsilon_{\max}^2\right) \quad \text{(compression)}$$
$$(7.85)$$

with the damage induced by compression being smaller than the damage induced by tension. An example of the model response is given in Fig. 7.10.

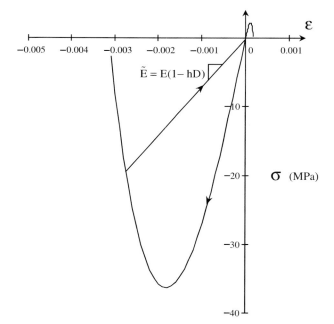

Fig. 7.10. Stress-stain curve for Marigo model with microdefects closure effect ($E = 30000$ MPa, linear law $D = Y_{\max}/S$, $S = 5 \cdot 10^{-4}$, $h = 0.057$)

The model applies to any state of stress. It is quite simple with material parameters E and ν for elasticity, Y_D as the damage threshold (which may be taken to be equal to zero for concrete), and S, s, and h as the damage parameters. We recommend using the simple damage evolution law (7.76)

$$D = \kappa^{-1}(Y_{\max}) = \left\langle \frac{Y_{\max} - Y_D}{S} \right\rangle^s. \tag{7.86}$$

7.4.1.3 Anisotropic Damage Model for Concrete

For concrete, it is sufficient to consider damage anisotropy with a quasi-unilateral effect acting on the hydrostatic stress only in the general thermodynamics framework of Sects. 1.2 and 1.5.

First, the Gibbs specific free enthalpy is represented by

$$\rho\psi^{\star} = \frac{1+\nu}{2E}\operatorname{tr}(\boldsymbol{H}\boldsymbol{\sigma}^{D}\boldsymbol{H}\boldsymbol{\sigma}^{D}) + \frac{1-2\nu}{6E}\left[\frac{\langle\operatorname{tr}\boldsymbol{\sigma}\rangle^{2}}{1-\operatorname{tr}\boldsymbol{D}} + \langle-\operatorname{tr}\boldsymbol{\sigma}\rangle^{2}\right], \tag{7.87}$$

with $\boldsymbol{H} = (\boldsymbol{1} - \boldsymbol{D})^{-1/2}$ so that the elasticity law becomes

$$\boldsymbol{\epsilon} = \rho\frac{\partial\psi^{\star}}{\partial\boldsymbol{\sigma}} = \frac{1+\nu}{E}\tilde{\boldsymbol{\sigma}} - \frac{\nu}{E}\operatorname{tr}\tilde{\boldsymbol{\sigma}}\,\boldsymbol{1} \tag{7.88}$$

and defines the symmetric effective stress as

$$\tilde{\boldsymbol{\sigma}} = (\boldsymbol{H}\boldsymbol{\sigma}^{D}\boldsymbol{H})^{D} + \frac{1}{3}\left[\frac{\langle\operatorname{tr}\boldsymbol{\sigma}\rangle}{1-\operatorname{tr}\boldsymbol{D}} - \langle-\operatorname{tr}\boldsymbol{\sigma}\rangle\right]\boldsymbol{1}. \tag{7.89}$$

The energy release rate tensor $\boldsymbol{Y} = \rho\frac{\partial\psi^{\star}}{\partial\boldsymbol{D}}$ is the thermodynamics variable associated with $\boldsymbol{D}$ (see Sect. 1.2.3).

Second, use the Mazars strain damage criterion ($f < 0 \longrightarrow$ elasticity) yielding

$$f = \epsilon^{\star} - \kappa(\operatorname{tr}\boldsymbol{D}) \quad \text{and} \quad \epsilon^{\star} = \sqrt{\langle\boldsymbol{\epsilon}\rangle_{+} : \langle\boldsymbol{\epsilon}\rangle_{+}}, \tag{7.90}$$

where $\kappa(\operatorname{tr} D)$ is chosen to ensure that at low cost (two damage parameters only), the damage rate in compression is lower than that in tension:

$$\kappa(\operatorname{tr}\boldsymbol{D}) = a \cdot \tan\left[\frac{\operatorname{tr}\boldsymbol{D}}{aA} + \arctan\left(\frac{\kappa_{0}}{a}\right)\right], \tag{7.91}$$

where κ_0 is the damage threshold. The terms A and a are the dimensionless damage parameters.

The anisotropic damage evolution law can be derived from the non-associated damage potential $F = F(\boldsymbol{Y};\boldsymbol{\epsilon}) = \boldsymbol{Y} : \langle\boldsymbol{\epsilon}\rangle_{+}^{2}$,

$$\dot{\boldsymbol{D}} = \dot{\mu}\frac{\partial F}{\partial\boldsymbol{Y}} = \dot{\mu}\langle\boldsymbol{\epsilon}\rangle_{+}^{2}. \tag{7.92}$$

The damage multiplier $\dot{\mu}$ is determined from the consistency condition $f = 0$ and $\dot{f} = 0$. The damage law finally reads as

$$\dot{\boldsymbol{D}} = \kappa^{-1\prime}(\epsilon^\star) \frac{\langle \boldsymbol{\epsilon} \rangle_+^2}{\epsilon^{\star 2}} \dot{\epsilon}^\star \qquad (7.93)$$

or

$$\boxed{\begin{array}{l} \dot{\boldsymbol{D}} = A \left[1 + \left(\dfrac{\epsilon^\star}{a}\right)^2 \right]^{-1} \dfrac{\langle \boldsymbol{\epsilon} \rangle_+^2}{\epsilon^{\star 2}} \dot{\epsilon}^\star \qquad \text{if} \quad \epsilon^\star > \kappa \\ \max D_{\mathrm{I}} = D_{\mathrm{c}} \longrightarrow \text{mesocrack initiation.} \end{array}} \qquad (7.94)$$

It ensures a damage rate that is proportional to the square of the positive part of the total strain tensor and in accordance with the damage anisotropy observed in concrete in tension and compression (see Sect. 1.4.5).

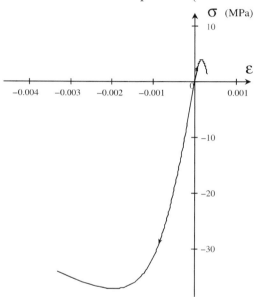

Fig. 7.11. Stress-stain curve for the anisotropic damage model ($E = 42000$ MPa, $\nu = 0.2$, $\kappa_0 = 5\ 10^{-5}$, $A = 5\ 10^3$, $a = 2.93\ 10^{-4}$, $D_{\mathrm{c}} = 0.92$)

A nonlocal anisotropic damage model is simply obtained by replacing $\epsilon^\star$ in the damage criterion with the nonlocal Mazars equivalent strain,

$$\epsilon^\star_{\mathrm{nl}} = \frac{1}{V_r} \int_V W(\boldsymbol{x} - \boldsymbol{s}) \epsilon^\star(\boldsymbol{s}) \mathrm{d}V , \qquad (7.95)$$

where $W(\boldsymbol{x} - \boldsymbol{s})$ is a nonlocal weight function (such as (2.113) or (2.114)) of Sect. 2.2.7) and where $V_r = \int_V W(\mathbf{x} - \mathbf{s}) \mathrm{d}V$. Consider then $f = \epsilon^\star_{\mathrm{nl}} - \kappa(\mathrm{tr}\ \boldsymbol{D})$, keeping $F = \boldsymbol{Y} : \langle \boldsymbol{\epsilon} \rangle_+^2$ unchanged so that the nonlocal anisotropic damage evolution law reads:

$$\boxed{\begin{array}{l} \dot{\boldsymbol{D}} = A \left[1 + \left(\dfrac{\epsilon^\star_{\mathrm{nl}}}{a}\right)^2 \right]^{-1} \dfrac{\langle \boldsymbol{\epsilon} \rangle_+^2}{\epsilon^{\star 2}} \dot{\epsilon}^\star_{\mathrm{nl}} \qquad \text{if} \quad \epsilon^\star_{\mathrm{nl}} > \kappa \\ \max D_{\mathrm{I}} = D_{\mathrm{c}} \longrightarrow \text{mesocrack initiation.} \end{array}} \qquad (7.96)$$

The local model has a total of 6 material parameters: E, ν for elasticity, κ_0 as damage threshold, and A, a, and D_c as damage parameters. An example of the concrete stress-strain curve is given in Fig. 7.11.

Finally, when permanent strains have a major effect, as in cyclic loading or in dynamics, use the Laborderie model (see Sect. 7.4.3) or others, taking into account the microdefects closure effect, damage anisotropy, and inelastic strains (A. Dragon and D. Halm 1998, F. Ragueneau 1999).

7.4.1.4 Damage Model for Composite Laminates Elementary Layers (P. Ladevèze 1986)

This model takes into account different damage mechanisms with different effects in tension and compression. Consider here laminates made of orthotropic layers or plies (long fiber reinforcement). In such composite materials, damage occurs at different levels:

- In the elementary plies with several degradation mechanisms, different scalar damage variables such as D_F for fiber breakage, D_T for transverse cracking, and D_S for shear damage are considered.
- Damage at the interfaces leads to the phenomenon of delamination.

The simple 2D quasi-brittle mesomodel for elementary plies described here is an extension of the Marigo model to orthotropy. The Gibbs energy is represented by

$$\rho\psi^\star = \frac{\langle\sigma_{11}\rangle^2}{2E_1(1-D_F)} + \frac{\phi(\langle-\sigma_{11}\rangle)}{2E_1} - \frac{1}{2}\left(\frac{\nu_{21}}{E_2} + \frac{\nu_{12}}{E_1}\right)\sigma_{11}\sigma_{22} \quad (7.97)$$
$$+ \frac{\langle\sigma_{22}\rangle^2}{2E_2(1-D_T)} + \frac{\langle-\sigma_{22}\rangle^2}{2E_2} + \frac{\sigma_{12}^2}{2G_{12}(1-D_S)},$$

with the usual symmetry $\nu_{21}/E_2 = \nu_{12}/E_1$, $\vec{e}_1$ in the fiber direction and ϕ as the microdefects closure function. The initial model considers $\phi = \langle-\sigma_{11}\rangle^2$ (instantaneous modulus recovery in compression). A thermodynamically consistent choice taking into account the nonlinear elastic response in compression is (B. Desmorat 2002)

$$\phi = -\frac{1}{b_D}\ln\left(1 - b_D\langle-\sigma_{11}\rangle^2\right), \quad (7.98)$$

with b_D as a material parameter. This leads to an effective Young's modulus $\tilde{E}_1^-$ in compression that is different from the effective modulus in tension $\tilde{E}_1^+$: $\tilde{E}_1^- = E_1$ for the instantaneous recovery; $\tilde{E}_1^- = E_1(1 - b_D\langle-\sigma_{11}\rangle^2))$ for the nonlinear response.

The elasticity law coupled with damage yields the following from the thermodynamics potential

$$\epsilon_{11} = \frac{\langle \sigma_{11} \rangle}{E_1(1-D_F)} - \frac{\langle -\sigma_{11} \rangle}{E_1^-} - \frac{\nu_{21}}{E_2}\sigma_{22},$$

$$\epsilon_{22} = \frac{\langle \sigma_{22} \rangle}{E_2(1-D_T)} - \frac{\langle -\sigma_{22} \rangle}{E_2^-} - \frac{\nu_{12}}{E_1}\sigma_{11}, \qquad (7.99)$$

$$\epsilon_{12} = \frac{\sigma_{12}}{2G_{12}(1-D_S)}.$$

The energy release rate densities associated with the damage variables are

$$Y_F = \frac{\langle \sigma_{11} \rangle^2}{2E_1(1-D_F)^2},$$

$$Y_T = \frac{\langle \sigma_{22} \rangle^2}{E_2(1-D_T)^2}, \qquad (7.100)$$

$$Y_S = \frac{\sigma_{12}^2}{2G_{12}(1-D_S)^2}.$$

Damage evolution laws such as (7.78) can be used for each damage variable (up to a critical damage D_c equal to 1). An adequate choice for organic matrix composites is to consider brittle fiber breakage, $D_F = 0$ or 1 depending on the fiber rupture stress σ_{RF}, and mixed evolution of damage for tranverse cracking and shear damage (P. Ladevèze and E. Le Dantec 1992):

$$\hat{Y} = \sup_{[0,t]} \{Y_S + b_T Y_T\},$$

$$D_T = \left\langle \frac{\sqrt{\hat{Y}} - \sqrt{Y_{TD}}}{S_T} \right\rangle, \qquad (7.101)$$

$$D_S = \left\langle \frac{\sqrt{\hat{Y}} - \sqrt{Y_{SD}}}{S_S} \right\rangle,$$

with b_T, S_T, Y_{TD}, S_S, and Y_{TS} as the damage parameters.

In tension, the stress-strain curves are similar to those obtained for the Marigo model. Note that the model can be completed to take into account the plastic-like behavior in the $\pm 45°$ directions. See Sect. 7.4.4 for an application mixing both layers degradation and interfacial delamination.

7.4.2 Failure of Pre-stressed Concrete 3D Structures♂♂
(F. Gatuingt and F. Ragueneau 2001, 2004)

Consider first concrete structures reinforced with passive bars and/or active (pre-stressed) cables. Mesh the concrete parts with 2D or 3D finite elements of size varying from 0.01 m (cement) or 0.1 m (concrete) to usually 0.3 m or more for large structures or uniformly loaded zones. Use a law of elasticity coupled with damage for the concrete elements. Model the bars and the cables with 1D elements with interpolation functions of the same type as those for the 2D or 3D connecting elements and use elasto-plasticity or elasto-plasticity coupled with isotropic damage for the steel reinforcements (see Sect. 1.5.2).

It is usually not necessary to model damage in bars and cables as the designed solutions tend to prevent it:

- For monotonic loading applications, the value of the uniaxial strain in the steel parts must remain lower than the damage threshold ϵ_{pD} or the rupture strain ϵ_{pR}.
- For cyclic or dynamic loadings, compute at least the accumulated plastic strain p or the stored energy w_s for comparison with the damage threshold in terms of plastic strains p_D or in terms of stored energy w_D (see Sect. 1.4.1).

Still, for the steels, use monotonic hardening only for monotonic applications; use kinematic hardening (or both kinematic and isotropic hardening) for cyclic and dynamics applications.

Finally, the numerical problem of the application of the pre-stresses in the active cables is not so simple. A possible procedure for damage computations with pre-stressed cables is as follows:

1. Estimate the cables tensile stress σ_0 and enter it at the Gauss points of the cables elements.
2. Use a finite element interpolation to transform it into elementary nodal loads $\{F_e\}$ (with the notations of Sect. 2.2.1) at the FE nodes of the cables elements,

$$\{F_e\} = \int_{V_e} [B]^T \{\sigma\} \, dV. \tag{7.102}$$

They represent the loading applied by the concrete body on the cables.
3. Change the sign of previous nodal loads to obtain the loading applied, by the cables on concrete. This defines the loading equivalent to the cables action.
4. The computation of the structure submitted to previous equivalent loading gives both initial deformation and damage states due to the application of the pre-stress. Do not model the cables at this stage.
5. Consider finally the cables as passive elements in further structure computations. To correctly model the cables plastification, do not forget to decrease, in the FE input file, the cables yield stress to $\sigma_{y0} = \sigma_y - \sigma_0$ (σ_y is the yield stress of the material). Start the computation from the previous initial deformation state.

The stress, strain, and damage fields numerically obtained take into account the cables pre-stress.

As an example, consider the short term behavior of the concrete structure in Fig. 7.12 is representative of a ring of a power plant containment vessel reinforced by half circular steel bars and pre-stressed by metallic cables. The

348   7 Failure of Brittle and Quasi-Brittle Materials

diameter of the ring is 46.8 m, its thickness is 0.9 m, and only a height of 0.4 m is meshed (see Fig. 7.13). A vertical compressive stress of 8.5 MPa is applied to model the weight of the upper part of the structure (not meshed). There are two symmetric anchoring parts and the steel reinforcement is made of 88 rectangular stirrups (not considered in the computations), 4 circular reinforcement bars and 2 pre-stressed half-circular cables. The finite element analysis of the ring is made in 3D with the bars and the cables modelled with 2-node bar elements and the concrete part with of 8-node bricks. The C.E.A. CASTEM 2000 computer code is used.

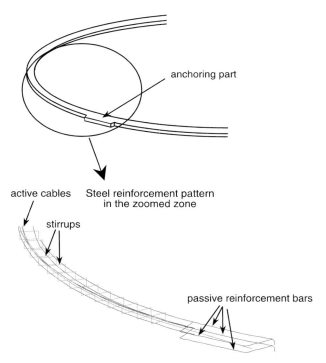

Fig. 7.12. Concrete ring and its steel reinforcement

Both the passive steel bars and the active metallic cables are assumed to be perfectly plastic. The concrete is modelled with the anisotropic damage model of the previous section and the corresponding material parameters are

- $E = 190000$ MPa, $\nu = 0.3$, and $\sigma_y = 500$ MPa for the steel bars of diameter 25 mm
- $E = 190000$ MPa, $\nu = 0.3$, and $\sigma_y = 1814$ MPa for the active cables with cross sections of 5143 mm^2

- $E = 42000$ MPa, $\nu = 0.2$, $\kappa_0 = 5\ 10^{-5}$, $A = 5\ 10^3$, and $a = 2.93\ 10^{-4}$ for concrete (with a rupture stress in compression $\sigma_u^- = 38$ MPa and a peak stress in tension $\sigma_u^+ = 4$ MPa)

The pre-stress is introduced using the above procedure to a tensile stress $\sigma_0 = 385$ MPa along the cables so that when the cables are considered passive, their yield stress is decreased from $\sigma_y = 1814$ MPa to $\sigma_{y0} = \sigma_y - \sigma_0 = 1429$ MPa in the CASTEM input data file. Note that enforcing the pre-stress leads to an initial cracking localized at the lower side of the anchoring parts (see Fig. 7.13 in which the displacements have an amplification factor of 724).

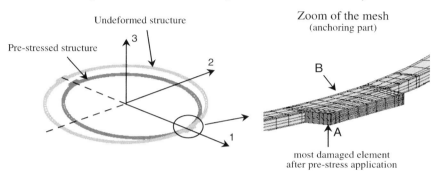

Fig. 7.13. Undeformed and pre-stressed structures

An internal pressure $P(t)$ is then applied. The damage fields are computed up to $P_{max} = 9\ 10^5$ Pa for which the average damage intensity corresponds to a structure that has already collapsed. Due to the steel reinforcement, a non-uniformly deformed mesh is obtained (see Fig. 7.14 in which the displacements have an amplification factor of 633).

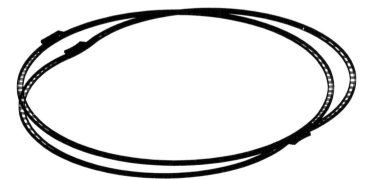

Fig. 7.14. Undeformed structure and deformed structure at $P_{max} = 9\ 10^5$ Pa

The evolution of the principal damages at point A and at the cross section center point B (Fig. 7.13) are plotted in Fig. 7.15 as functions of the cables pre-stressing tension and then as functions of the pressure intensity P.

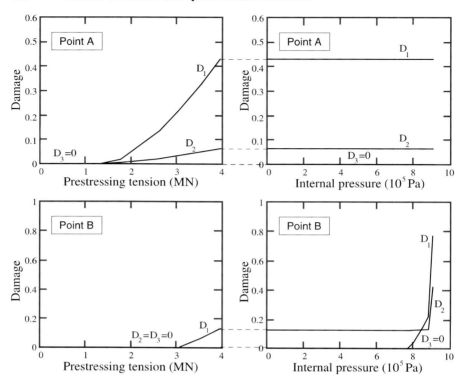

Fig. 7.15. Principal damages at points A and B during pre-stress application and during pressure loading

This simple test shows the possibility of using 3D anisotropic damage computations of reinforced concrete structures up to a very high damage level and up to plastification of the reinforcement steels. It illustrates the fact that initial cracking patterns often exist in such structures. The steel reinforcements prevent any localization modes and there is no need to introduce strain localization limiters.

Coupled with the consideration of a safety criterion $D < D_{\text{given}}$, these computations can be used to design civil engineering structures at their most loaded parts.

7.4.3 Seismic Response of Reinforced Concrete Structures[66]
(F. Ragueneau and J. Mazars 2001)

Dynamics and seismic FE analyses need efficient and robust modelling for both space and time discretizations. Explicit schemes are often used (such as the Newmark scheme) and some include numerical damping for computation reasons. Seismic loading looks like a random fatigue loading, but quick, so that it is given as the seism acceleration history (also called seism accelerogram).

7.4 Numerical Failure Analysis

Geometric simplifications are welcome and are sometimes possible for civil engineering buildings made of walls and flats naturally meshed by beams and plates. For reinforced concrete parts, a homogeinized behavior must be considered. A possibility consists of the choice of a multilayered beam (or plate) element which averages concrete and steel nonlinear constitutive equations (see Fig. 7.16). Using beam or plate kinematics or multilayered beam modelling conserves the steel reinforcement ratios as well as their location through their effect on the beam inertia.

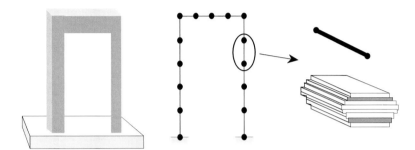

Fig. 7.16. Example of a frame frame discretized by multilayered beam elements

For seismic civil engineering applications, consider a damage model with permanent strains for concrete and use elasto-plasticity for steels (up to damage initiation, see Sect. 7.4.2) or elasto-plasticity coupled with isotropic damage if computations up to a total collapse are performed. Note that the main advantage of such multifiber type finite elements is the uniaxial implementation of each layer's constitutive law, leading to efficient computations in dynamics. As illustrated next, an accurate modelling of concrete behavior is necessary, the major role of the dynamics force variation enhances the effect of the microdefects closure modelling on the seismic response of the structure.

Table 7.1. Steel reinforcements of the walls

	Sa, Sb (mm^2)	Sc (mm^2)
Level 5	15.9	78.4
Level 4	28.3	78.4
Level 3	94.4	110.2
Level 2	188.9	138
Level 1	289.4	138

As an example, consider the $1/3^{rd}$ mock-up of Fig. 7.17 on the "C.E.A." shaking table ("CAMUS" experiment) computed for the "Nice S1" accelerogram that is representative of earthquake far field and for the "San Francisco 1957" accelerogram representative of earthquake near field. It is composed of 2 parallel walls linked by 6 square slabs and is anchored on the shaking table

352   7 Failure of Brittle and Quasi-Brittle Materials

by a highly reinforced footing. Table 7.1 lists the steel reinforcement cross sections for each wall.

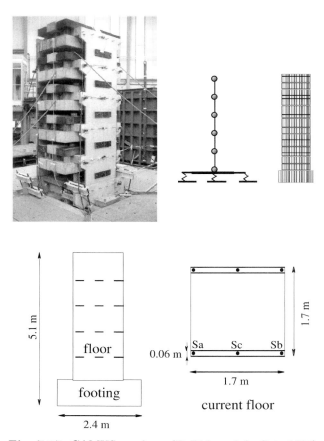

Fig. 7.17. CAMUS mock-up (P. Bish and A. Coin 1998)

The mock-up is loaded by a horizontal displacement parallel to the walls. Steel bar bracing systems disposed perpendicularly to the loading direction prevent torsional modes. The complete experimental sequence is an accelerogram made of Nice 0.24 g, followed by San Francisco 1.13 g, Nice 0.40 g, and Nice 0.71 g (Fig. 7.18) modified in time with a ratio $1/\sqrt{3}$ to take into account the similarity rules.

7.4.3.1 FE Modelling and Eigenmode Adjustment

The choice of a multilayered beam modelling is made with all the constituents restricted to beam kinematics, but with nonlinear material behavior.

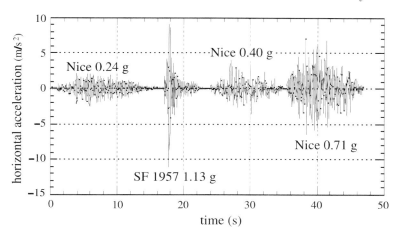

Fig. 7.18. Earthquake accelerograms

- The Laborderie model is used for concrete where only the uniaxial constitutive equations are needed. They introduce 11 material parameters: E and ν for elasticity; α_t, α_c, and σ_{clos} for the microdefects closure effect; Y_{tD} and Y_{cD} as damage thresholds in tension and compression; and S_t, s_t, S_c, and s_c for damage.
Elasticity with permanent strains due to damage is represented by

$$\epsilon = \frac{\langle \sigma \rangle}{E(1-D^+)} - \frac{\langle -\sigma \rangle}{E(1-D^-)} + \frac{\alpha_t D^+}{E(1-D^+)} g'(\sigma) + \frac{\alpha_c D^-}{E(1-D^-)} \quad (7.103)$$

and the damage evolution laws are

$$D^+ = 1 - \frac{1}{1 + \left\langle \dfrac{Y_{\max}^+ - Y_{tD}}{S_t} \right\rangle^{s_t}}$$

$$D^- = 1 - \frac{1}{1 + \left\langle \dfrac{Y_{\max}^- - Y_{cD}}{S_c} \right\rangle^{s_c}}, \quad (7.104)$$

where

$$Y^+ = \frac{\langle \sigma \rangle^2}{2E(1-D^+)^2} + \frac{\alpha_t g(\sigma)}{E(1-D^+)^2}$$

$$Y^- = \frac{\langle -\sigma \rangle^2}{2E(1-D^-)^2} + \frac{\alpha_c \sigma}{E(1-D^-)^2}, \quad (7.105)$$

and where the microcrack closure function $g(\sigma)$ is defined as follows

$$g(\sigma) = \sigma + \frac{\langle -\sigma \rangle^2}{2\sigma_{\text{clos}}} \quad \text{if} \quad \sigma > -\sigma_{\text{clos}},$$
$$g(\sigma) = -\frac{\sigma_{\text{clos}}}{2} \quad \text{if} \quad \sigma \leq -\sigma_{\text{clos}}.$$
(7.106)

The set of material parameters, here for a concrete with a maximum compressive strength of 35 MPa and tensile strength of 3 MPa, are: $E = 22 \cdot 10^9$ Pa, $Y_{\text{tD}} = 127$ Pa, $Y_{\text{cD}} = 6 \cdot 10^4$ Pa, $S_t = 1.11 \cdot 10^{-4}$ Pa, $s_t = 1.2$, $S_c = 1.89 \cdot 10^5$ Pa, $s_c = 1.4$, $\alpha_t = 10^6$ Pa, $\alpha_c = -40 \cdot 10^6$ Pa, and $\sigma_{\text{clos}} = 1.3 \cdot 10^6$ Pa. The stress-strain response of the model in tension-compression is given in Fig. 7.19.

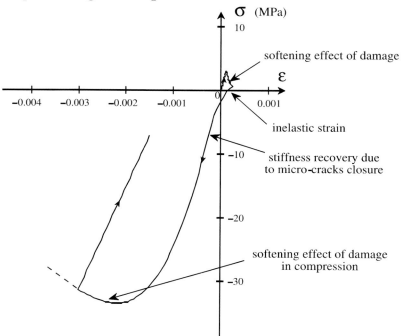

Fig. 7.19. Stress-strain curve for the Laborderie model

- The steels are assumed to be elasto-plastic (no damage). They have a Young's modulus $E = 200 \cdot 10^9$ Pa, a yield stress $\sigma_y = 414 \cdot 10^6$ Pa, and an ultimate stress $\sigma_u = 480 \cdot 10^6$ Pa.

The LMT-Cachan computer code EFICOS (J.F. Dubé, C. Laborderie, J. Mazars 1994) is used with the Newmark scheme for the time discretization.

Figure 7.17 shows the mock-up modelling as well as the finite element mesh. The maximum element size is limited to 30 cm in the less loaded zones with a total of 22 elements, each made of 20 layers. The additional mass and the weight load of each floor are concentrated at each storey (black circles). The stiffness of the springs below the shaking table is identified so as to fit the first two eigenmodes measured on the non-damaged (virgin) structure.

7.4.3.2 Seismic Structure Response

Figure 7.20 gives both the experimental and numerical responses of the structure. The computed horizontal mock-up top displacement compares well with the measured one.

Shocks are induced as microcracks close when the vertical mode is activated. Experimental observations show a vertical acceleration of the shaking table. The vertical force variations are computed (Fig. 7.21) and a comparison with the experiment (summarized in Table 7.2) demonstrates that the Laborderie model for concrete, representing permanent strains due to damage and the strong difference of behavior in tension and in compression, is relevant to seismic analyses.

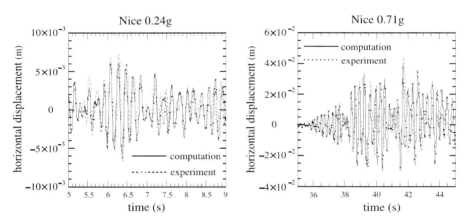

Fig. 7.20. Horizontal top displacement

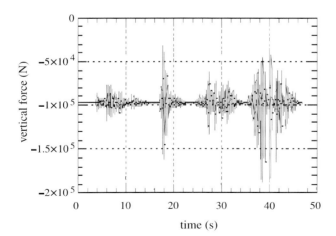

Fig. 7.21. Vertical force variations

A small change in the microcrack closure function through the value of σ_{clos}, responsible for the gradual stiffness recovery in compression, has a big influence on the results on a structural level (it activates the second vibration mode here). In conclusion, it is of first importance in dynamics to properly model the microdefects closure effect for loaded concrete structures.

Table 7.2. Maximum dynamics forces (kN)

	Model	Experiments
Nice 0.24 g	119	138
San Francisco 1.13 g	160	198
Nice 0.40 g	132	146
Nice 0.71 g	190	248

7.4.4 Damage and Delamination in Composite Structures♂♂♂
(O. Allix and P. Ladevèze 1992–98)

Damage of composite laminates is complex, with different damage mechanisms in each ply of the stacking structure but also with delamination and crack propagation at the interfaces (Fig. 7.22). Based on the damage models of Sect. 7.4.1 for the orthotropic layers, a mesomodel can be built, allowing for the computation of damage and delamination in laminate structures.

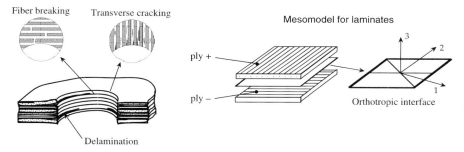

Fig. 7.22. Damage mechanism in laminates (*left*). Laminate modelling (*right*)

For engineering applications, perform first the whole structural computation in elasticity using laminated plates theory (with no damage and with perfect interfaces). Then reanalyze the most loaded zones by meshing every layer and interface using damage models.

For 3D reanalyses, the Gibbs energy for a single layer (7.97) must be completed with normal stresses terms:

$$\rho\psi^\star = \frac{\langle\sigma_{11}\rangle^2}{2E_1(1-D_F)} + \frac{\phi(\langle-\sigma_{11}\rangle)}{2E_1} + \frac{\langle\sigma_{22}\rangle^2}{2E_2(1-D_T)} + \frac{\langle-\sigma_{22}\rangle^2}{2E_2} - \frac{\nu_{12}}{E_1}\sigma_{11}\sigma_{22}$$
$$+ \frac{\langle\sigma_{33}\rangle^2}{2E_3(1-D_T)} + \frac{\langle-\sigma_{33}\rangle^2}{2E_2} - \frac{\nu_{13}}{E_3}\sigma_{11}\sigma_{33} - \frac{\nu_{23}}{E_2}\sigma_{22}\sigma_{33}$$
$$+ \frac{\sigma_{12}^2}{2G_{12}(1-D_S)} + \frac{\sigma_{23}^2}{2G_{23}(1-D_S)} + \frac{\sigma_{13}^2}{2G_{13}(1-D_S)}\,, \tag{7.107}$$

with ϕ as a nonlinear function such as (7.98) and where the subscripts 1, 2, and 3 designate the fibers direction, the transverse direction inside the layer, and the normal direction, respectively, and $\langle.\rangle$ stands for the positive part. The elasticity coupled with damage laws is derived from the Gibbs potential. A delay effect is introduced here for regularization purpose within the damage evolution laws: for the damage variables D_S and D_T, one considers

$$\dot{D}_{S\ or\ T} = \dot{D}_\infty \left[1 - \exp\left(-a\langle\kappa_{\text{Lad}}^{-1}(\hat{Y}) - D_{S\ or\ T}\rangle\right)\right], \tag{7.108}$$

which recovers the laws $D_{S\ or\ T} = \kappa_{\text{Lad}}^{-1}(\hat{Y}) = \left\langle \frac{\sqrt{\hat{Y}} - \sqrt{Y_D}}{S_{S\ or\ T}} \right\rangle$ of Sect. 7.4.1 for low strain rates and which bounds the damage rates to $\dot{D}_\infty$ at high strain rates (with $\dot{D}_\infty$ and a as material parameters). Note that the inelastic strains in the layers (due to the microcracks sliding with friction) are taken into account through the consideration of a plasticity-like yield function, such as $f < 0$ corresponding to elasticity and $f = 0$ to internal sliding:

$$f = \sqrt{\tilde{\sigma}_{12}^2 + \tilde{\sigma}_{23}^2 + \tilde{\sigma}_{13}^2 + \chi^2(\tilde{\sigma}_{22}^2 + \tilde{\sigma}_{33}^2)} - R - \tau_y \leq 0, \tag{7.109}$$

with R as the hardening function, τ_y as the yield stress and χ as a material parameter. The effective stresses are then

$$\begin{aligned}\tilde{\sigma}_{11} &= \sigma_{11}, \\ \tilde{\sigma}_{22} &= \frac{\langle\sigma_{22}\rangle}{1-D_T} - \langle-\sigma_{22}\rangle, \\ \tilde{\sigma}_{33} &= \frac{\langle\sigma_{33}\rangle}{1-D_T} - \langle-\sigma_{33}\rangle, \\ \tilde{\sigma}_{ij} &= \frac{\sigma_{ij}}{1-D_S} \quad (i \neq j).\end{aligned} \tag{7.110}$$

Next, mesh the imperfect interfaces by using interface elements (isoparametric elements here) with the interface damage model developed by O. Allix and P. Ladevèze (1992). It considers an interface energy per unit area function of the components σ_{i3} of the normal stress vector (constant across the interface) and 3 damage variables d_1, d_2, and d_3 (anisotropic interface):

$$\Psi^\star = \frac{\langle\sigma_{33}\rangle^2}{2k_3(1-d_3)} + \frac{\langle-\sigma_{33}\rangle^2}{2k_3} + \frac{\sigma_{13}^2}{2k_1(1-d_1)} + \frac{\sigma_{23}^2}{2k_2(1-d_2)}, \tag{7.111}$$

where $\vec{n} = \vec{e}_3$ is the interface normal (see also Fig. 7.4 of Sect. 7.2.2). The interface stiffness terms k_i are material parameters which depend on the thickness estimated for the physical interface (1 µm here).

The variables associated with the stresses σ_{i3} are the displacement discontinuities $[\![u_i]\!]$ at the interface and the interface elasticity law coupled with damage derived from the thermodynamics potential $\Psi^\star$:

$$[\![u_1]\!] = \frac{\partial \Psi^\star}{\partial \sigma_{13}} = \frac{\sigma_{13}}{k_1(1-d_1)},$$

$$[\![u_2]\!] = \frac{\partial \Psi^\star}{\partial \sigma_{23}} = \frac{\sigma_{23}}{k_2(1-d_2)}, \quad (7.112)$$

$$[\![u_3]\!] = \frac{\partial \Psi^\star}{\partial \sigma_{33}} = \frac{\langle \sigma_{33} \rangle}{k_3(1-d_3)} - \frac{\langle -\sigma_{33} \rangle}{k_3},$$

as do the strain energy release rates,

$$Y_1 = \frac{\partial \Psi^\star}{\partial d_1} = \frac{\sigma_{13}}{2k_1(1-d_1)^2},$$

$$Y_2 = \frac{\partial \Psi^\star}{\partial d_2} = \frac{\sigma_{23}}{2k_2(1-d_2)^2}, \quad (7.113)$$

$$Y_3 = \frac{\partial \Psi^\star}{\partial d_3} = \frac{\langle \sigma_{33} \rangle}{2k_3(1-d_3)^2}.$$

The interface damage evolution law is written here in a delayed form bounding the damage rate (with S_I, s_I, $\dot{\delta}_\infty$ and a_I the interface damage parameters), as represented by

$$\dot{\delta} = \dot{\delta}_\infty \left(1 - \exp\left\{-a_\mathrm{I}\left[\left(\frac{\hat{Y}_\mathrm{I}}{S_\mathrm{I}}\right)^{s_\mathrm{I}} - \delta\right]\right\}\right) \quad (7.114)$$

$$\hat{Y}_\mathrm{I} = [Y_3^{\gamma_0} + (\gamma_1 Y_1)^{\gamma_0} + (\gamma_2 Y_2)^{\gamma_0}]^{1/\gamma_0},$$

where $d_1 = d_2 = d_3 = \delta$ here.

Perform then the damage post-processing of the zone of interest such as around perforations, holes, or rivets. Use as boundary conditions the displacement field along the multilayered plate medium plane computed in the elastic reference calculation. Be careful to apply this field on a single medium line along the border of the reanalyzed zone: imposing displacements all over the thickness nodes would lead to edges effects and non-physical damaged zones localized along this border.

As an example, consider the M55J/M18 high modulus carbon-fiber/epoxy-resin laminate with stacking sequence $[0_3/\pm 45_2/90]_S$ and which is loaded in tension (Fig. 7.23). The testing specimen is 50 mm in width, 150 mm in gauge length, and has a 10 mm diameter hole. The ply thickness is 0.125 mm. The tensile test is performed at a fixed displacement rate of 0.5 mm·min^{-1}. Figure 7.24 shows the evolution of the X-ray revealed damage map near the hole for an increasing applied load. Rupture is obtained for a load of 430 MPa.

Fig. 7.23. Carbon-fiber/epoxy-resin M55J/M18 holed specimen

At 55% of the rupture load, both transverse cracking (in 90°-plies near the hole) and matrix cracking (in the 0°-plies near the hole in the fiber direction, phenomenon also called "splitting") occur. Delamination initiates later, at about 80% of the rupture load. The delaminated area develops and remains located between the two splitting lines. At 97% of the rupture load, its length is about the equivalent of two hole diameters.

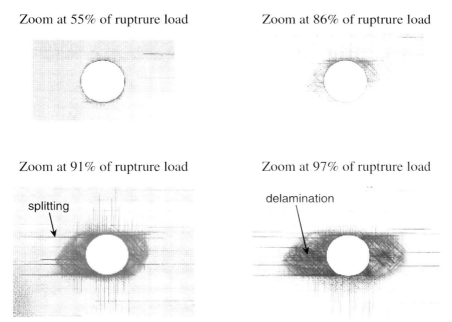

Fig. 7.24. X-ray photographs of the M55J/M18 holed specimen (D. Levêque 1998)

The damage reanalysis of the circular zone around the hole is performed by post-processing an initial elastic multilayered plate analysis. The DSDM LMT-Cachan computer code is used with the following material parameters:

- Single ply in the orthotropy framework ($\vec{e}_1$: fibers direction):
  $E_1 = 311000$ MPa, $E_2 = 6350$ MPa, $E_3 = 20000$ MPa, $\nu_{12} = \nu_{13} = 0.35$, $\nu_{23} = 0.48$, $G_{12} = G_{13} = G_{23} = 4870$ MPa, $\sigma_{RF} = 1760$ MPa, $b_T = 0$,

$Y_{SD} = Y_{TD} = 0$, $S_S = 2.55$ MPa$^{1/2}$, $S_T = 1.96$ MPa$^{1/2}$, $\dot{D}_\infty = 500$ s^{-1}, $a = 1$, $\chi = 1.3$, $\tau_y = 17$ MPa, $R(p) = 1445 \cdot p^{1/2}$ MPa.

- 0°-0° interface: $k_1 = k_2 = k_3 = 4 \cdot 10^5$ N/mm^{-3}, $\gamma_0 = 1.6$, $\gamma_1 = \gamma_2 = 0.4$, $a_I = 1$, $\dot{\delta}_\infty = 1000$ s^{-1}, $S_I = 0.66$ N/mm^{-1}, $s_I = 0.2$.
- $\pm\theta$ interfaces: $k_1 = k_2 = 3 \cdot 10^4$ N/mm^{-3}, $k_3 = 4 \cdot 10^4$ N/mm^{-3}, $\gamma_0 = 1.2$, $\gamma_1 = \gamma_2 = 0.4$, $a_I = 1$, $\dot{\delta}_\infty = 1000$ s^{-1}, $S_I = 0.54$ N/mm^{-1}, $s_I = 0.5$.

The damage maps computed in the layers and interfaces are given in Figs. 7.25 and 7.26. The delaminated areas (in black) correspond to the zones where the damage $d_3 = 1$. Note that when the 0°-fibers break near the hole, the local load is transferred by shear in the matrix in the adjacent fibers. The shear damage D_S is then the indicator for splitting.

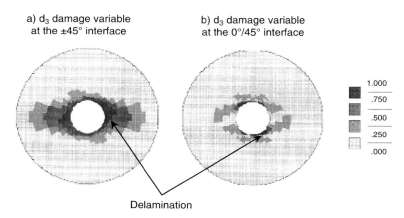

Fig. 7.25. Computed damage maps at the interfaces (at rupture load)

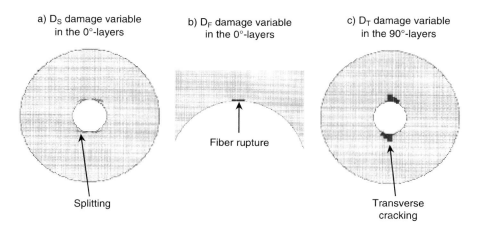

Fig. 7.26. Computed damage maps in the layers (at rupture load)

In conclusion, damage mechanics is an adequate tool to predict the damage pattern in composite materials and structures but it needs an important material database. It models interface cracking and delamination phenomenon and can therefore be used to optimize the layers sequence. Finally, note that in such quasi-static fracture the material constants $\dot{D}_\infty$, a, $\dot{\delta}_\infty$, and a_I do not play an important role but they become of most importance in dynamic cases such as chock problems.

7.4.5 Failure of CMC Structures♂♂
(F. Hild, P.-L. Larsson, F.A. Leckie 1994)

Due to long fiber reinforcement, ceramic matrix composites (CMC) are orthotropic and quasi-brittle. The strength of structures made of CMC can be checked either by using the Weibull analysis of Sect. 7.2.3 or Continuum Damage Mechanics. The aim of this section is to illustrate the second possibility with the orthotropic damage model for composite layers (see Sect. 7.4.1) extended to transverse isotropy.

The 3D elasticity coupled with damage model is obtained by completing the thermodynamics potential (7.97) with terms ensuring that the directions 2 and 3 play the same role. Perform then the damage computations in a fully coupled FE analysis, bearing in mind that once strain localization takes place, a mesocrack should be initiated and the FE solutions become mesh-dependent.

As an example, consider a spinning disk and a spinning ring made of CMC fiber reinforced in the hoop direction (Fig. 7.27). Both the disk and the ring have an outer radius $a = 0.3$ m. The inner radius of the ring is $b = a/2 = 0.15$ m. The two structures are thin enough for the plane stress assumption to apply and the cylindrical coordinates system (r, φ) is chosen. The loading is due to a rotation at the angular velocity ω and ρ denotes the material density.

The constitutive model is elasticity coupled with damage and is locally orthotropic (tangential to the fibers direction). The damage variable is defined as the percentage of broken fibers and is given by the corresponding failure probability (Eq. (7.37) with $\sigma_{\varphi\varphi} = E_\varphi(1 - D_\varphi)\epsilon_{\varphi\varphi}$). Thus

$$D_\varphi = 1 - \exp\left\{-\frac{L}{L_0}\left[\frac{2E_\varphi Y_\varphi}{\phi^2 \sigma_\mathrm{w}^2}\right]^{\frac{m_\mathrm{w}}{2}}\right\} \quad \text{and} \quad Y_\varphi = \frac{\sigma_{\varphi\varphi}^2}{2E_\varphi(1-D_\varphi)^2} \quad (7.115)$$

if $\epsilon_{\varphi\varphi} > 0$ and $\dot{\epsilon}_{\varphi\varphi} > 0$ and where the Weibull parameters m_w and σ_w are identified from tensile tests made on fibers of length L_0 and ϕ is the fiber volume fraction in the hoop direction.

Due to distributed pull-out in these spinning structures, the evolution of fiber breakage is not dictated by the length of the fibers but by the strength over which, once a fiber is broken, the tensile stress field recovers its original

7 Failure of Brittle and Quasi-Brittle Materials

level. This length is a function of the fiber radius R_f and the interfacial shear stress τ_I which is assumed to be constant along the interface cracks. For weak interfaces (τ_I < 10 to 15 MPa here), a shear lag analysis using the global load sharing assumption gives (R.B. Henstenburg and S.L. Phoenix 1989, W.A. Curtin 1989)

$$L = \frac{R_f \sigma_{\varphi\varphi}}{\phi \tau_I (1 - D_\varphi)} \tag{7.116}$$

and leads to a damage law representing the gradual fiber multifragmentation which is less conservative than the initial Weibull law:

$$D_\varphi = 1 - \exp\left\{-\left[\frac{2E_\varphi Y_\varphi}{\phi^2 \sigma_c^2}\right]^{\frac{m_w+1}{2}}\right\} \quad \text{and} \quad \sigma_c = \left(\frac{\sigma_w^{m_w} L_0 \tau_I}{R_f}\right)^{\frac{1}{m_w+1}}. \tag{7.117}$$

Fully coupled computations are performed with the ABAQUS code under a plane stress assumption. Satisfactory results are obtained with only 20 elements. The material parameters are:

- $E_\varphi = 140\ 10^9$ Pa, $E_r = 20\ 10^9$ Pa, $G_{r\varphi} = 13\ 10^9$ Pa, and $\nu_{r\varphi} = 0.214$ for elasticity
- $m_w = 4$, $\phi\sigma_w = 1450\ 10^6$ Pa, $L_0 = 12.6\ 10^{-3}$ m, and $\phi\sigma_c = 1300\ 10^6$ Pa for damage

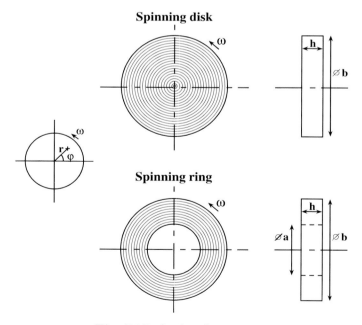

Fig. 7.27. Analyzed structures

Figure 7.28 compares the stress field in the hoop direction of the ring obtained for elasticity coupled with damage to the stress field obtained in pure elasticity. There is a stress redistribution and the global stiffness of the damaged structure softens as the applied load increases.

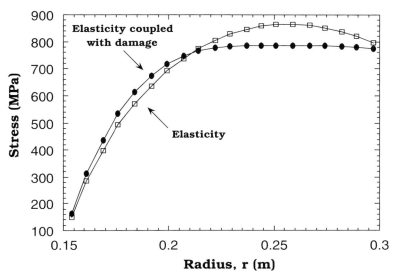

Fig. 7.28. Hoop stress in the ring $\sigma_{\varphi\varphi}$ as a function of the radius ($\rho\omega^2 = \rho\omega_{\text{loc}}^2 = 2.26 \cdot 10^{10}$ kg·m^{-2}·s^{-2})

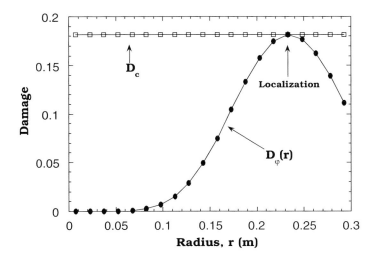

Fig. 7.29. Damage in the disk as a function of the radius ($\rho\omega^2 = \rho\omega_{\text{loc}}^2 = 1.87 \cdot 10^{10}$ kg·m^{-2}·s^{-2})

The damage field computed in the disk is given in Fig. 7.29 where the critical damage $D_c = 0.18$ corresponds to strain localization occurrence. Note that the model predicts strain localization and therefore, mesocrack initiation inside the structure. There is also a difference in the computed load levels at localization $\rho\omega_{\text{loc}}^2$ between the two structures studied: $\rho\omega_{\text{loc}}^2 = 1.87 \cdot 10^{10}$ kg·m^{-2}·s^{-2} for the disk and $\rho\omega_{\text{loc}}^2 = 2.26 \cdot 10^{10}$ kg·m^{-2}·s^{-2} for the ring. These values are found to be 20 to 25% larger than the values obtained with the initial Weibull law (7.115) which corresponds to a more conservative design, i.e., heaviest mechanical components.

To conclude, FE damage analyses of quasi-brittle CMC structures can be performed up to strain-damage localization (the failure criterion considered here). They give the location of the fiber reinforcement weakness in the ring and can be used to design non-uniformly reinforced disks and rings and to optimize the fiber density distribution.

7.4.6 Single and Multifragmentation of Brittle Materials♂♂
(C. Denoual and F. Hild 1997)

Pure brittle materials may break either by single cracking or multifragmentation depending on the stress rate. This is the case for glass, concrete, ceramics, and rocks when impacted by a projectile.

At low stress rate, failure occurs as a result of a single crack propagation with a high degree of scatter. A classical analysis using the $\sigma^\star$ criterion (see Sect. 7.2.1) or Weibull analysis (see Sect. 7.2.3) gives the failure conditions. In dynamics, fragmentation occurs with a more deterministic behavior and the microcrack pattern can be computed using Continuum Damage Mechanics. In order to make proper predictions, it is important to know a priori the transition between the two mechanisms.

The Weibull model corresponds to the initiation and immediate growth of a crack in a brittle material. In dynamics, the degradation mechanism is complex because only a part of the unstable defects or flaws lead to microcrack growth. This is due to a stress relaxation (or obscured) zone which develops around each initiated microcrack during the propagation of the stress wave. To illustrate this phenomenon, the microcrack initiation is represented on the space-time graph of Fig. 7.30 where the first initiation occurs at time t_1 at the space location M_1 for the stress $\sigma^\star(t_1)$. This produces an obscured zone of volume V_{obs} as a function of $t - t_1$ in which no flaws become unstable.

Assuming self similar relaxation (or interaction) zones of size governed by a constant fraction kc_L of the longitudinal stress wave velocity c_L, we have

$$V_{\text{obs}} = V_{\text{obs}}(t - t_1) = \kappa \left[kc_L \cdot (t - t_1) \right]^3 , \qquad (7.118)$$

where κ is a shape parameter. The quantity $kc_L \cdot (t - t_1)$ is a representative length of the relaxation zone around a microcrack. At time t_2 corresponding to $\sigma^\star(t_2) > \sigma^\star(t_1)$, a second microcrack starts to propagate and produces its own obscured zone (Fig. 7.30). The third and fourth defects do not nucleate here as they are obscured by the first two microcracks.

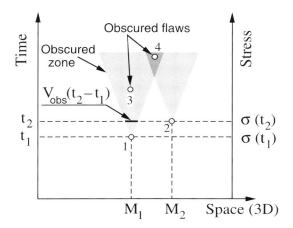

Fig. 7.30. Defects obscuration in a space-time graph

To model this dynamic process consider, as in a Weibull analysis, a flaw distribution λ_{tot} described by a Poisson process of intensity,

$$\lambda_{tot}(\sigma^\star) = \lambda_0 \left[\frac{\sigma^\star}{\sigma_0}\right]^{m_w}, \qquad (7.119)$$

with m_w as the Weibull modulus and σ_0 as the scale parameter relative to a reference flaw density λ_0. The quantity λ_{tot} represents the mean number of flaws that may break in a unit volume for a local stress less than or equal to $\sigma^\star$.

At the beginning of the loading, no interactions occur and the crack density (denoted by λ_{crack}) is equal to the total density of the flaws able to break, λ_{tot}. Due to the dynamic relaxation process, λ_{crack} becomes smaller than λ_{tot} following the law

$$\frac{d\lambda_{crack}}{dt} = \frac{d\lambda_{tot}}{dt} \cdot [1 - P_{obs}], \qquad (7.120)$$

with $\lambda_{crack}(t=0) = 0$ and where P_{obs} is the probability of defects obscuration or relaxation which defines the damage variable,

$$D \equiv P_{obs} = 1 - \exp\left(-\int_0^t \frac{d\lambda_{tot}}{d\tau} V_{obs}(t-\tau)d\tau\right). \qquad (7.121)$$

Knowing the local stress history $\sigma^\star(t)$ gives the flaw distribution history $\lambda_{tot}(t)$ (7.119), then the probability of defects obscuration (7.121), and finally by time integration of (7.120) the crack density.

7.4.6.1 Quasi-Static Weibull Model Recovered

The Weibull model corresponds to the case of failure due to a single defect (weakest link hypothesis). Setting $V_{\text{obs}} = const = V_{\text{eff}}$ as the effective volume (7.27), the probability of failure is equal to P_{obs}, i.e.,

$$P_{\text{F}} = 1 - \exp\left(-\int_0^t \frac{\mathrm{d}\lambda_{\text{tot}}}{\mathrm{d}\tau} V_{\text{eff}} \mathrm{d}\tau\right) = 1 - \exp\left(-\lambda_{\text{tot}} V_{\text{eff}}\right), \quad (7.122)$$

which recovers Weibull failure probability (7.28), or

$$P_{\text{F}} = 1 - \exp\left(-\frac{V_{\text{eff}}}{V_0}\left[\frac{\sigma^\star}{\sigma_{\text{w}}}\right]^{m_{\text{w}}}\right) \quad (7.123)$$

for the Poisson process (7.119).

7.4.6.2 Fragmentation to Multifragmentation Transition

A closed-form solution of previous set of equations (7.119)–(7.122) can be derived in the case of the impact loading $\sigma^\star = \dot{\sigma} t$ with constant stress rate $\dot{\sigma}$:

$$\lambda_{\text{tot}} = \lambda_0 \left(\frac{\dot{\sigma}}{\sigma_0}\right)^{m_{\text{w}}} t^{m_{\text{w}}} = \lambda_{\text{c}} \hat{t}^{m_{\text{w}}},$$

$$\lambda_{\text{crack}} = \lambda_{\text{c}} \frac{m_{\text{w}}}{m_{\text{w}} + 3} \left[\frac{(m_{\text{w}} + 3)!}{6 m_{\text{w}}!}\right]^{\frac{m_{\text{w}}}{m_{\text{w}}+3}} \gamma\left(\frac{m_{\text{w}}}{m_{\text{w}} + 3}, \frac{6 m_{\text{w}}!}{(m_{\text{w}} + 3)!} \hat{t}^{m_{\text{w}}+3}\right),$$

$$D = P_{\text{obs}} = 1 - \exp\left(-\frac{6 m_{\text{w}}!}{(m_{\text{w}} + 3)!} \hat{t}^{m_{\text{w}}}\right).$$

$$(7.124)$$

The function of the dimensionless time $\hat{t} = t/t_{\text{c}}$ introduces the characteristic time t_{c}, stress rate $\dot{\sigma}_{\text{c}}$, and microcrack density λ_{c}. The term γ is the incomplete gamma function, $\gamma(a, x) = \int_0^x t^{a-1} \exp(-t) \mathrm{d}t$, such that

$$t_{\text{c}} = \left(\frac{\sigma_0^{m_{\text{w}}}}{\kappa (k c_{\text{L}})^3 \lambda_0 \dot{\sigma}^{m_{\text{w}}}}\right)^{\frac{1}{m_{\text{w}}+3}} \quad \text{and} \quad \lambda_{\text{c}} = \frac{1}{V_{\text{c}}} = \left(\frac{\lambda_0^{\frac{1}{m_{\text{w}}}} \dot{\sigma}}{\kappa^{\frac{1}{3}} k c_{\text{L}} \sigma_0}\right)^{\frac{3 m_{\text{w}}}{m_{\text{w}}+3}}.$$

$$(7.125)$$

The characteristic volume V_{c} contains, on average, one flaw that may break at the characteristic time t_{c} ($\lambda_{\text{c}} V_{\text{c}} = 1$).

The applied stress denoted here, Σ, is related to the local or effective stress via $\sigma^\star = \Sigma/(1 - D)$ so that the stress-rate-dependent tensile strength σ_{u} is given by $\mathrm{d}\Sigma/\mathrm{d}\sigma = 0$ or

$$\sigma_{\text{u}} = \sigma_{\text{c}} \left(\frac{(m_{\text{w}} + 2)!}{6 e \cdot m_{\text{w}}!}\right)^{\frac{1}{m_{\text{w}}+3}} \quad \text{where} \quad \sigma_{\text{c}} = \left(\frac{\sigma_0^{m_{\text{w}}} \dot{\sigma}^3}{\kappa (k c_{\text{L}})^3 \lambda_0}\right)^{\frac{1}{m_{\text{w}}+3}}. \quad (7.126)$$

This is the dashed line (stress-rate-dependent) in the dimensionless tensile strength versus dimensionless stress rate diagram (Fig. 7.31) in which the solid circles (average) and bars (standard deviation) are given by Monte-Carlo simulations (500 realizations for each point). Figure 7.31 also shows the stress-rate-independent results for a classical Weibull analysis from which the mean failure stress is (Γ is the Euler Γ-function)

$$\overline{\sigma} = \frac{\sigma_0}{(\lambda_0 V_{\text{eff}})^{\frac{1}{m_w}}} \Gamma\left(1 + \frac{1}{m_w}\right) \tag{7.127}$$

and the corresponding standard deviation is

$$\overline{\overline{\sigma}} = \frac{\sigma_0}{(\lambda_0 V_{\text{eff}})^{\frac{1}{m_w}}} \sqrt{\Gamma\left(1 + \frac{1}{m_w}\right) - \Gamma^2\left(1 + \frac{1}{m_w}\right)}. \tag{7.128}$$

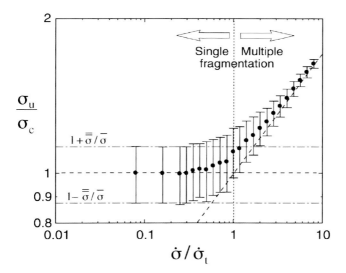

Fig. 7.31. Normalized tensile strength σ_u/σ_c vs normalized stress rate $\dot{\sigma}/\dot{\sigma}_t$ for $m_w = 10$ and $V_{\text{eff}} = 1/\lambda_0$ (C. Denoual and F. Hild 1997)

The stress-rate-independent regime $\dot{\sigma} < \dot{\sigma}_t$ corresponds to single fragmentation. It is obtained from the classical Weibull analysis when the weakest link hypothesis is made. The regime $\dot{\sigma} > \dot{\sigma}_t$ corresponds to multifragmentation. It is dependent on the stress rate and is obtained from the dynamic obscuration mechanism. The transition between both regimes occurs at the stress rate $\dot{\sigma} = \dot{\sigma}_t$, defined as the intersection between "quasi-static" and "dynamic" strengths,

$$\sigma_{\mathrm{u}}(\dot{\sigma} = \dot{\sigma}_{\mathrm{t}}) = \overline{\sigma}, \tag{7.129}$$

giving

$$\boxed{\dot{\sigma}_{\mathrm{t}} = \sigma_0 k c_{\mathrm{L}} (\lambda_0 \kappa)^{\frac{1}{3}} (\lambda_0 V_{\mathrm{eff}})^{\frac{m_{\mathrm{w}}+3}{3 m_{\mathrm{w}}}} \left[\frac{6 e \cdot m_{\mathrm{w}}!}{(m_{\mathrm{w}}+2)!} \Gamma^{m_{\mathrm{w}}+3} \left(1 + \frac{1}{m_{\mathrm{w}}}\right) \right]^{\frac{1}{3}}.}$$
(7.130)

Remark — When $\dot{\sigma} > \dot{\sigma}_{\mathrm{t}}$, Continuum Damage Mechanics can be used to model the multifragmented microcrack pattern. According to the definition of damage as the probability of obscuration,

$$\ln(1 - D) = -\int_0^t \frac{\mathrm{d}\lambda_{\mathrm{tot}}}{\mathrm{d}\tau} V_{\mathrm{obs}}(t - \tau) \mathrm{d}\tau = -\int_0^t \frac{\mathrm{d}\lambda_{\mathrm{tot}}}{\mathrm{d}\tau} \kappa \left[k c_{\mathrm{L}} \cdot (t - \tau)\right]^3 \mathrm{d}\tau$$

and a damage evolution law for high stress rates ($> \dot{\sigma}_{\mathrm{t}}$) is obtained by eliminating the explicit dependency upon time:

$$-\frac{\mathrm{d}^3}{\mathrm{d}t^3} \ln(1 - D) = 6\kappa (k c_{\mathrm{L}})^3 \lambda_{\mathrm{tot}}(\sigma^*),$$

with $D(t=0) = 0$, $\dot{D}(t=0) = 0$, and $\ddot{D}(t=0) = 0$ due to $\lambda_{\mathrm{tot}}(t=0) = 0$. Using $Y = \sigma^{\star 2}/2E$, it can be written as a function of the strain energy release rate,

$$\frac{\mathrm{d}^2}{\mathrm{d}t^2}\left(\frac{1}{1-D}\frac{\mathrm{d}D}{\mathrm{d}t}\right) = \frac{6\kappa(kc_{\mathrm{L}})^3}{V_0}\left[\frac{2EY}{\sigma_{\mathrm{w}}^2}\right]^{\frac{m_{\mathrm{w}}}{2}}.$$

To conclude, the damage law at low stress rate has to be of the form $D = D(Y)$ when a differential equation in time is needed at high stress rate.

7.4.7 Hierarchic Approach up to Homogenized Behavior^{♂,♂♂♂}

Failure of brittle or quasi-brittle components is mainly governed by initial defects that are random in their size and space distribution. Therefore, the accuracy of prediction is often poor for at least two reasons: high scatter of basic tests results to identify the parameters and ignorance of the state of initial defects. Nevertheless, speaking of probabilities gives a sense to these uncertainties.

For brittle materials or interfaces:

- In absence of any information other than an ultimate stress, use the damage equivalent stress criterion (and not the von Mises stress!),

$$\sigma^\star = \sigma_{\mathrm{eq}} R_\nu^{1/2} = \sigma_{\mathrm{u}} \quad \text{with} \quad R_\nu = \frac{2}{3}(1+\nu) + 3(1-2\nu)\left(\frac{\sigma_{\mathrm{H}}}{\sigma_{\mathrm{eq}}}\right)^{1/2}. \tag{7.131}$$

Introducing the microdefects closure is an improvement if some compression occurs.

- The Mazars damage equivalent strain is better for concrete:
$$\epsilon^\star = \sqrt{\langle\epsilon\rangle^+ : \langle\epsilon\rangle^+} = \frac{\sigma_u^+}{E} = \nu\sqrt{2}\frac{\sigma_u^-}{E}. \quad (7.132)$$

- For interfaces a debonding criterion which needs two material parameters may be applied:
$$\left\langle \sigma_{33}|\sigma_{33}| + \left(\frac{\sigma_R^I}{\tau_R^I}\right)^2 (\sigma_{13}^2 + \sigma_{23}^2) \right\rangle^{1/2} = 1. \quad (7.133)$$

- The Weibull model should be used to characterize the probability of rupture but the material parameters need 10 to 20 specimens for their identification:
$$P_F = 1 - \exp\left\{-\frac{V_{\text{eff}}}{V_0}\left[\frac{\sigma^\star}{\sigma_w}\right]^{m_w}\right\}. \quad (7.134)$$

Using the two-scale damage model is a way to predict the rupture of quasi-brittle materials that occurs when a dissipation prior to cracking exists (if some fatigue results complete the identification database!) but several other damage models have their specific applications:

- The Marigo model with or without microdefects closure effects is a general model valid for most materials while the anisotropic damage model of Sect. 7.4.1 is suitable for concrete
- The Laborderie model with permanent strains and microdefects closure effects in dynamics (seismic effects on civil engineering structures)
- Mesomodels for composites where three damage variables are considered
- Probabilistic models for ceramics and fragmentation in dynamics

Finally, elasticity and damage models for reinforced concrete are of main importance in civil engineering. Due to the size of the structures, it is interesting in FE computations to avoid meshing the steel bars and the concrete body separately. Homogenization procedures give the equivalent elastic properties of heterogeneous materials. They apply to undamaged reinforced concrete and uniaxial bar reinforcement, yielding orthotropic elasticity characteristics. For example, if the steel spacing is the same in the two tranverse directions, tranverse isotropy is obtained with longitudinal and transverse Young's moduli E_L and E_T, Poisson ratios ν_{LT} and ν_T and shear moduli G_{LT} and G_{TT}:

$$\begin{aligned}
E_L &= E_L(\phi, E_c, E_s, \ldots), \\
E_T &= E_T(\phi, E_c, E_s, \ldots), \\
\nu_{LT} &= \nu_{LT}(\phi, E_c, E_s, \nu_c, \nu_s), \\
\nu_T &= \nu_T(\phi, E_c, E_s, \nu_c, \nu_s), \\
G_{LT} &= G_{LT}(\phi, E_c, E_s, \nu_c, \nu_s), \\
G_{TT} &= \frac{E_T}{2(1+\nu_T)},
\end{aligned} \quad (7.135)$$

whose specific expressions depend on the homogenization procedure. They are functions of steel volume fraction (ϕ), the elastic properties of concrete (E_c, ν_c) and steel (E_s, ν_s).

The question of coupling with damage arises then and one needs to take into account the microcrack closure effect in concrete. One of the simplest possible models uses the anisotropic and damage framework of Sect. 7.4.1, but extended to tranverse anisotropy,

$$\rho\psi^\star = \frac{\langle\sigma_{11}\rangle^2}{2\tilde{E}_L^+} + \frac{\langle-\sigma_{11}\rangle^2}{2\tilde{E}_L^-} + \frac{\langle\sigma_{22}\rangle^2 + \langle\sigma_{33}\rangle^2}{2\tilde{E}_T^+} + \frac{\langle-\sigma_{22}\rangle^2 + \langle-\sigma_{33}\rangle^2}{2\tilde{E}_T^-}$$
$$- \frac{\nu_{LT}}{E_L}(\sigma_{11}\sigma_{22} + \sigma_{11}\sigma_{33}) - \frac{\nu_{TT}}{E_T}\sigma_{22}\sigma_{33} + \frac{\sigma_{12}^2 + \sigma_{13}^2}{2G_{LT}} + \frac{\sigma_{23}^2}{2G_{TT}}, \quad (7.136)$$

where the coupling with damage is reduced to the minimum by making the damage variable D_L act on the longitudinal modulus, the damage variable D_T on the tranverse modulus, and by neglecting the coupling of the ratios ν_{ij}/E_j and the shear moduli with damage:

$$\begin{aligned}\tilde{E}_L^+ &= E_L(\phi, E_c(1-D_L), E_s, \ldots), \\ \tilde{E}_L^- &= E_L(\phi, E_c(1-hD_L), E_s, \ldots), \\ \tilde{E}_T^+ &= E_T(\phi, E_c(1-D_T), E_s, \ldots), \\ \tilde{E}_T^- &= E_T(\phi, E_c(1-hD_T), E_s, \ldots),\end{aligned} \quad (7.137)$$

where h is the microdefects closure parameter.

The damage evolution laws are written as $D = D(Y)$ laws (such as Eqs. (7.76) to (7.79))

$$D_L = D_L(Y_{L\,max}) \quad \text{and} \quad D_T = D_T(Y_{T\,max}), \quad (7.138)$$

with $Y_{L\,max}$ and $Y_{T\,max}$ as the maximum values reached during the loading of the longitudinal and transverse strain energy release rates

$$Y_L = \rho\frac{\partial\psi^\star}{\partial D_L} \quad \text{and} \quad Y_T = \rho\frac{\partial\psi^\star}{\partial D_T}. \quad (7.139)$$

Using the simple mixture laws

$$E_L = \phi E_s + (1-\phi)E_c \quad \text{and} \quad E_T = \left(\frac{\phi}{E_s} + \frac{1-\phi}{E_c}\right)^{-1} \quad (7.140)$$

gives:

$$\begin{aligned}Y_L &= \frac{(1-\phi)E_c}{2}\left[\frac{\langle\sigma_{11}\rangle^2}{\tilde{E}_L^{+\,2}} + h\frac{\langle-\sigma_{11}\rangle^2}{\tilde{E}_L^{-\,2}}\right] \\ Y_T &\approx \frac{1-\phi}{2E_c}\left[\frac{\langle\sigma_{22}\rangle^2 + \langle\sigma_{33}\rangle^2}{(1-D_T)^2} + h\frac{\langle-\sigma_{22}\rangle^2 + \langle-\sigma_{33}\rangle^2}{(1-hD_T)^2}\right]\end{aligned} \quad (7.141)$$

which take into account (through h) the much lower damage growth for concrete in compression than in tension.

...
Congratulations to all readers who read up to this last page!
...

Bibliography

1. *Introduction to Continuum Damage Mechanics*, L.M. Kachanov (Martinus Nijhoff Dortrecht, The Netherlands 1986)
2. *Continuum Damage Mechanics Theory and Applications*, D. Krajcinovic, J. Lemaitre, CISM Lectures (Springer, Berlin Heidelberg New York 1987)
3. *A Course on Damage mechanics*, J. Lemaitre (Springer, Berlin Heidelberg New York 1992–1996)
4. *Damage Mechanics*, D. Krajcinovic, North Holland Series in Appl. Math and Mech (Elsevier, 1996)
5. *Modeling of Material Damage and Failure of Structures*, J. Skrzypek, A. Ganczarski (Springer, Berlin Heidelberg New York 1999)
6. *Advances in Damage Mechanics: Metals and Metal Matrix Composites*, G.Z. Voyiadjis, P.I. Kattan (Elsevier, 1999)
7. *Continuous Damage and Fracture*, ed. by A. Benallal (Elsevier, 2000)
8. *Handbook of Materials Behavior Models*, ed. by J. Lemaitre (Academic Press, New York 2001)
9. *Damage Mechanics with Finite Elements*, P.I. Kattan, G.Z. Voyiadjis (Springer, Berlin Heidelberg New York 2002)
10. *Formulas for Stress and Strain*, R.J. Roark (Mc Graw Hill, New York 1965)
11. *Stress Concentration Factors*, R.E. Peterson (Wiley, New York 1974)
12. *Handbook of Formulas for Stress and Strain*, W. Griffel (Frederick Ungar, 1976)

Index

ABAQUS, 175, 187, 261, 264, 362
accumulated plastic strain, 10
accumulation of damage, 196
adaptation, 211
adiabatic, 255
Aifantis, 113, 317
Albert, 277
Allix, 2, 356
aluminum, 175, 190
aluminum alloy, 267
Amar, 135
analysis
   fully coupled, 213
   post-processing, 214
   stochastic, 323
Andrade, 233
anisotropy, 21
   composites, 344
   damage, 4
   induced, 44, 51, 111
   plastic, 189
Armstrong–Frederick, 46, 229
Arrhenius, 255
ASTREE, 26, 219

Børvik, 269
Bažant, 112–114, 321
Babuska, 112
ballistic limit, 272
Batdorf, 321
Bathias, 282
Bauschinger, 277
Belytschko, 112, 114
Benallal, 65, 69, 77, 97, 230, 258
Bennani, 267
Berstad, 269
Berthaud, 18, 44
Besson, 97, 103, 105, 172, 214
BFGS method, 125
bifurcation, 66
Billardon, 22, 65, 77, 88, 97, 117, 174, 184, 187, 258
Bish, 352
Boussinesq, 315
Broyden, 125
buckling, 158
   post-$\sim$, 255
Burlet, 230

Cailletaud, 230
Calloch, 64, 190, 230
Cano, 70
Cantournet, 221
carbon-fiber/epoxy-resin, 358
Carmet, 303
CASTEM, 348
ceramic, 364
ceramic matrix composites, 361
Cerruti, 315
Chaboche, 4, 10, 233, 281
Chaudonneret, 245
Chow, 4
Chranowski, 233
Chrysochoos, 28
Chu, 177
Claise, 23
Clausius–Duhem inequality, 9

Coffin, 191
Coin, 352
Coleman, 330
composites, 2, 13, 339, 346, 361
concrete, 13, 343, 346, 364
condition
  consistency, 10, 43, 46, 340
  Kuhn–Tucker, 10
  quasi-unilateral, 12, 33
constitutive equations, 8, 45, 46
convergence, 99, 103, 111
Cordebois, 6, 172
CORRELI, 23
Cough, 277
Cox, 315
crack
  arrest, 152, 208, 336
  initiation, 27, 33, 65
  orientation, 73
crash, 267
creep, 237
  -damage, 33, 233, 240, 255
  -fatigue, 237, 239, 251, 262
Crisfield, 94
criterion
  damage $\sim$, 339
  Dang Van, 282
  Hill, 268
  of debonding, 327
  rupture, 324
  Sines, 282
  von Mises, 10, 45
  yield, 45, 83
Croix, 267
cross identification, 132, 166, 178
Curtin, 362
curve
  isochronous, 236
  Manson–Coffin, 192, 193
  R-$\sim$, 153
  Wöhler, 38, 64, 193
cyclic plasticity, 202

damage
  accumulation, 196, 217, 287, 302
  anisotropic, 5, 15, 18, 34, 40, 57, 59, 75, 80, 171, 343
  criterion, 65, 171, 339, 344
  critical, 31, 65

  definition, 3
  equivalent strain, 326
  equivalent stress, 11, 79, 157, 166, 188, 324, 328
  hydrostatic, 6
  interface, 358
  isotropic, 5, 10, 52, 53, 55, 75, 80
  multiplier, 340
  non-local $\sim$ model, 114, 344
  parameters, 47, 78, 138, 340, 343
  quasi-brittle, 33, 183, 339
  tensorial, 4
  threshold, 27, 43, 76, 130, 161, 300
  two-scale $\sim$ model, 283, 332
DAMAGE 2000, 119, 305, 311
Dang Van, 315
de Borst, 114
deep drawing, 170
delamination, 345, 356
Denoual, 364, 367
Deperrois, 315
Desmorat B., 345
Desmorat R., 14, 34, 97, 105, 172, 214, 221, 263
digital image correlation, 23
dissipation, 43, 60, 70, 222
distribution, 26, 278, 309, 365
  Gaussian, 113, 323
Doghri, 65, 77, 88, 97, 117, 174
Dolbow, 112
Doudard, 64
Dragon, 345
Dubé, 354
Dufailly, 17, 22, 48, 50, 135, 143
dynamic plasticity, 142, 234, 248, 255

earthquake, 351
effective volume, 329, 366
EFICOS, 354
elastic predictior, 117
elastic predictor, 95
elasticity change, 17, 54, 178
elasto-(visco-)plasticity, 27, 45, 53, 55, 57, 59, 78, 267
elastomers, 221, 274
elementary layer, 345
Elshelby, 120
energy density, 218
  effective elastic $\sim$, 12, 34, 85

release rate, 10, 12, 32, 80
   strain ∼, 41
energy, stored, 8, 9, 27, 38, 41, 197
Engel, 2
Eshelby, 62, 116, 316
Eshelby–Kröner law, 62, 116, 284, 315
Evans, 321
exponential law, 98
extrusion, 172

Fairbarn, 277
fatigue
   asymptotic ∼ limit, 29, 37, 61, 64, 278, 282
   creep-∼, 237, 251, 262
   engineering ∼ limit, 36, 278, 282
   mean stress effect, 62, 81, 287, 295
   multilevel, 216
   non-proportional, 305
fiber bundle, 330
Findley, 317
finite strains, 167, 221
Flórez-López, 181
Fletcher, 125
Forest, 114
FORGE 2, 174
forming limits, 170
fragmentation, 366
frames, 181, 271, 351
François, 190, 230
friction, 221, 223
fully coupled analysis, 213
function
   shape ∼, 91
   triaxiality ∼, 11

Galtier, 64, 312
Ganczarski, 233
Gatuingt, 346
Gauss–Newton method, 124
Gaussian distribution, 113, 323
geomaterials, 176
Geymonat, 65
Gibbs
   energy, 345, 356
   potential, 8, 357
   specific free enthalpy, 343
glass, 364
Golfarb, 125

Goodman diagram, 280
gradient effect, 316
Grange, 173, 305
Green–Lagrange, 167, 221
Griffith, 321
Gupta, 190
Gurson, 176

Halford, 191
Halm, 345
hardening
   isotropic, 8, 45
   kinematic, 8, 45, 82
Hart–Smith, 221
Hayhurst, 77, 233
heat capacity, 8, 70
Helmholtz potential, 8, 340
Hencky–Mises, 150, 161, 203
Henstenburg, 362
Hessian, 123, 125
Hild, 23, 321, 322, 329, 361, 364, 367
Hill, 66, 189, 268
homogenization, 61, 317, 369
Hopperstad, 269
Hult, 1, 233
hydrostatic sensitivity parameter, 6
hyperelasticity, 221, 275

identification, 47, 64, 120, 127, 135
   cross ∼, 178, 181
image correlation, 23, 138, 185
impact, 269, 366
impact loading, 366
implicit scheme, 96, 117
IN 100 stainless alloy, 236, 262
Inconel alloy, 48
initial defect, 307, 322
interface damage, 358
internal
   friction, 41
   sliding, 41, 43, 223, 275
intrusion-extrusion, 281

Jacobian, 124
   matrix, 100, 107
   terms, 101
Jaumann, 168
Johnson–Cook, 270
jump procedure, 88

378 Index

Kachanov, 1, 3, 233, 235
Kirchhoff–Love, 304
Klingele, 2
Kröner, 62, 116, 120
Krajcinovic, 4, 330
Kuhn–Tucker condition, 10
Kurtyka, 190, 230

Laborderie, 341, 353, 354
Ladevèze, 3, 11, 14, 168, 213, 341, 345
Lambert–Diani, 221
laminates, 345, 356
Langseth, 269
Lapra, 224
large scale yielding, 166, 245
Larsson, 361
Lauro, 267
law
    conduction, 70
    elasticity, 10, 14, 70
    evolution, 70
    exponential, 51, 98
    Hencky–Mises, 150, 161
    Norton, 51, 98, 244, 245, 248, 254, 255, 258
    Paris, 293
    state, 9, 70
    unified damage, 32, 37, 41
Lechatelier, 277
Leckie, 4, 77, 233, 261, 361
Le Dantec, 341, 346
Lemaitre, 1, 8, 11, 14, 33, 77, 120, 263
Levèque, 359
Levenberg–Marquardt method, 125
limit analysis, 166, 181
limit curve, 172
    ballistic ∼, 272
    localization ∼, 171
    strain ∼, 164
    stress ∼, 165
Lin, 114
Lin–Taylor, 120
localization
    Eshelby–Kröner law, 62, 116, 284, 315
    limiters, 112, 350
    strain ∼, 175
    strain ∼, 65, 114, 170
    tensor, 316

Lorentz, 114
LS-DYNA, 271
Luong, 64

Mac Clintoch, 141
Manson, 191
Manson–Coffin curve, 192, 193
maraging steel, 281
Marigo, 339
Mariott, 261
Marquis, 137, 190, 230
matrix
    ceramic ∼ composites, 361
    Jacobian, 100, 107
    sensitivity, 135, 146, 200, 242, 290, 334
Maugin, 168
Mazars, 44, 73, 321, 326, 350, 354
McClintock, 321
Melenk, 112
mesocrack initiation, 27, 31, 34, 65
method
    θ-, 98, 106
    BFGS, 125
    Gauss–Newton, 124
    Levenberg–Marquardt, 125
    Monte Carlo, 309
    Neuber, 147, 150, 152, 155, 201, 244
    Newton, 94, 124
    SQP, 123, 125
    strain energy density ∼, 201, 203
Meyers, 190
microdefects closure, 12, 171, 288, 324
    effect, 55, 59, 341
    parameter, 13
microhardness, 22, 186
microstress, 41, 62, 221
Miner rule, 197
Mises
    signed von ∼ stress, 82
    von ∼ criterion, 10, 45
    von ∼ equivalent stress, 11
Moës, 112
Molinari, 69
Monte Carlo method, 308
Mooney, 221
Moreau, 91
Moret–Bailly, 258
Morrow, 191

Moussy, 172
Mroz, 112, 230
Mulhaus, 114
multifragmentation, 362, 364
Murakami, 4

Needleman, 141, 176
negative part, 13, 15
Neuber method, 147, 150, 152, 155, 201, 244
Newton–Raphson method, 94, 124
Nguyen Q.S., 91
Nocedal, 123
non-local damage model, 114, 344
nonlinear accumulation, 280
normality rule, 9, 177, 222
Norton law, 43, 98, 233, 244, 245, 248, 254, 255, 258

Ožbolt, 113
Oh, 112
Ohno, 230
Onat, 4
Ortiz, 270

Palmgreen–Miner rule, 195, 215, 229, 280, 302
PAMCRASH 2000, 268
Panoskaltsis, 317
Papadopoulos, 317
parameter
  necking, 35
  viscosity, 50, 128, 265
Paris law, 293
perturbation, 69, 103
Phoenix, 362
Pietruszczak, 112
Pijaudier-Cabot, 113
Pilvin, 122
Pineau, 191, 215, 321
Piola–Kirchhoff, 167, 221
plastic
  corrector, 92, 96
  hinges, 160, 181, 255
Poisson process, 365
polymers, 13, 75, 167, 275
porosity, 176
positive part, 13, 15
post-buckling, 158, 255

post-processing analysis, 78, 85, 115, 214
post-processor DAMAGE 2000, 119, 301, 305, 311
potential
  Gibbs $\sim$, 8, 343, 357
  Helmholtz $\sim$, 8, 70, 339
  of dissipation, 9, 26, 42, 45, 222
  state $\sim$, 8, 15, 41, 221
powders, 176
pre-stress, 234, 346
predamaged, 184, 227, 272, 312
predeformed, 184, 227, 272, 312
principle
  of strain equivalence, 8, 10, 11, 42
  second, 9
probabilistic safety margin, 335
proportional loading, 82, 160

quasi-brittle materials, 321, 339

R-curve, 153
Rabotnov, 5, 233
Ragueneau, 345, 346, 350
Ramtani, 44
random loading, 297, 314
Representative Volume Element, 75
Rey, 221
Rice, 66, 141
Riks, 94
rocks, 364
Rousselier, 112, 176
Rousset, 190
Roux, 23
rubber, 86, 223
Rudnicki, 66

Saanouni, 172
safety factor, 152, 208, 247, 292, 334
Sauzay, 21, 34, 303, 315
scale effect, 281
scatter, 247, 292, 299, 307
SED method, 203, 204, 216
seismic loading, 350
sensitivity, 135, 146, 200, 242, 290, 334
Sermage, 130, 134, 218, 262, 263, 300, 313
Shah, 2
shakedown, 210

Shanno, 125
shape function, 91
shear band, 67, 112
Shi, 112
SIDOLO, 120, 132
Sidoroff, 6
Simo, 91, 103
Sines fatigue criterion, 282
Skrzypek, 233
sliding, 41, 43, 223, 275
small scale yielding, 86, 148, 201
Socie, 191
SQP method, 123, 125
stabilized cycle, 49, 88, 131, 211, 226
steel, 21, 39, 64, 75, 171, 256
  2-1/4 CrMo, 129, 132, 217, 265
  316 stainless, 249
  35 NCD 16, 280
  dual phase, 64
  ferritic, 36, 158
  low carbon, 187
  SOLDUR 355, 21, 214
  Weldox 460E, 271
stirrups, 348
stochastic analysis, 292, 297, 309, 323
strain
  accumulated plastic $\sim$, 10
  energy density, 11, 41, 218
  energy release rate, 292
  equivalence principle, 8
  finite, 167
strains
  finite, 221
stress
  back, 45, 62, 231
  concentration, 147, 201, 208, 244
    coefficient, 148, 154
  effective, 5, 8, 12, 13, 31, 34, 42, 104, 116, 119, 189, 322
  hydrostatic, 11
  interface equivalent $\sim$, 327
  mean, 301
  nominal, 148
  Piola–Kirchhoff, 221
  signed von Mises $\sim$, 82
  ultimate, 36
  viscous, 45
  Weibull, 329
  yield, 248

Taira rule, 237, 274
Tanaka, 230
tangent operator, 66, 72
  consistent $\sim$, 92, 111
Taylor, 91, 103
thermodynamics, 7
Tison, 267
toughness, 74, 154, 292, 335
Tracey, 141
triaxiality, 11, 34, 145, 150
  effective $\sim$ function, 81
  function, 11, 86
Truesdell, 168
Tvergaard, 141, 176, 177

UDIMET 700 alloy, 193
ultrasonic waves, 17
unified damage law, 32, 37, 41

variables of state, 7
Vincent, 190
Virely, 218
virtual work principle, 91
VISCOENDO, 261, 262, 264
Voigt, 91
Voyiadjis, 321

Wöhler curve, 38, 64, 193, 279, 301, 313
Wang, 230
weakest link hypothesis, 329, 366
Weibull, 307, 321
  model, 328, 366
  modulus, 328, 361
  stress, 329
Wempner, 94
Wright, 123

yield criterion, 45, 83
yield stress, 22, 248

ZeBuLon, 171, 214
Zyczkowski, 190

Printing: Krips bv, Meppel
Binding: Litges & Dopf, Heppenheim